OXFORD REVISION GUIDES

AS & A Level PHYSICS *through diagrams*

Stephen Pople

Great Clarendon Street, Oxford OX2 6DP

Oxford University Press is a department of the
University of Oxford. It furthers the University's objective
of excellence in research, scholarship, and education by
publishing worldwide in

Oxford New York

Auckland Cape Town Dar es Salaam Hong Kong Karachi
Kuala Lumpur Madrid Melbourne Mexico City Nairobi
New Delhi Shanghai Taipei Toronto

With offices in

Argentina Austria Brazil Chile Czech Republic France Greece
Guatemala Hungary Italy Japan Poland Portugal Singapore
South Korea Switzerland Thailand Turkey Ukraine Vietnam

Oxford is a registered trade mark of Oxford University Press
in the UK and in certain other countries

First published 2000
Second edition 2001

British Library Cataloguing in Publication Data

Data available

ISBN-13: 978-0-19-915078-6
ISBN-10: 0-19-915078-8

10 9 8 7 6 5 4 3 2 1

Designed and typeset in Optima
by Hardlines, Charlbury, Oxfordshire UK
Printed in Great Britain by Bell & Bain Ltd, Glasgow

CONTENTS

How to use this book

- If you are studying for an AS or A level in physics, start here! (If you are not aiming for one of these qualifications, you can use this book as a general reference for physics up to advanced level: there is an index to help you find the topic(s) you require.)
- Obtain a copy of the specification you are going to be examined on. Specifications are available from the exam boards' websites: www.aqa.org.uk; www.edexcel.org.uk; www.ocr.org.uk.
- With the table below as a starting point, make your own summary of the content of the specification you will be following.
- Use the pathways on pages 6 and 7 to help match the material in this book with that required by your specification.
- Find out the requirements for any coursework and the dates of your exams and plan your revision accordingly. Page 8 has some helpful advice.
- Begin revising! The self-assessment questions on pages 146–151 will help you to check your progress.

Note:

- This book covers AS and A2 material for all the main specifications and therefore contains some sections that you will not require.
- The material in this book is not divided up into AS and A2 because the level required may vary from one specification to another.
- If your specification is not listed, most of the material you need will still be included in this book, but you will have to construct your own route through the book.

Specification structures

This table summarizes the six main AS and A level specifications. Satisfactory assessment in units 1–3 corresponds to an AS level pass. Satisfactory assessment in the AS units 1–3 and the A2 units 4–6 corresponds to an A level pass. In each column are listed the unit names and main subdivisions as given in the specification. The method of assessment in each unit is listed, together with the percentage of marks assigned to the entire AS or A level. Do check your specification for the latest information.

		AQA Physics A	AQA Physics B	Edexcel Physics
AS units	Unit 1	**Particles, radiation, and quantum phenomena (Module 1)** *1h30m written exam on Module 1 (short structured questions)* *AS 30% A 15%*	**Foundation physics (Module 1)** *1h30m written exam on Module 1 (short answer & structured questions)* *AS 35% A 17.5%*	**Mechanics and radioactivity** *1h20m written exam (short & long structured questions)* *AS 30% A 15%*
	Unit 2	**Mechanics and molecular kinetic theory (Module 2)** *1h30m written exam on Module 2 (short structured questions)* *AS 30% A 15%*	**Waves and nuclear physics (Module 2)** *1h30m written exam on Module 2 (short answer & structured questions)* *AS 35% A 17.5%*	**Electricity and thermal physics** *1h20m written exam (short & long structured questions)* *AS 30% A 15%*
	Unit 3	**Current electricity and elastic properties of solids (Module 3)** *1h30m written exam on Module 3 (short structured questions)* *AS 25% A 12.5%* *1h30m practical exam OR Coursework* *AS 15% A 7.5% AS 15% A 7.5%*	**Experimental work (Module 3)** *2h practical exam* *AS 30% A 15%*	**Topics** One of: Astrophysics Solid materials Nuclear and particle physics Medical physics *1h20m written exam (structured questions)* *AS 20% A 10%* *45m practical exam* *AS 20% A 10%*
A2 units	Unit 4	**Waves, fields, and nuclear energy (Module 4)** *1h30m written exam on Module 4 (multiple-choice and structured questions)* *A 15%*	**Further physics (Module 4)** *1h30m written exam on Module 1 (short answer & structured questions)* *A 15%*	**Waves and our Universe** *1h20m written exam (short & long structured questions)* *A 15%*
	Unit 5	**Nuclear instability (Module 5)** **Options (Module 6)** One of: Astrophysics Medical physics Applied physics Turning points in physics Electronics *1h30m written exam on Modules 5 & 6 (structured questions)* *A 10%* *1h30m practical exam OR Coursework* *A 5% A 5%*	**Fields and their applications (Module 5)** *2h written exam (synoptic assessment: structured questions & comprehension question)* *A 20%*	**Fields and forces** *1h written exam* *A 7.5%* *1h30m practical exam* *A 7.5%*
	Unit 6	*2h written exam on Modules 1–5 (structured synoptic questions)* *A 20%*	**Experimental work (Module 6)** *3h practical exam & synoptic assessment in a practical context* *A 15%*	**Synthesis** *2h written exam (synoptic assessment: passage analysis & long structured questions)* *A 20%*

		Edexcel Physics (Salters Horners)	OCR Physics A	OCR Physics B (Advancing Physics)
AS units	Unit 1	**Physics at work, rest, and play** The sound of music Technology in space Higher, faster, stronger *1h30m written exam* *AS 33.3% A 16.7%*	**Forces and motion** *1h30m written exam* *AS 30% A 15%*	**Physics in action** Communication Designer materials *1h30m written exam* *AS 33.4% A 16.7%*
	Unit 2	**Physics for life** Good enough to eat Digging up the past Spare part surgery *1h30m written exam* *AS 33.3% A 16.7%*	**Electrons and photons** *1h30m written exam* *AS 30% A 15%*	**Understanding processes** *1h30m written exam* *AS 36.6 A 18.3%*
	Unit 3	**Working with physics** Two laboratory practical activities and an out-of-school visit. *Coursework* *AS 33.3% A 16.7%*	**Wave properties/experimental skills** *1h written exam* *AS 20% A 10%* *1h 30m practical exam OR Coursework* *AS 20% A10% AS 20% A10%*	**Physics in practice** *Coursework* *AS 30% A 15%*
A2 units	Unit 4	**Moving with physics** Transport on track The medium is the message Probing the heart of matter *1h30m written exam* *A 15%*	**Forces, fields, and energy** *1h30m written exam* *A 15%*	**Rise and fall of the clockwork Universe** Models and rules Matter in extremes *1h20m written exam* *A 10.8%* Practical investigation *Coursework* *A 7.5%*
	Unit 5	**Physics from creation to collapse** Two-week individual practical project *Coursework* *A 10%* Reach for the stars Build or bust? *1h written exam* *A 10%*	**Options in physics** One of: Cosmology Health physics Materials Nuclear and particle physics Telecommunications *1h30m written exam* *A 15%*	**Field and particle pictures** Fields Fundamental particles *1h10m written exam* *A 10.8%* Research report *Coursework* *A 7.5%*
	Unit 6	**Exploring physics** *1h30m written exam (synoptic questions)* *A 15%*	**Unifying concepts in physics/experimental skills** *1h written exam* *A 10%* *Coursework* *A 10%* *1h 30m practical exam* *A 10%*	**Advances in physics** *1h30m written exam* *A 15%*

What are...

...short-answer questions?
These questions will require just a few words or sentences as answers.

...structured questions?
This type of question is broken up into smaller parts. Some parts will ask you to define or show you understand a given term; explain a phenomenon or describe an experiment; plot sketch graphs or obtain information from given graphs; draw labelled diagrams or indicate particular features on a given diagram. Other parts will lead you to the solution of a complex problem by asking you to solve it in stages.

...comprehension questions?
In these questions you will be given a passage (short or extended) on a topic and then tested on your understanding of the topic and the scientific concepts in it.

...data-analysis questions?
In this type of question you will be given data in a variety of forms: graphs, tables, in text, as a list. You will then be asked to analyse the data to derive new results or information and may be asked to link the results with explanations of the scientific principles involved.

...synoptic questions?
When answering these you will have to apply physics principles or skills in contexts that are likely to be unfamiliar to you. Some questions will require you to show that you understand how different aspects of physics relate to one another or are used to explain different aspects of a particular application. Questions of this type will require you to draw on the knowledge, understanding, and skills developed during your study of the whole course. 20% of the A level marks are allocated to synoptic questions.

Pathways

The following pathways identify the main sections in the book that relate to the topics required by each specification.
Note:

- You will not necessarily need all the material that is given in any section.
- There may be material in other sections (e.g. applications) that you need to know.
- You should identify the relevant material by referring to the specification you are following.
- If this is your own copy of the book, highlight all the relevant topics throughout the book.

AQA Physics A

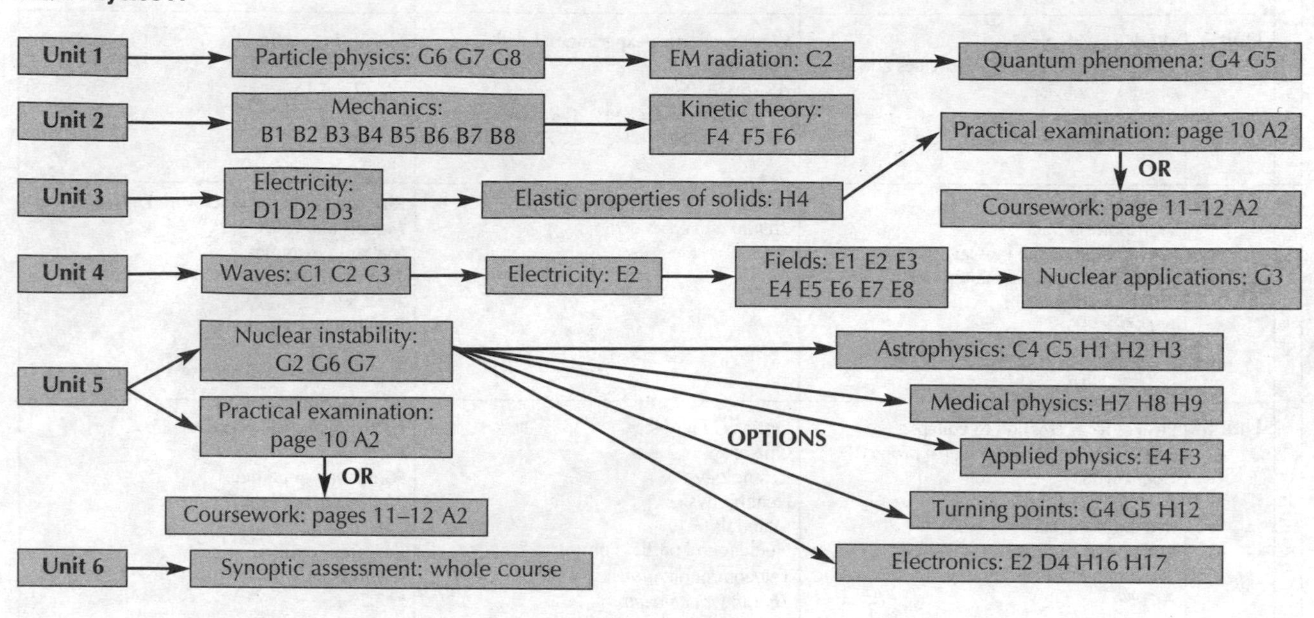

AQA Physics B

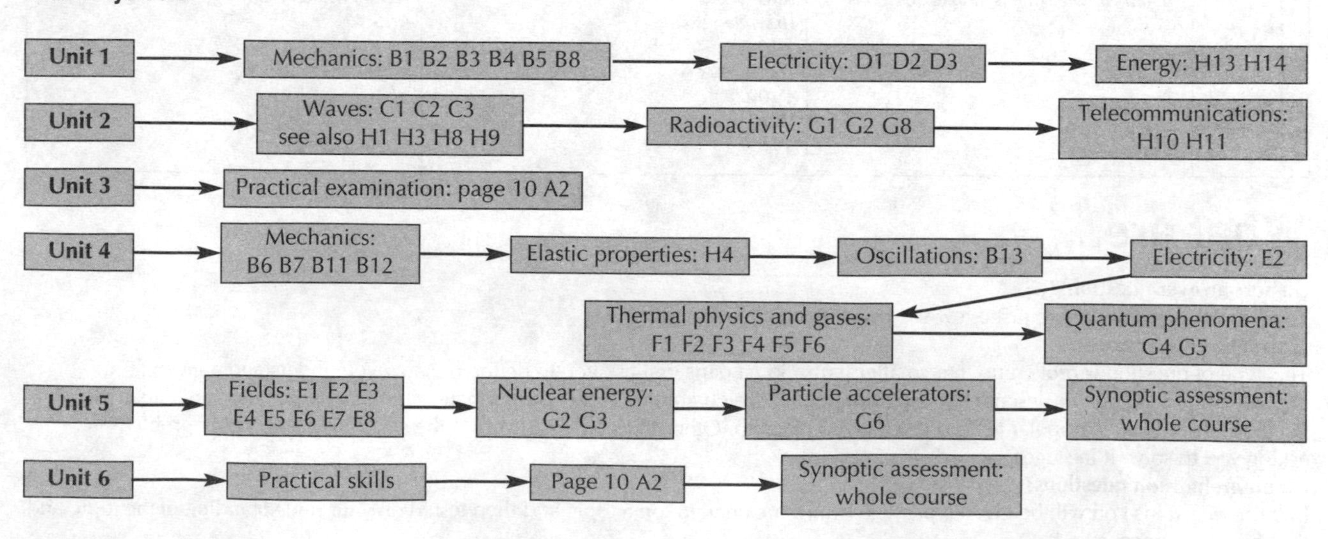

Edexcel Physics B (Salters Horners)

The Edexcel Salters Horners course structure is thematic. Concepts are covered as they are required for explanations within a given theme. It is therefore not possible to summarize the content in the same way as the other specifications.

If you are following this course you should:

- use the index and the Salters Horners specification to link the learning outcomes required to the pages on which the topics appear
- note the sections where relevant information appears as you cover them in the modules
- highlight the relevant material if this copy of the book is your own property.

Edexcel Specification A

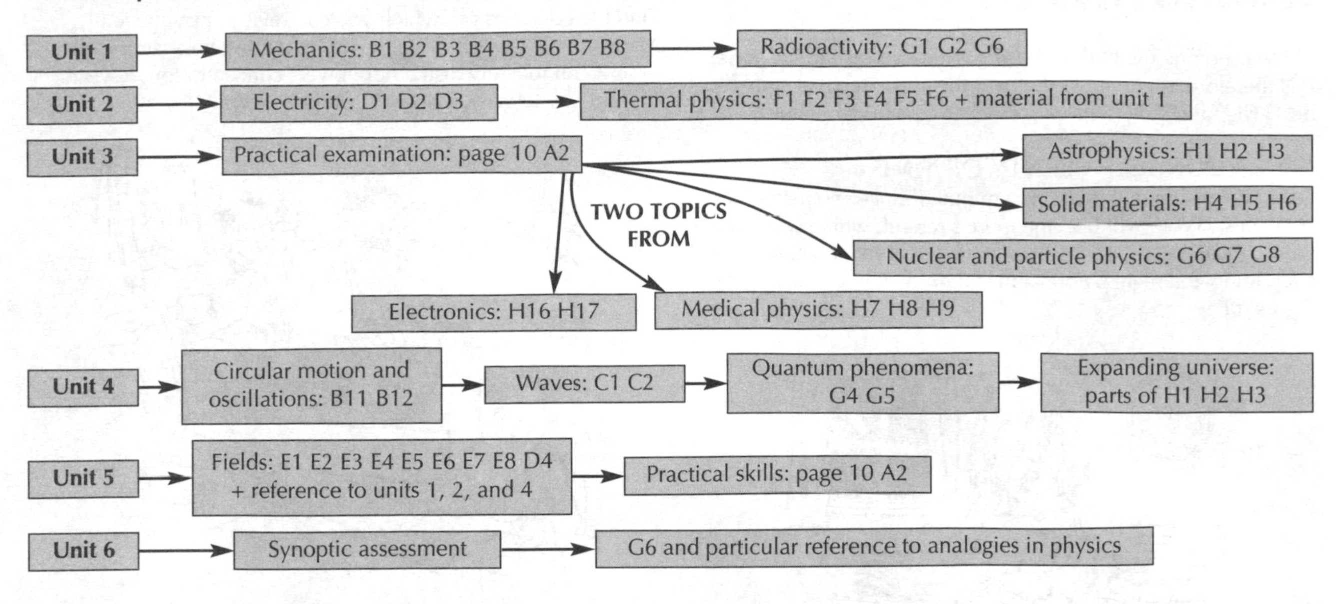

OCR Physics A

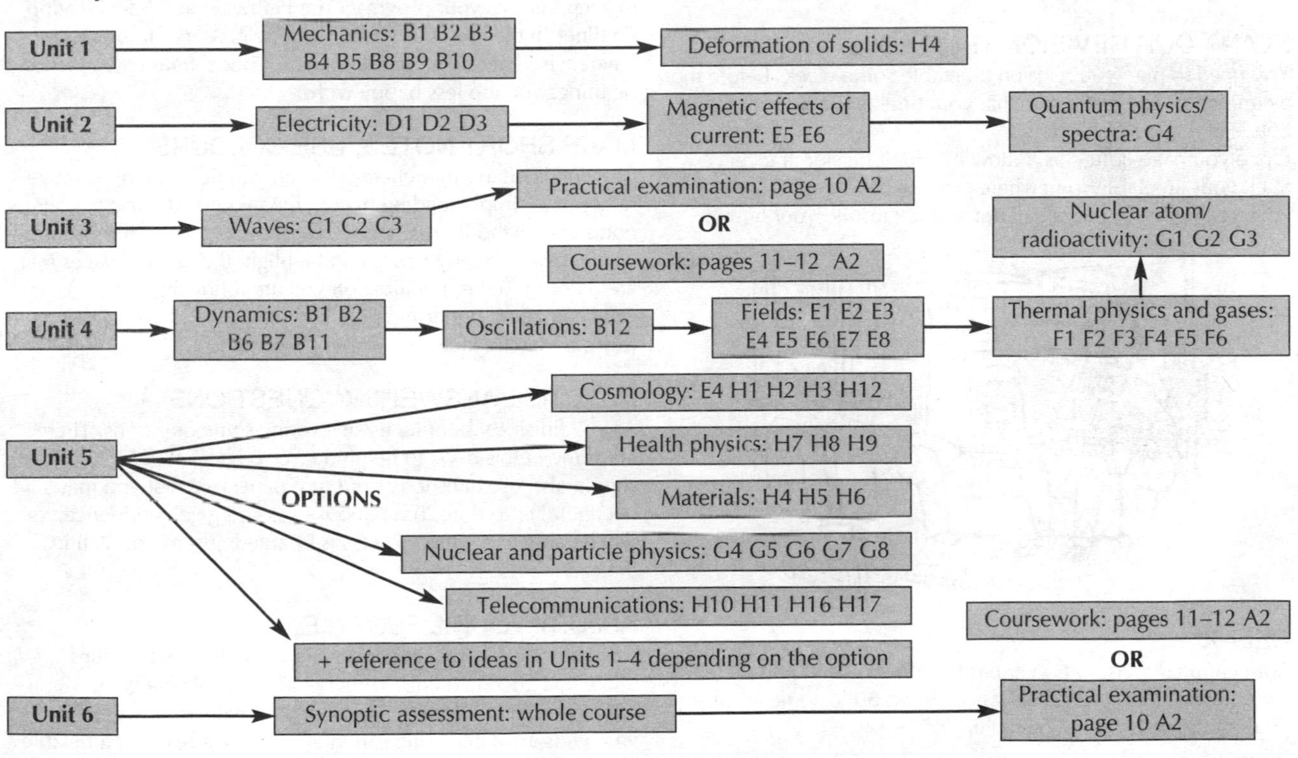

OCR Physics B (Advancing Physics)

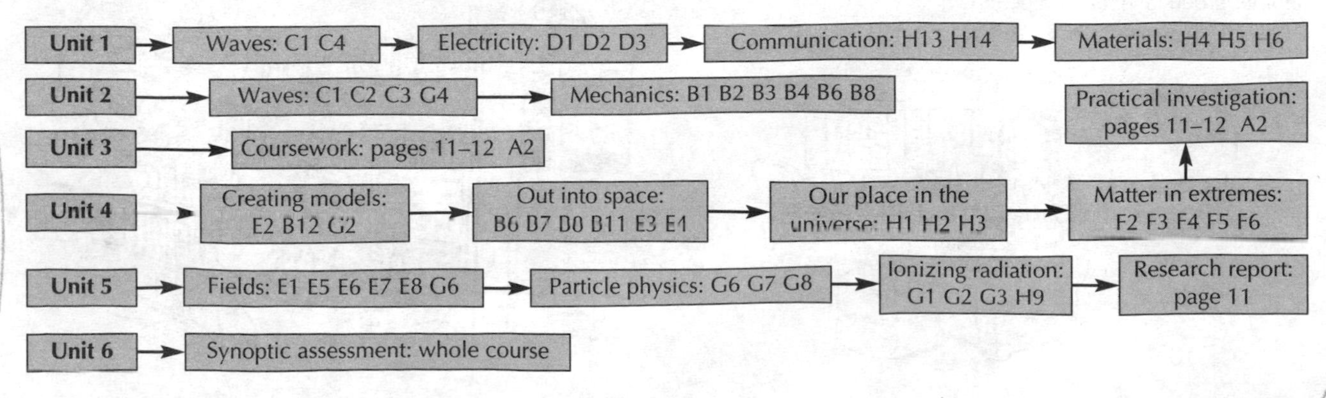

How to revise

There is no one method of revising which works for everyone. It is therefore important to discover the approach that suits you best. The following rules may serve as general guidelines.

GIVE YOURSELF PLENTY OF TIME

Leaving everything until the last minute reduces your chances of success. Work will become more stressful, which will reduce your concentration. There are very few people who can revise everything 'the night before' and still do well in an examination the next day.

PLAN YOUR REVISION TIMETABLE

You need to plan you revision timetable some weeks before the examination and make sure that your time is shared suitably between all your subjects.
Once you have done this, follow it – don't be side-tracked. Stick your timetable somewhere prominent where you will keep seeing it – or better still put several around your home!

RELAX

Concentrated revision is very hard work. It is as important to give yourself time to relax as it is to work. Build some leisure time into your revision timetable.

GIVE YOURSELF A BREAK

When you are working, work for about an hour and then take a short tea or coffee break for 15 to 20 minutes. Then go back to another productive revision period.

FIND A QUIET CORNER

Find the conditions in which you can revise most efficiently. Many people think they can revise in a noisy busy atmosphere – most cannot! Any distraction lowers concentration. Revising in front of a television doesn't generally work!

KEEP TRACK

Use checklists and the relevant examination board specification to keep track of your progress. The Pathways and Specification Outlines in the previous section will help. Mark off topics you have revised and feel confident with. Concentrate your revision on things you are less happy with.

MAKE SHORT NOTES, USE COLOURS

Revision is often more effective when you do something active rather than simply reading material. As you read through your notes and textbooks make brief notes on key ideas. If this book is your own property you could highlight the parts of pages that are relevant to the specification you are following.
Concentrate on understanding the ideas rather than just memorizing the facts.

PRACTISE ANSWERING QUESTIONS

As you finish each topic, try answering some questions. There are some in this book to help you (see pages 146–151). You should also use questions from past papers. At first you may need to refer to notes or textbooks. As you gain confidence you will be able to attempt questions unaided, just as you will in the exam.

ADJUST YOUR LIFESTYLE

Make sure that any paid employment and leisure activities allow you adequate time to revise. There is often a great temptation to increase the time spent in paid employment when it is available. This can interfere with a revision timetable and make you too tired to revise thoroughly. Consider carefully whether the short-term gains of paid employment are preferable to the long-term rewards of examination success.

Success in examinations

EXAMINATION TECHNIQUE

The following are some points to note when taking an examination.

- Read the question carefully. Make sure you understand exactly what is required.
- If you find that you are unable to do a part of a question, do not give up. The next part may be easier and may provide a clue to what you might have done in the part you found difficult.
- Note the number of marks per question as a guide to the depth of response needed (see below).
- Underline or note the key words that tell you what is required (see opposite).
- Underline or note data as you read the question.
- Structure your answers carefully.
- Show all steps in calculations. Include equations you use and show the substitution of data. Remember to work in SI units.
- Make sure your answers are to suitable significant figures (usually 2 or 3) and include a unit.
- Consider whether the magnitude of a numerical answer is reasonable for the context. If it is not, check your working.
- Draw diagrams and graphs carefully.
- Read data from graphs carefully; note scales and prefixes on axes.
- Keep your eye on the clock but don't panic.
- If you have time at the end, use it. Check that your descriptions and explanations make sense. Consider whether there is anything you could add to an explanation or description. Repeat calculations to ensure that you have not made a mistake.

DEPTH OF RESPONSE

Look at the **marks allocated to the question.**
This is usually a good guide to the depth of the answer required. It also gives you an idea how long to spend on the question. If there are 60 marks available in a 90 minute exam, your 1 mark should be earned in 1.5 minutes.

Explanations and descriptions

If a *4 mark* question requires an explanation or description, you will need to make *four* distinct relevant points.
You should note, however, that simply mentioning the four points will not necessarily earn full marks. The points need to be made in a coherent way that makes sense and fits the context and demands of the questions.

Calculations

In calculation questions marks will be awarded for method and the final answer.

In a *3 mark* calculation question you *may* obtain all three marks if the final answer is correct, even if you show no working. However, you should *always show your working* because

- sometimes the working is a requirement for full marks
- if you make an error in the calculation you cannot gain any method marks unless you have shown your working.

In general in a *3 mark* calculation you earn

1 mark for quoting a relevant equation or using a suitable method

1 mark for correct substitution of data or some progress toward the final answer

1 mark for a correct final answer given to suitable significant figures with a correct unit.

Errors carried forward

If you make a mistake in a calculation and need to use this incorrect answer in a subsequent part of the question, you can still gain full marks. Do not give up if you think you have gone wrong. Press on using the data you have.

KEY WORDS

How you respond to a question can be helped by studying the following, which are the more common key words used in examination questions.

Name: The answer is usually a technical term consisting of one or two words.

List: You need to write down a number of points (often a single word) with no elaboration.

Define: The answer is a formal meaning of a particular term.

What is meant by...? This is often used instead of 'define'.

State: The answer is a concise word or phrase with no elaboration.

Describe: The answer is a description of an effect, experiment, or (e.g.) graph shape. No explanations are required.

Suggest: In your answer you will need to use your knowledge and understanding of topics in the specification to deduce or explain an effect that may be in a novel context. There may be no single correct answer to the question.

Calculate: A numerical answer is to be obtained, usually from data given in the question. Remember to give your answer to a suitable number of significant figures and give a unit.

Determine: Often used instead of 'calculate'. You may need to obtain data from graphs, tables, or measurements.

Explain: The answer will be extended prose. You will need to use your knowledge and understanding of scientific phenomena or theories to elaborate on a statement that has been made in the question or earlier in your answer. A question often asks you to 'state and explain...'.

Justify: Similar to 'explain'. You will have made a statement and now have to provide a reason for giving that statement.

Draw: Simply draw a diagram. If labelling or a scale drawing is needed, you will usually be asked for this, but it is sensible to provide labelling even if it is not asked for.

Sketch: This usually relates to a graph. You need to draw the general shape of the graph on labelled axes. You should include enough quantitative detail to show relevant intercepts and/or whether the graph is exponential or some inverse function, for example.

Plot: The answer will be an accurate plot of a graph on graph paper. Often it is followed by a question asking you to 'determine some quantity from the graph' or to 'explain its shape'.

Estimate: You may need to use your knowledge and/or your experience to deduce the magnitude of some quantities to arrive at the order of magnitude for some other quantity defined in the question.

Discuss: This will require an extended response in which you demonstrate your knowledge and understanding of a given topic.

Show that: You will have been given either a set of data and a final value (that may be approximate) or an algebraic equation. You need to show clearly all basic equations that you use and all the steps that lead to the final answer.

REVISION NOTE

In your revision remember to

- learn the formulae that are not on your formula sheet
- make sure that you know what is represented by all the symbols in equations on your formula sheet.

Practical assessment

Your practical skills will be assessed at both AS and A level. Make sure you know how your practical skills are going to be assessed.
You may be assessed by

- **coursework**
- **practical examination**

The method of assessment will depend on the specification you are following and the choice of your school/college. You may be required to take

- two practical examinations (one at AS and one at A level)
- two coursework assessments
- one practical examination and one coursework assessment.

PRACTISING THE SKILLS

Whichever assessment type is used, you need to learn and practise the skills during your course.

Specific skills

You will learn specific skills associated with particular topics as a natural part of your learning during the course. Make sure that you have hands-on experience of all the apparatus that is used. You need to have a good theoretical background of the topics on your course so that you can

- devise a sensible hypothesis
- identify all variables in an experiment
- control variables
- choose suitable magnitudes for variables
- select and use apparatus correctly and safely
- tackle analysis confidently
- make judgements about the outcome.

PRACTICAL EXAMINATION

The form of the examination varies from one examination board to another, so make sure you know what your board requires you to do. Questions generally fall into three types which fit broadly into the following categories:
You may be required to

- examine a novel situation, create a hypothesis, consider variables, and design an experiment to test the hypothesis
- examine a situation, analyse data that may be given to you, and evaluate the experiment that led to the data
- obtain and analyse data in an experiment which has been devised by the examination board.

In any experiment you may be required to determine uncertainties in raw data, derived data, and the final result.

Designing experiments and making hypotheses

Remember that you can only gain marks for what you write, so take nothing for granted. Be thorough. A description that is too long is better than one that leaves out important detail.

Remember to

- use your knowledge of AS and A level physics to support your reasoning
- give quantitative reasoning wherever possible
- draw clear labelled diagrams of apparatus
- provide full details of measurements made, equipment used, and experimental procedures
- be prepared to state the obvious.

A good test of a sufficiently detailed account is to ask yourself whether it would be possible to do the experiment you describe without needing any further infomation.

PRACTICAL SKILLS

There are four basic skill areas:

Planning
Implementing
Analysing
Evaluating

The same skills are assessed in both practical examinations and coursework.

GENERAL ASSESSMENT CRITERIA

You will be assessed on your ability to

- identify what is to be investigated
- devise a hypothesis or theory of the expected outcome
- devise a suitable experiment, use appropriate resources, and plan the procedure
- carry out the experiment or research
- describe precisely what you have done
- present your data or information in an appropriate way
- draw conclusions from your results or other data
- evaluate the uncertainties in your experiment
- evaluate the success or otherwise of the experiment and suggest how it might have been improved.

GENERAL SKILLS

The general skills you need to practise are

- the accurate reporting of experimental procedures
- presentation of data in tables (possibly using spreadsheets)
- graph drawing (possibly using IT software)
- analysis of graphical and other data
- critical evaluation of experiments

Carrying out experiments

When **making observations** and **tabulating data** remember to

- consider carefully the range and intervals at which you make your observations
- consider the accuracy to which it is reasonable to quote your observations (how many significant figures are reasonable)
- repeat all readings and remember to average
- be consistent when quoting data
- tabulate all data (including repeats and averages) remembering to give units for all columns
- make sure figures are not ambiguous.

When **deriving data** remember to

- work out an appropriate unit
- make sure that the precision is consistent with your raw data.

When **drawing graphs** remember to

- choose a suitable scale that uses the graph paper fully
- label the axes with quantity and unit
- mark plotted points carefully with a cross using a sharp pencil
- draw the best straight line or curve through the points so that the points are scattered evenly about the line.

When **analysing data** remember to

- use a large gradient triangle in graph analysis to improve accuracy
- set out your working so that it can be followed easily
- ensure that any quantitative result is quoted to an accuracy that is consisted with your data and analysis methods
- include a unit for any result you obtain.

Carrying out investigations

Keep a notebook

Record

- all your measurements
- any problems you have met
- details of your procedures
- any decisions you have made about apparatus or procedures including those considered and discarded
- relevant things you have read or thoughts you have about the problem.

Define the problem

Write down the aim of your experiment or investigation. Note the variables in the experiment. Define those that you will keep constant and those that will vary.

Suggest a hypothesis

You should be able to suggest the expected outcome of the investigation on the basis of your knowledge and understanding of science. Try to make this as quantitative as you can, justifying your suggestion with equations wherever possible.

Do rough trials

Before commencing the investigation in detail do some rough tests to help you decide on

- suitable apparatus
- suitable procedures
- the range and intervals at which you will take measurements
- consider carefully how you will conduct the experiment in a way that will ensure safety to persons and to equipment.

Remember to consider alternative apparatus and procedures and justify your final decision.

Carry out the experiment

Remember all the skills you have learnt during your course:

- note all readings that you make
- take repeats and average whenever possible
- use instruments that provide suitably accurate data
- consider the accuracy of the measurements you are making
- analyse data as you go along so that you can modify the approach or check doubtful data.

Presentation of data

Tabulate all your observations, remembering to

- include the quantity, any prefix, and the unit for the quantity at the head of each column
- include any derived quantities that are suggested by your hypothesis
- quote measurements and derived data to an accuracy/significant figures consistent with your measuring instruments and techniques, and be consistent
- make sure figures are not ambiguous.

Graph drawing

Remember to

- label your axes with quantity and unit
- use a scale that is easy to use and fills the graph paper effectively
- plot points clearly (you may wish to include 'error bars')
- draw the best line through your plotted points
- consider whether the gradient and area under your graph have significance.

Analysing data

This may include

- the calculation of a result
- drawing of a graph
- statistical analysis of data
- analysis of uncertainties in the original readings, derived quantities, and results.

Make sure that the stages in the processing of your data are clearly set out.

Evaluation of the investigation

The evaluation should include the following points:

- draw conclusions from the experiment
- identify any systematic errors in the experiment
- comment on your analysis of the uncertainties in the investigation
- review the strengths and weaknesses in the way the experiment was conducted
- suggest alternative approaches that might have improved the experiment in the light of experience.

Use of information technology (IT)

You may have used data capture techniques when making measurements or used IT in your analysis of data. In your analysis you should consider how well this has performed. You might include answers to the following questions.

- What advantages were gained by the use of IT?
- Did the data capture equipment perform better than you could have achieved by a non-IT approach?
- How well has the data analysis software performed in representing your data graphically, for example?

THE REPORT

Remember that your report will be read by an assessor who will not have watched you doing the experiment. For the most part the assessor will only know what you did by what you write, so do not leave out important information.

If you write a good report, it should be possible for the reader to repeat what you have done should they wish to check your work.

A **word-processed report** is worth considering. This makes the report much easier to revise if you discover some aspect you have omitted. It will also make it easier for the assessor to read.

Note:
The report may be used as portfolio evidence for assessment of Application of Number, Communication, and IT Key Skills.

Use subheadings

These help break up the report and make it more readable. As a guide, the subheadings could be the main sections of the investigation: aims, diagram of apparatus, procedure, etc.

Coping with coursework

TYPES OF COURSEWORK

Coursework takes different forms with different specifications. You may undertake

- short experiments as a routine part of your course
- long practical tasks prescribed by your teacher/lecturer
- a long investigation of a problem decided by you and agreed with your teacher
- a research and analysis exercise using book, IT, and other resources.

A short experiment
This may take one or two laboratory sessions to complete and will usually have a specific objective that is closely linked to the topic you are studying at the time.
You may only be assessed on one or two of the skills in any one assessment.

A long investigation
This may take 5 to 10 hours of class time plus associated homework time.
You will probably be assessed on all the skills in a long investigation.

Research and analysis task
This may take a similar amount of time but is likely to be spread over a longer period. This is to give you time to obtain information from a variety of sources.
You will be assessed on

- the planning of the research
- the use of a variety of sources of information
- your understanding of what you have discovered
- your ability to identify and evaluate relevant information
- the communication of your findings in writing or in an oral presentation.

Make sure you know in detail what is expected of you in the course you are following. Consult the Pathways and Specification outlines on pages 4–7.

STUDY THE CRITERIA

Each examination board produces criteria for the assessment of coursework. The practical skills assessed are common to all boards, but the way each skill is rewarded is different for each specification. Ensure that you have a copy of the assessment criteria so that you know what you are trying to achieve and how your work will be marked.

PLAN YOUR TIME

Meeting the deadline is often a major problem in coping with coursework.

Do not leave all the writing up to the end
Using a word processor you can draft the report as you go along. You can then go back and tidy it up at the end.

Draw up an initial plan
Include the following elements:

The aim of the project
What are you going to investigate practically?
or
What is the topic of your research?

A list of resources
What are your first thoughts on apparatus?
or
Where are you going to look for information?
(Books; CD ROMs; Internet)
or
Is there some organization you could write to for information?

Theoretical ideas
What does theory suggest will be the outcome?
or
What are the main theoretical ideas that are linked with your investigation or research project?

Timetable
What is the deadline?
What is your timetable for?

Laboratory tasks
How many lab sessions are there?
Initial thoughts on how they are to be used

Non-laboratory tasks
Initial analysis of data
Writing up or word-processing part of your final report
Making good diagrams of your apparatus
Revising your time plan
Evaluating your data or procedures

Key Skills

What are Key Skills?

These are skills that are not specific to any subject but are general skills that enable you to operate competently and flexibly in your chosen career. Visit the Key Skills website (www.keyskillssupport.net) or phone the Key Skills help line to obtain full, up-to-date information.
While studying your AS or A level courses you should be able to gather evidence to demonstrate that you have achieved competence in the Key Skills areas of

- ***Communication***
- ***Application of Number***
- ***Information Technology.***

You may also be able to prove competence in three other key skills areas:

- ***Working with Others***
- ***Improving your own Learning***
- ***Problem Solving.***

Only the first three will be considered here and only an outline of what you must do is included. You should obtain details of what you need to know and be able to do. You should be able to obtain these from your examination centre.

Communication

You must be able to

- create opportunities for others to contribute to group discussions about complex subjects
- make a presentation using a range of techniques to engage the audience
- read and synthesize information from extended documents about a complex subject
- organize information coherently, selecting a form and style of writing appropriate to complex subject matter.

Application of Number

You must be able to plan and carry through a substantial and complex activity that requires you to

- plan your approach to obtaining and using information, choose appropriate methods for obtaining the results you need and justify your choice
- carry out multistage calculations including use of a large data set (over 50 items) and re arrangement of formulae
- justify the choice of presentation methods and explain the results of your calculations.

Information Technology

You must be able to plan and carry through a substantial activity that requires you to

- plan and use different sources and appropriate techniques to search for and select information based on judgement of relevance and quality
- automated routines to enter and bring together information, and create and use appropriate methods to explore, develop, and exchange information
- develop the structure and content of your presentation, using others' views to guide refinements, and information from difference sources.

A complex subject is one in which there are a number of ideas, some of which may be abstract and very detailed. Lines of reasoning may not be immediately clear. There is a requirement to come to terms with specialized vocabulary.

A substantial activity is one that includes a number of related tasks. The result of one task will affect the carrying out of others. You will need to obtain and interpret information and use this to perform calculations and draw conclusions.

What standard should you aim for?

Key Skills are awarded at four levels (1–4). In your A level courses you will have opportunities to show that you have reached level 3, but you could produce evidence that demonstrates that you are competent at a higher level.
You may achieve a different level in each Key Skill area.

What do you have to do?

You need to show that you have the necessary underpinning knowledge in the Key Skills area and produce evidence that you are able to apply this in your day-to-day work.
You do this by producing a portfolio that contains

- evidence in the form of reports when it is possible to provide written evidence
- evidence in the form of assessments made by your teacher when evidence is gained by observation of your performance in the classroom or laboratory.

The evidence may come from only one subject that you are studying, but it is more likely that you will use evidence from all of your subjects.
It is up to you to produce the best evidence that you can.

The specifications you are working with in your AS or A level studies will include some ideas about the activities that form part of your course and can be used to provide this evidence. Some general ideas are summarized below, but refer to the specification for more detail.

Communication: in science you could achieve this by

- undertaking a long practical or research investigation on a complex topic (e.g. use of nuclear radiation in medicine)
- writing a report based on your experimentation or research using a variety of sources (books, magazines, CD-ROMs, Internet, newspapers)
- making a presentation to your fellow students
- using a presentation style that promotes discussion or criticism of your findings, enabling others to contribute to a discussion that you lead.

Application of Number: in science you could achieve this by

- undertaking a long investigation or research project that requires detailed planning of methodology
- considering alternative approaches to the work and justifying the chosen approach
- gathering sufficient data to enable analysis by statistical and graphical methods
- explaining why you analysed the data as you did
- drawing the conclusions reached as a result of your investigation.

Information Technology: in science you could achieve this by

- using CD-ROMs and the Internet to research a topic
- identifying those sources which are relevant
- identifying where there is contradictory information and identifying which is most probably correct
- using a word processor to present your report, drawing in relevant quotes from the information you have gathered
- using a spreadsheet to analyse data that you have collected
- using data capture techniques to gather information and mathematics software to analyse the data.

Answering the question

This section contains some examples of types of questions with model answers showing how the marks are obtained. You may like to try the questions and then compare your answers with the model answers given.

MARKS FOR QUALITY OF WRITTEN COMMUNICATION

In questions that require long descriptive answers or explanations, marks may be reserved for the quality of language used in your answers.

2 marks if your answer
- uses scientific terms correctly
- is written fluently and/or is well argued
- contains only a few spelling or grammatical errors.

1 mark if your answer
- generally uses scientific terms correctly
- generally makes sense but lacks coherence
- contains poor spelling and grammar.

An answer that is scientifically inaccurate, is disjointed, and contains many spelling and grammatical errors loses both these marks.

The message is: **do not let your communication skills let you down**.

ALWAYS SHOW YOUR WORKING

In calculation questions one examination board might expect to see the working for all marks to be gained. Another might sometimes give both marks if you give the correct final answer. It is wise always to show your working. If you make a mistake in processing the data you could still gain the earlier marks for the method you use.

Question 1

Description and explanation question

(a) Describe the nuclear model of an atom that was proposed by Rutherford following observations made in Geiger and Marsden's alpha-particle scattering experiment. *(4 marks)*

(b) Explain why when gold foil is bombarded by alpha particles

(i) some of the alpha particles are deviated through large angles that are greater than 90°; *(3 marks)*

(ii) most of the alpha particles pass through without deviation and lose little energy while passing through the foil. *(2 marks)*

Note: In explanations or descriptive questions there are often alternative relevant statements that would earn marks. For example in part **(a)** you could earn credit for stating that electrons have small mass or negative charge.

Question 2

Calculation question

The supply in the following circuit has an EMF of 12.0 V and negligible internal resistance.

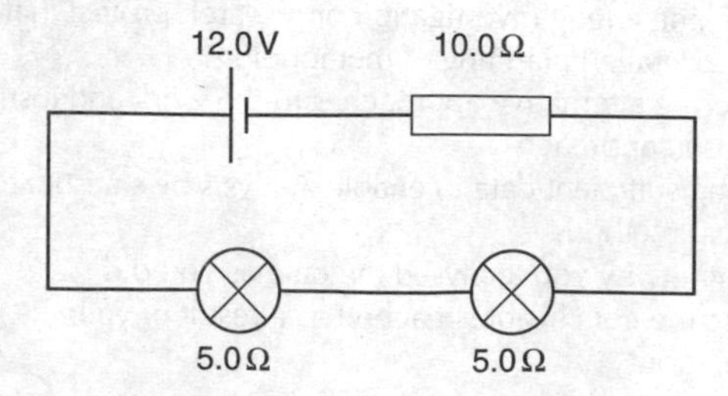

(a) Calculate

(i) the current through each lamp; *(2 marks)*

(ii) the power dissipated in each lamp; *(2 marks)*

(iii) the potential difference across the 10.0 Ω resistor. *(1 mark)*

(b) A student wants to produce the same potential difference across the 10.0 Ω resistor using two similar resistors in parallel.

(i) Sketch the circuit the student uses. *(1 mark)*

(ii) Determine the value of each of the series resistors used. Show your reasoning. *(3 marks)*

Answer to question 1

(a) The atom consists of a small nucleus (✓) which contains most of the mass (✓) of the atom. The nucleus is positive (✓). Electrons orbit the nucleus (✓).

(b) (i) A few alpha particles pass close to a nucleus (✓). There is a repelling force between the alpha particle and the gold nucleus because they are both positively charged (✓). This causes deflection of the alpha particle. Because the alpha particle is much less massive than the gold nucleus it may deviate through a large angle (✓).

(ii) Few alpha particles collide with a nucleus since most of matter is empty space occupied only by electrons (✓). The alpha particles deviate only a little and lose very little energy because an electron has a very small mass compared to that of an alpha particle (✓).

Answer to question 2

(a) (i) Current in circuit = EMF/total resistance (✓)
=12.0/20.0
Current in circuit = 0.60 A (✓)

(ii) Power $= I^2R$ (✓)
$= 0.60^2 \times 5.0$
Power = 1.8 W (✓)

(iii) PD $= IR = 0.60 \times 10.0 = 6.0$ V (✓)

(b) (i)

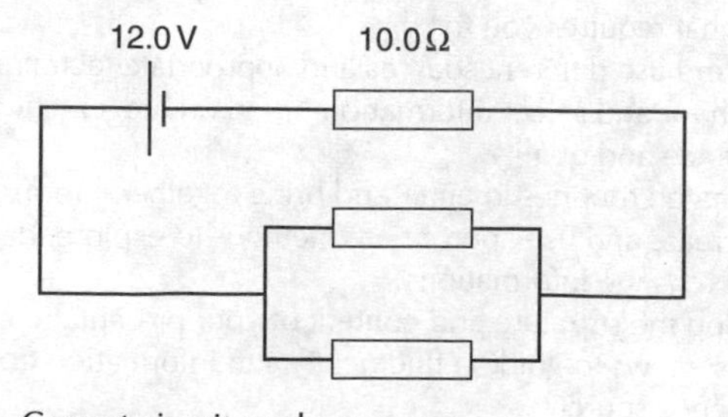

Correct circuit as above. (✓)

(ii) Parallel combination must be 10.0 Ω (✓)
Two similar parallel resistors have total resistance equal to half that of one resistor. (✓)
(*or* $\frac{1}{10} = \frac{1}{R} + \frac{1}{R}$)
Each resistor = 20 Ω (✓)

Question 3

Graph interpretation and graph sketching

The diagram shows how the pressure p varies with the volume V for a fixed mass of gas.

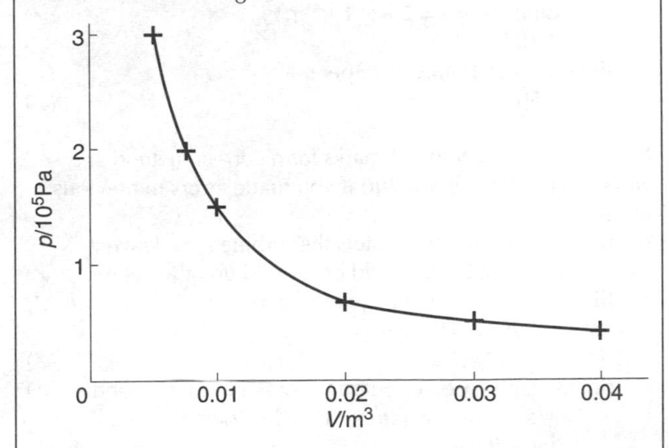

(a) Use data from the graph to show that the changes take place at constant temperature. *(3 marks)*

(b) Sketch a graph to show how the pressure varies with $1/V$ for this gas. *(2 marks)*

Question 4

Experiment description

The fundamental frequency f of a stretched string is given by the equation $f=\frac{1}{2l}+\sqrt{\frac{T}{\mu}}$, where T is the tension and μ is the mass per unit length of the string.

(a) Sketch the apparatus you would use to test the relationship between f and T. *(2 marks)*

(b) State the quantities that are kept constant in the experiment. *(2 marks)*

(c) Describe how you obtain data using the apparatus you have drawn and how you would use the data to test the relationship. *(7 marks)*

Synoptic Questions

Application type (AEB 1994 part question)

Figure 1 shows the principle of the operation of a hydro-electric power station. The water which drives the turbine comes from a reservoir high in the mountains.

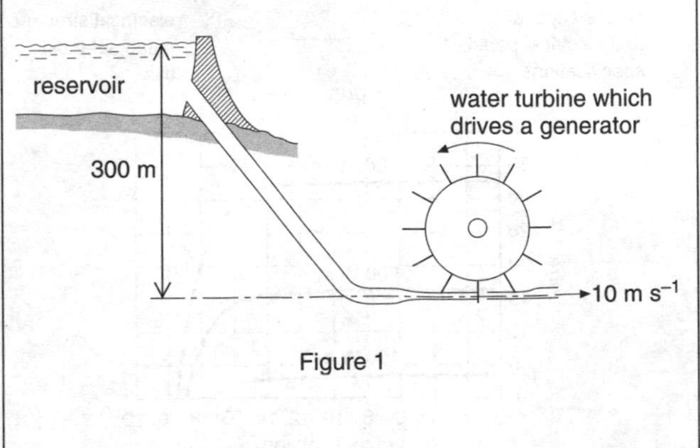

Figure 1

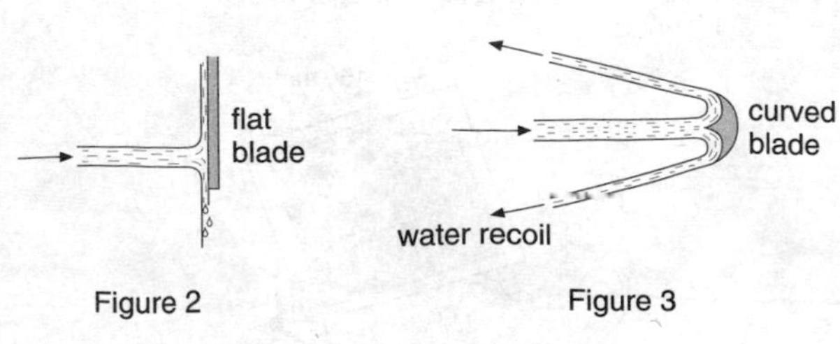

Figure 2

Figure 3

Answer to question 3

(a) For a change at constant temperature, pV = constant (✓).
Use co-ordinates from three points A, B, and C on the graph (✓) (NB using only two would lose this mark).
e.g. units (m³,10⁵ Pa). A (0.005, 3) B (0.01,1.5) C (0.03, 0.5).
Product in each case is $0.015 \times 10^5\,m^3\,Pa$.
The product pV is constant within limits of experimental uncertainties, so the changes take place at constant temperature (✓).

(b) Straight line through the origin (✓).
pV for the line is consistent with data in given graph (✓).

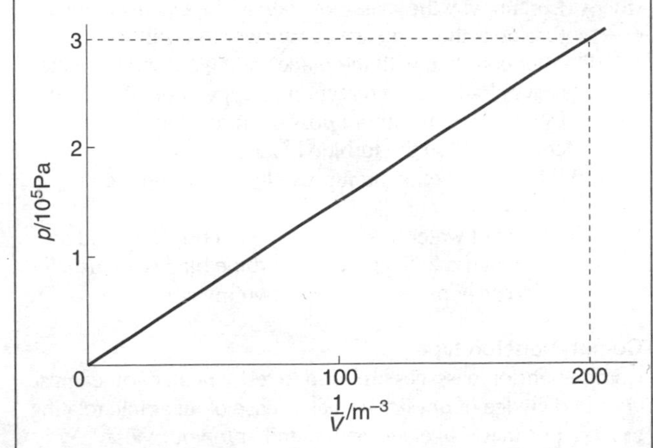

Answer to question 4

(a)

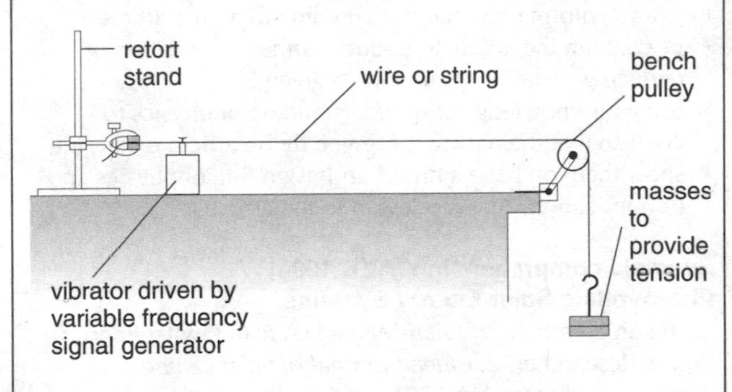

Means of determining frequency. (✓)
Sensible arrangement with means of changing tension. (✓)

(b) The constant quantities are:

- The mass per unit length of the wire. The material and the diameter must not be changed. (✓)
- The length of the wire used. (✓)

(c) A suitable tension is produced by adding masses at the end of the wire. The tension is noted (✓). When the mass used to tension the wire is m the tension is mg (✓). The oscillator frequency drives the vibrator which causes the wire to vibrate (✓). The oscillator frequency is adjusted until the wire vibrates at its fundamental frequency (i.e. a single loop is observed) (✓). The output frequency of the oscillator is noted (✓). The tension is changed and the new frequency at which the wire vibrates with one loop is determined (✓). A graph is plotted of frequency f against the square root of the tension, $\sqrt{T}$ (✓). If $f \propto \sqrt{T}$ the graph should be a straight line through the origin (✓).

The water level in the reservoir is 300 m above the nozzle which directs the water onto the blades of the turbine. The diameter of the water jet emerging from the nozzle is 0.060 m. The density of the water is 100 kg m^{-3} and the acceleration of free fall, *g*, is 9.8 m s^{-2}.

(a) Assuming that the kinetic energy of the water leaving the nozzle is equal to the potential energy of the water at the surface of the reservoir, estimate
  (i) the speed of the water as it leaves the nozzle;
  (ii) the mass of water flowing from the nozzle in 1.00 s;
  (iii) the power input to the turbine. (*6 marks*)

(b) **(i)** Explain why the mass flow rate at the exit from the turbine is the same as your answer to **(a)(ii)**.
  (ii) After colliding with the blades of the turbine the water moves in the same direction at a speed of 10.0 m s^{-1}. Estimate the maximum possible force that the water could exert on the turbine blades.
  (iii) Estimate the maximum possible power imparted to the turbine.

(c) When a jet of water hits a flat blade it tends to spread as shown in Figure 2. Suggest why turbine blades are usually shaped to give the recoil flow shown in Figure 3.

Answer to application question

(a) (i) $\frac{1}{2}mv^2 = mgh$ or $\frac{1}{2}v^2 = 9.8 \times 300$
velocity = 77 m s^{-1} (✓)
(ii) Mass (per s) = volume (per s) × density (✓)
Volume per s = 2.8×10^{-3} m^3 (✓)
= 220 kg s^{-1} (✓)
(iii) Power available = KE per s (✓)
= 650 kW (✓)

Note: You could gain full marks for a correct method and workings in parts **(ii)** and **(iii)** if you made errors in previous parts.

(b) (i) All the water that enters the turbine must leave it otherwise there would be a build up of water. (✓)
(ii) Force = rate of change of momentum (✓)
OR 220 × (77 – 10)
Force = 14.7 kN (✓)
(iii) Maximum power output = loss of KE per second (✓)
$= \frac{1}{2} \times$ mass flow rate × {(initial velocity)^2 – (final velocity)^2} (✓)
OR
$= \frac{1}{2} \times 220 \times \{77^2 - 10^2\}$
= 640 kW (✓)

Note: Throughout part **(b)** errors may be carried forward from earlier answers.

(c) Using the system in Figure 3 the *change in momentum* is greater. (✓)
This results in a greater force on the wheel. (✓)

Comprehension type

Comprehension passages are used to test whether you can use your knowledge of physics to make sense of an article relating to a context that is likely to be unfamiliar to you. Most comprehension questions also include some data analysis.

Questions may require you to
- extract information that is given directly in the article
- use data in the article to deduce further information or deduce whether it agrees with a given law
- use your knowledge and understanding of physics to confirm that the data that is given in the article is sensible
- show that you have a broad understanding of physics and its applications that is relevant to the article.

Example comprehension (AEB 1994)
Photovoltaic Solar Energy Systems

Based on an article by Gian-Mattia Schucan (Switzerland), Young Researcher, *European Journal of Science and Technology*, September 1991.

1. One means of converting the Sun's energy directly into electrical energy is by photovoltaic cells.
2. In 1989 photovoltaic installations in Switzerland provided approximately 4.0×10^5 kW h of electricity, sufficient for 100 households. It is hoped that 3.0×10^9 kW h of electrical energy per year will be produced by photovoltaic installations by the year 2025. This is about seven per cent of Switzerland's present annual energy consumption.
3. The yield (output) of a photovoltaic installation is determined by technical and environmental influences. The technical factors are summarised in Figure 1.
4. Single solar cells are interconnected electrically to form a solar panel. A typical panel has an area of $\frac{1}{3}$ m^2 and an output of 50 W under standard test conditions which correspond to 1000 W m^{-2} of solar radiation and 25 °C cell temperature. The electrical characteristics of a larger panel are given in Figure 2.
5. Panels are connected together in series and parallel to form a Solar Cell Field, and a Maximum Power Tracker adjusts the Field to its optimum operating point. In order to change the direct current from the solar panels into alternating current for use in the country's power transmission system a device known as an inverter is used.
6. Figure 3 shows a weatherproof photovoltaic solar module suitable for experiments in schools and colleges. Its nominal output is 6 V, 0.3 W, rising to a maximum of about 8 V, 0.5 W.

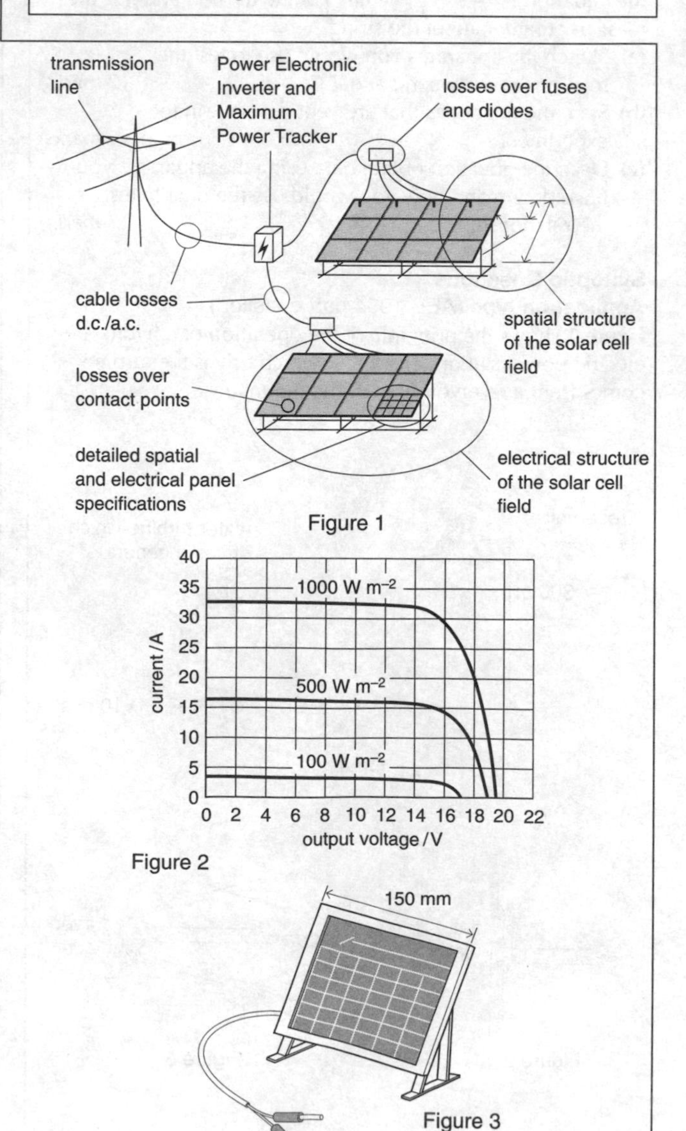

Figure 1

Figure 2

Figure 3

Questions

1 Using the information in Paragraph 2 estimate:
(a) the annual energy consumption in kW h in Switzerland in 1991; *(2 marks)*
(b) the number of Swiss households which could be powered by energy generated from photovoltaic installations in the year 2025. State any assumptions made. *(3 marks)*

2 Using data in Figure 2 determine whether the output current is directly proportional to the solar irradiation in W m^{-2}, for a photovoltaic solar panel operating up to 14 V. *(4 marks)*

3 This question is about the characteristic A in Figure 2.
(a) (i) What is the current when the output voltage is 12.0 V? *(1 mark)*
(ii) What is the output power when the output voltage is 12.0 V? *(2 marks)*
(iii) Draw up a table showing the output power and corresponding output voltages, for output voltages between 12.0 V and 18.0 V. *(2 marks)*
(iv) Plot a graph of output power (*y-axis*) against output voltage (*x-axis*). *(6 marks)*
(v) Use your graph to determine the maximum output power and the corresponding output voltages. *(2 marks)*
(b) From the information given in Paragraph 4, estimate the area of the solar panel which was used for producing Figure 2. *(3 marks)*
(c) What is the maximum efficiency of this panel? *(3 marks)*

4 Why is alternating current used in power transmission systems? *(3 marks)*

5 Suggest three environmental factors which will affect the power output from a particular panel. *(3 marks)*

6 Draw a circuit which would enable you to measure the output power, on a hot summer's day, of the module shown in Figure 3 and described in Paragraph 6. Give the ranges of any meters used and the values of any components in your circuit, showing all relevant calculations. *(6 marks)*

Useful tips for comprehension passage

- Read the passage carefully.
- Questions frequently refer to particular lines in the passage. When answering a question highlight or underline such references.
- Data is not always easy to keep in mind when in a long sentence. Make a note of any data you consider relevant to the question in a form that is easier to use. Make a list.
- Use number of marks per question to judge the detail required in an answer.

Answers to comprehension questions

1 (a) 7% of 1991 consumption = 3.0×10^9 kW h (✓)
1991 consumption = $(100/7) \times 3.0 \times 10^9$
= 4.3×10^{10} W (✓)
(b) Each household uses $4.0 \times 10^5/100$ = 4000 kW h (✓)
3.0×10^9 kW h supplies $3.0 \times 10^9/4000$
= 750 000 households (✓)
assuming average electricity use per house is same in 2025 as in 1991. (✓)

2 Check whether I/P is constant: (✓)
For 100 W, $I = 3$ A $I/P = 0.030$
For 500 W $I = 15$ A $I/P = 0.030$
For 1000 W $I = 32$ A $I/P = 0.032$ (✓)
Within uncertainties reading from the graph I/P is constant and I is therefore proportional to P. (✓)

Note: This could also be shown by plotting a graph of I against P. This would produce a straight line through the origin.

3 (a) (i) 32 or 33 A (✓)
(ii) $P = VI$ (✓)
384 W or 396 W (✓)

Note: Strictly this should be rounded off to 2 significant figures.

(iii)

V/V	12	13	14	15	16	17	18
I/A	32	32	32	31	29	27	19
P_{out}/W	380	420	450	470	460	460	340

(✓✓) for complete table
(✓) e.g. only even voltages used
(iv) Sketch graph shown is general shape. This should be drawn accurately on graph paper.

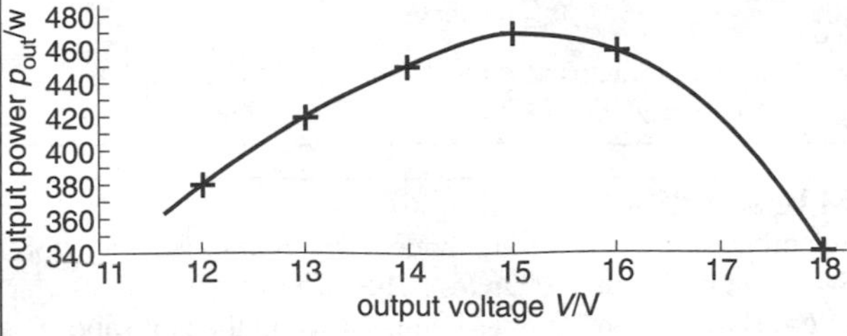

Axis labelled with units (✓)
Axis with scales shown (✓)
Good scale (✓)
Correct plotting (✓)
Smooth curve (✓)
(v) Peak around 470 W (✓)
when output voltage is about 15 V (✓)
(b) 50 W output corresponds to a panel of area 1/3 m^2. (✓)
470 W output requires a panel of area
$(470/50) \times 1/3$ (✓)
= 3.1 m^2 (✓)

Note: If you obtained an incorrect power in **(a)(v)** you could still gain full marks here if you use the correct method and working.

(c) Efficiency = output power/input power (✓)
Input power = 3.1×1000 W = 3200 W (✓)
Efficiency = 0.15 or 15% (✓)

Note: Again you could gain full marks even if you determined the area of the solar panel incorrectly in **(b)**.

4 You could give any three of the following or some other sensible comment that is relevant: (✓✓✓)
AC is easy to transform
Power loss in cables can be reduced by transforming
Currents in cable can be reduced
Power loss in cables = I^2R

5 You could give any three of the following or some other sensible comment that is relevant: (✓✓✓)
Weather conditions (rain cloud)
Shading by buildings or trees
Pollution in atmosphere
Dirt on panel

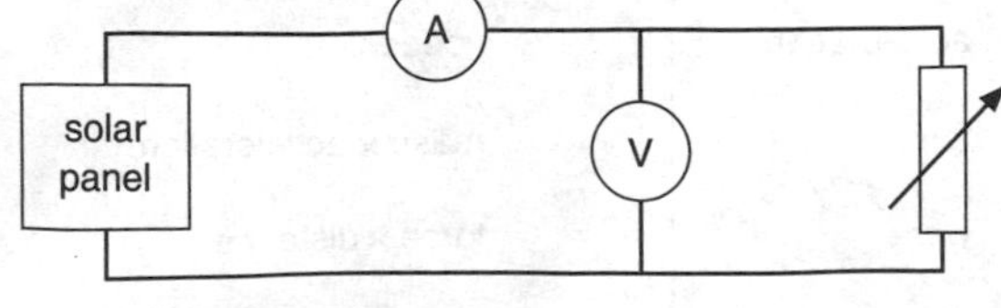

On diagram
Load resistor (✓)
Ammeter in series with load (✓)
Voltmeter across cell (or across load) (✓)
Clearly stated
Voltmeter range 0–10 V (✓)
Ammeter range 0–100 mA (✓)
Maximum current = 0.5/8 = 62 mA (✓)
Load resistance required about 130 Ω

Note: You would need to show at least one calculation (of load or current) to gain full marks.

A1 Units and dimensions

Physical quantity

Say a plank is 2 metres long. This measurement is called a ***physical quantity***. In this case, it is a length. It is made up of two parts:

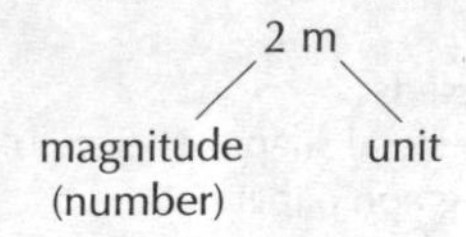

Note:

- '2 m' really means '2 × metre', just as, in algebra, $2y$ means '$2 \times y$'.

Physical quantity	Unit	
	Name	Symbol
length	metre	m
mass	kilogram	kg
time	second	s
current	ampere	A
temperature	kelvin	K
amount*	mole	mol

SI base units

Scientific measurements are made using SI units (standing for Système International d'Unités). The system starts with a series of ***base units***, the main ones being shown in the table above right. Other units are derived from these.

SI base units have been carefully defined so that they can be accurately reproduced using equipment available to national laboratories throughout the world.

* In science, 'amount' is a measurement based on the number of particles (atoms, ions or molecules) present. One mole is 6.02×10^{23} particles, a number which gives a simple link with the total mass. For example, 1 mole (6.02×10^{23} atoms) of carbon-12 has a mass of 12 grams. 6.02×10^{23} is called the ***Avogadro constant***.

SI derived units

There is no SI base unit for speed. However, speed is defined by an equation (see B1). If an object travels 12 m in 3 s,

$$\text{speed} = \frac{\text{distance travelled}}{\text{time taken}} = \frac{12\ \text{m}}{3\ \text{s}} = 4\,\frac{\text{m}}{\text{s}}$$

The units m and s have been included in the working above and treated like any other numbers or algebraic quantities. To save space, the final answer can be written as 4 m/s, or 4 m s^{-1}. (Remember, in maths, $1/x = x^{-1}$ etc.)

The unit m s^{-1} is an example of a ***derived SI unit***. It comes from a defining equation. There are other examples below. Some derived units are based on other derived units. And some derived units have special names. For example, 1 joule per second (J s^{-1}) is called 1 watt (W).

Prefixes

Prefixes can be added to SI base and derived units to make larger or smaller units.

Prefix	Symbol	Value	Prefix	Symbol	Value
pico	p	10^{-12}	kilo	k	10^{3}
nano	n	10^{-9}	mega	M	10^{6}
micro	µ	10^{-6}	giga	G	10^{9}
milli	m	10^{-3}	tera	T	10^{12}

For example,

1 mm = 10^{-3} m   1 km = 10^{3} m

Note:

- 1 gram (10^{-3} kg) is written '1 g' and not '1 mkg'.

Physical quantity	Defining equation (simplified)	Derived unit	Special symbol (and name)
speed	distance/time	m s^{-1}	–
acceleration	speed/time	m s^{-2}	–
force	mass × acceleration	kg m s^{-2}	N (newton)
work	force × distance	N m	J (joule)
power	work/time	J s^{-1}	W (watt)
pressure	force/area	N m^{-2}	Pa (pascal)
density	mass/volume	kg m^{-3}	–
charge	current × time	A s	C (coulomb)
voltage	energy/charge	J C^{-1}	V (volt)
resistance	voltage/current	V A^{-1}	Ω (ohm)

Dimensions

Here are three measurements:

length = 10 m   area = 6 m^2 volume = 4 m^3

These three quantities have ***dimensions*** of length, length squared, and length cubed.

Starting with three basic dimensions – length [L], mass [M], and time [T] – it is possible to work out the dimensions of many other physical quantities from their defining equations. There are examples on the right and below.

Example 1

$$\text{speed} = \frac{\text{distance travelled}}{\text{time taken}} = \frac{[L]}{[T]} = [LT^{-1}]$$

So the dimensions of speed are $[LT^{-1}]$.

Example 2

$$\text{density} = \frac{\text{mass}}{\text{volume}} = \frac{[M]}{[L^3]} = [ML^{-3}]$$

So the dimensions of density are $[ML^{-3}]$.

Physical quantity	Defining equation (simplified)	Dimensions		In terms of base units
		from equation	reduced form	
length	–	–	[L]	m
mass	–	–	[M]	kg
time	–	–	[T]	s
speed	$\frac{\text{distance}}{\text{time}}$	$\frac{[L]}{[T]}$	$[LT^{-1}]$	m s^{-1}
acceleration	$\frac{\text{speed}}{\text{time}}$	$\frac{[LT^{-1}]}{[T]}$	$[LT^{-2}]$	m s^{-2}
force	mass × acceleration	$[M] \times [LT^{-2}]$	$[MLT^{-2}]$	kg m s^{-2}
work	force × distance	$[MLT^{-2}] \times [L]$	$[ML^2T^{-2}]$	kg m^2 s^{-2}
power	$\frac{\text{work}}{\text{time}}$	$\frac{[ML^2T^{-2}]}{[T]}$	$[ML^2T^{-3}]$	kg m^2 s^{-3}
pressure	$\frac{\text{force}}{\text{area}}$	$\frac{[MLT^{-2}]}{[L^2]}$	$[ML^{-1}T^{-2}]$	kg m^{-1} s^{-2}

Using dimensions or base units to check equations

Each term in the two sides of an equation must always have the same units or dimensions. For example,

$$\text{work} = \text{force} \times \text{distance moved}$$
$$[ML^2T^{-2}] = [MLT^{-2}] \times [L]$$
$$= [ML^2T^{-2}]$$

An equation cannot be accurate if the dimensions on both sides do not match. It would be like claiming that '6 apples equals 6 oranges'.

Dimensions are a useful way of checking that an equation is reasonable.

Example *Check whether the equation PE = mgh is dimensionally correct.*

To do this, start by working out the dimensions of the right-hand side:

$$mgh = [M] \times [LT^{-2}] \times [L] = [ML^2T^{-2}]$$

These are the dimensions of work, and therefore of energy. So the equation is dimensionally correct.

Note:

- A dimensions check cannot tell you whether an equation is accurate. For example, both of the following are dimensionally correct, but only one is right:

  PE = mgh PE = $2mgh$

Dimensionless numbers

A pure number, such as 6, has no dimensions. Here are two consequences of this fact.

Dimensions and units of frequency The frequency of a vibrating source is defined as follows:

$$\text{frequency} = \frac{\text{number of vibrations}}{\text{time taken}}$$

As number is dimensionless, the dimensions of frequency are $[T^{-1}]$. The SI unit of frequency in the hertz (Hz):

1 Hz = 1 s^{-1}

Dimensions and units of angle

On the right, the angle θ in ***radians*** is defined like this:

$$\theta = \frac{s}{r}$$

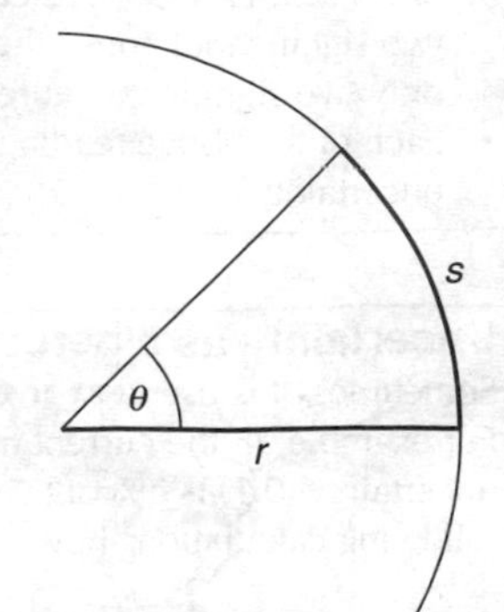

s/r has no dimensions because $[L] \times [L^{-1}] = 1$. However, when measuring an angle in radians, a unit is often included for clarity: 2 rad, for example.

A2 Measurements, uncertainties, and graphs

Scientific notation

The average distance from the Earth to the Sun is 150 000 000 km.

There are two problems with quoting a measurement in the above form:

- the inconvenience of writing so many noughts,
- uncertainty about which figures are important (i.e. How approximate is the value? How many of the figures are significant?).

These problems are overcome if the distance is written in the form 1.50×10^8 km.

'1.50×10^8' tells you that there are three significant figures – 1, 5, and 0. The last of these is the least significant and, therefore, the most uncertain. The only function of the other zeros in 150 000 000 is to show how big the number is. If the distance were known less accurately, to two significant figures, then it would be written as 1.5×10^8 km.

Numbers written using powers of 10 are in ***scientific notation*** or ***standard form***. This is also used for small numbers. For example, 0.002 can be written as 2×10^{-3}.

Uncertainty

When making any measurement, there is always some ***uncertainty*** in the reading. As a result, the measured value may differ from the true value. In science, an uncertainty is sometimes called an ***error***. However, it is important to remember that it is *not* the same thing as a mistake.

In experiments, there are two types of uncertainty.

Systematic uncertainties These occur because of some inaccuracy in the measuring system or in how it is being used. For example, a timer might run slow, or the zero on an ammeter might not be set correctly.

There are techniques for eliminating some systematic uncertainties. However, this spread will concentrate on dealing with uncertainties of the random kind.

Random uncertainties These can occur because there is a limit to the sensitivity of the measuring instrument or to how accurately you can read it. For example, the following readings might be obtained if the same current was measured repeatedly using one ammeter:

2.4 2.5 2.4 2.6 2.5 2.6 2.6 2.5

Because of the uncertainty, there is variation in the last figure. To arrive at a single value for the current, you could find the mean of the above readings, and then include an estimation of the uncertainty:

current = 2.5 ± 0.1

(2.5 = mean; 0.1 = uncertainty)

Writing '2.5 ± 0.1' indicates that the value could lie anywhere between 2.4 and 2.6.

Note:

- On a calculator, the mean of the above readings works out at 2.5125. However, as each reading was made to only two significant figures, the mean should also be given to only two significant figures i.e. 2.5.
- Each of the above readings may also include a systematic uncertainty.

Uncertainty as a percentage

Sometimes, it is useful to give an uncertainty as a percentage. For example, in the current measurement above, the uncertainty (0.1) is 4% of the mean value (2.5), as the following calculation shows:

$$\text{percentage uncertainty} = \frac{0.1}{2.5} \times 100 = 4$$

So the current reading could be written as 2.5 ± 4%.

Combining uncertainties

Sums and differences Say you have to *add* two length readings, *A* and *B*, to find a total, *C*. If $A = 3.0 \pm 0.1$ and $B = 2.0 \pm 0.1$, then the minimum possible value of *C* is 4.8 and the maximum is 5.2. So $C = 5.0 \pm 0.2$.

Now say you have to subtract *B* from *A*. This time, the minimum possible value of *C* is 0.8 and the maximum is 1.2 . So $C = 1.0 \pm 0.2$, and the uncertainty is the same as before.

If $C = A + B$ or $C = A - B$, then

uncertainty in *C* = uncertainty in *A* + uncertainty in *B*

The same principle applies when several quantities are added or subtracted: $C = A + B - F - G$, for example.

Products and quotients If $C = A \times B$ or $C = A/B$, then

% uncertainty in *C* = % uncertainty in *A* + % uncertainty in *B*

For example, say you measure a current *I*, a voltage *V*, and calculate a resistance *R* using the equation $R = V/I$. If there is a 3% uncertainty in *V* and a 4% uncertainty in *I*, then there is a 7% uncertainty in your calculated value of *R*.

Note:

- The above equation is only an approximation – and a poor one for uncertainties greater than about 10%.
- To check that the equation works, try calculating the maximum and minimum values of *C* if, say, *A* is 100 ± 3 and *B* is 100 ± 4. You should find that $A \times B$ is 10 000 ± approximately 700 (i.e. 7%).
- The principle of adding % uncertainties can be applied to more complex equations: $C = A^2B/FG$, for example. As $A^2 = A \times A$, the % uncertainty in A^2 is twice that in *A*.

Calculated results

Say you have to calculate a resistance from the following readings:

voltage = 3.3 V (uncertainty ± 0.1 V, or ± 3%)
current = 2.5 A (uncertainty ± 0.1 A, or ± 4%)

Dividing the voltage by the current on a calculator gives a resistance of 1.32 Ω. However, as the combined uncertainty is ±7%, or ± 0.1 Ω, the calculated value of the resistance should be written as 1.3 Ω. As a general guideline, a calculated result should have no more significant figures than any of the measurements used in the calculation. (However, if the result is to be used in further calculations, it is best to leave any rounding up or down until the end.)

Choosing a graph

The general equation for a straight-line graph is

$$y = mx + c$$

In this equation, m and c are ***constants***, as shown below. y and x are ***variables*** because they can take different values. x is the ***independent variable***. y is the ***dependent variable***: its value depends on the value of x.

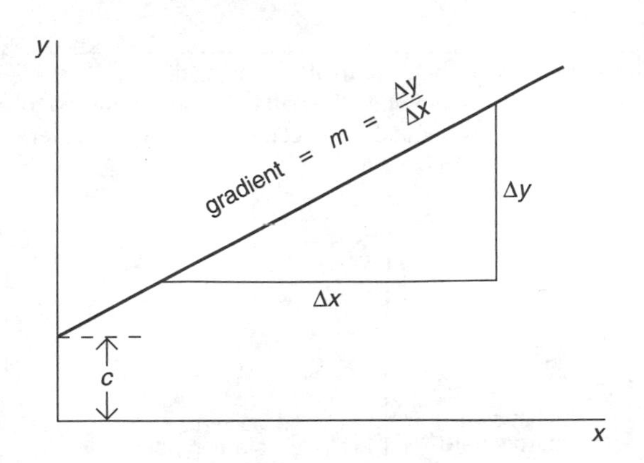

In experimental work, straight-line graphs are especially useful because the values of constants can be found from them. Here is an example.

Problem Theoretical analysis shows that the period T (time per swing) of a simple pendulum is linked to its length l, and the Earth's gravitational field strength g by the equation $T = 2\pi\sqrt{l/g}$. If, by experiment, you have corresponding values of l and T, what graph should you plot in order to work out a value for g from it?

Answer First, rearrange the equation so that it is in the form $y = mx + c$. Here is one way of doing this:

$$\underbrace{T^2}_{y} = \underbrace{\frac{4\pi^2}{g}}_{m} \underbrace{l}_{x} + \underbrace{0}_{c}$$

So, if you plot a graph of T^2 against l, the result should be a straight line through the origin (as $c = 0$). The gradient (m) is $4\pi^2/g$, from which a value of g can be calculated.

Showing uncertainties on graphs

In an experiment, a wire is kept at a constant temperature. You apply different voltages across the wire and measure the current through it each time. Then you use the readings to plot a graph of current against voltage.

The general direction of the points suggests that the graph is a straight line. However, before reaching this conclusion, you must be sure that the points' scatter is due to random uncertainty in the current readings. To check this, you could estimate the uncertainty and show this on the graph using short, vertical lines called uncertainty bars. The ends of each bar represent the likely maximum and minimum value for that reading. In the example below, the ***uncertainty bars*** show that, despite the points' scatter, it is reasonable to draw a straight line through the origin.

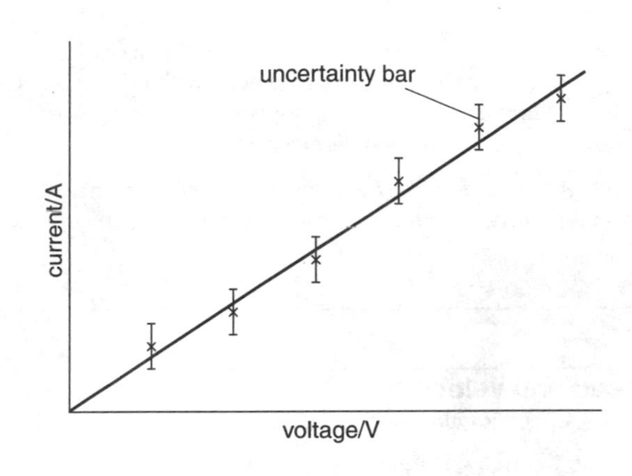

Labelling graph axes Strictly speaking, the scales on the graph's axes are pure, unitless numbers and not voltages or currents. Take a typical reading:

voltage = 10 V

This can be treated as an equation and rearranged to give:

voltage/V = 10

That is why the graph axes are labelled 'voltage/V' and 'current/A'. The values of these are pure numbers.

Reading a micrometer

The length of a small object can be measured using a micrometer screw gauge. You take the reading on the gauge like this:

Reading a vernier

Some measuring instruments have a vernier scale on them for measuring small distances (or angles). You take the reading like this:

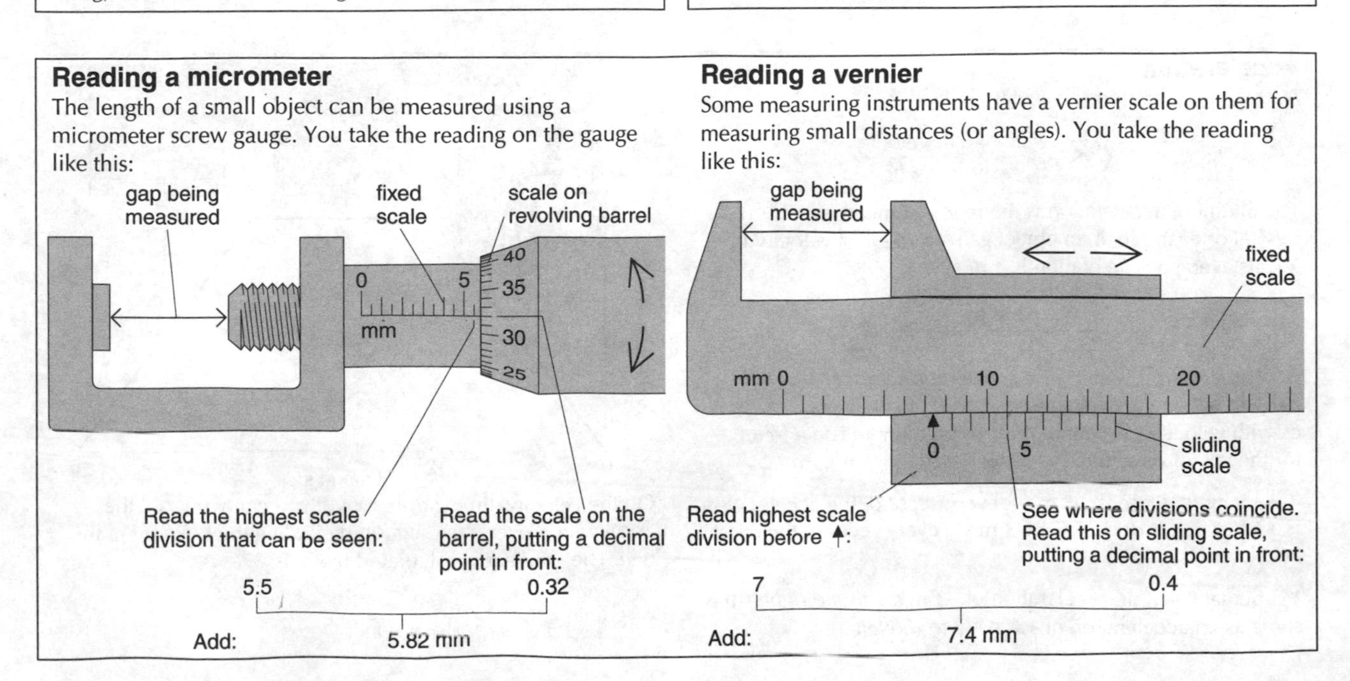

B1 Motion, mass, and forces

Units of measurement

Scientists make measurements using SI units such as the metre, kilogram, second, and newton. These and their abbreviations are covered in detail in A1. However, you may find it easier to appreciate the links between different units after you have studied the whole of section A.

For simplicity, units will be excluded from some stages of the calculations in this book, as in this example:

total length = 2 + 3 = 5 m

Strictly speaking, this should be written

total length = 2 m + 3 m = 5 m

Displacement

Displacement is distance moved in a particular direction. The SI unit of displacement is the ***metre*** (m).

Quantities, such as displacement, that have both magnitude (size) and direction are called ***vectors***.

A ——12 m——> B

The arrow above represents the displacement of a particle which moves 12 m from A to B. However, with horizontal or vertical motion, it is often more convenient to use a '+' or '–' to show the vector direction. For example,

Movement of 12 m *to the right*: displacement = +12 m
Movement of 12 m *to the left*: displacement = –12 m

Displacement is not necessarily the same as distance travelled. For example, when the ball below has returned to its starting point, its vertical displacement is zero. However, the distance travelled is 10 m.

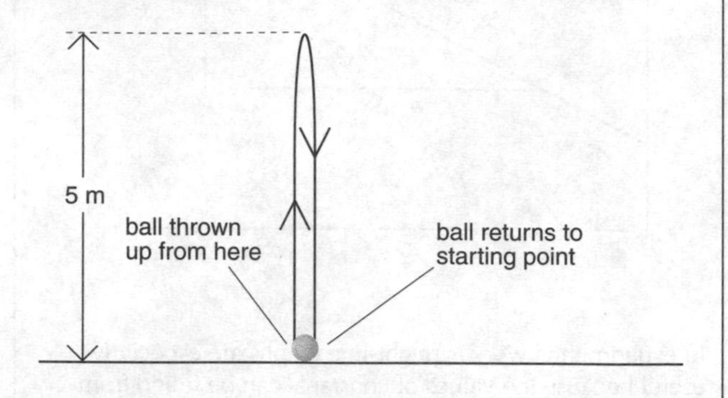

Speed and velocity

Average speed is calculated like this:

$$\text{average speed} = \frac{\text{distance travelled}}{\text{time taken}}$$

The SI unit of speed is the metre/second, abbreviated as m s^{-1}. For example, if an object travels 12 m in 2 s, its average speed is 6 m s^{-1}.

Average velocity is calculated like this:

$$\text{average velocity} = \frac{\text{displacement}}{\text{time taken}}$$

The SI unit of velocity is also the m s^{-1}. But unlike speed, velocity is a vector.

——6 m s^{-1}——>

The velocity vector above is for a particle moving to the right at 6 m s^{-1}. However, as with displacement, it is often more convenient to use a '+' or '–' for the vector direction.

Average velocity is not necessarily the same as average speed. For example, if a ball is thrown upwards and travels a total distance of 10 m before returning to its starting point 2 s later, its average speed is 5 m s^{-1}. But its average velocity is zero, because its displacement is zero.

Acceleration

Average acceleration is calculated like this:

$$\text{average acceleration} = \frac{\text{change in velocity}}{\text{time taken}}$$

The SI unit of acceleration is the m s^{-2} (sometimes written m/s^2). For example, if an object gains 6 m s^{-1} of velocity in 2 s, its average acceleration is 3 m s^{-2}.

——3 m s^{-2}——>>

Acceleration is a vector. The acceleration vector above is for a particle with an acceleration of 3 m s^{-2} to the right. However, as with velocity, it is often more convenient to use a '+' or '–' for the vector direction.

If velocity *increases* by 3 m s^{-1} every second, the acceleration is +3 m s^{-2}. If it *decreases* by 3 m s^{-1} every second, the acceleration is –3 m s^{-2}.

Mathematically, an acceleration of –3 m s^{-2} *to the right* is the same as an acceleration of +3 m s^{-2} *to the left.*

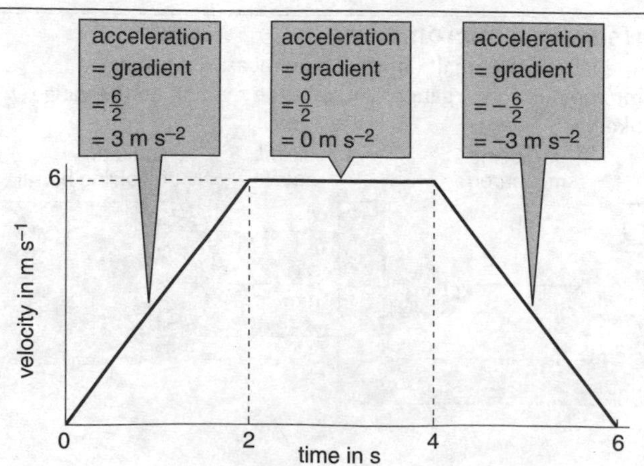

On the velocity–time graph above, you can work out the acceleration over each section by finding the *gradient* of the line. The gradient is calculated like this:

$$\text{gradient} = \frac{\text{gain along } y\text{-axis}}{\text{gain along } x\text{-axis}}$$

Force

Force is a vector. The SI unit is the ***newton*** (**N**).

If two or more forces act on something, their combined effect is called the ***resultant force***. Two simple examples are shown below. In the right-hand example, the resultant force is zero because the forces are ***balanced***.

A resultant force acting on a mass causes an acceleration. The force, mass, and acceleration are linked like this:

resultant force = mass × acceleration $F = ma$

For example, a 1 N resultant force gives a 1 kg mass an acceleration of 1 m s^{-2}. (The newton is defined in this way.)

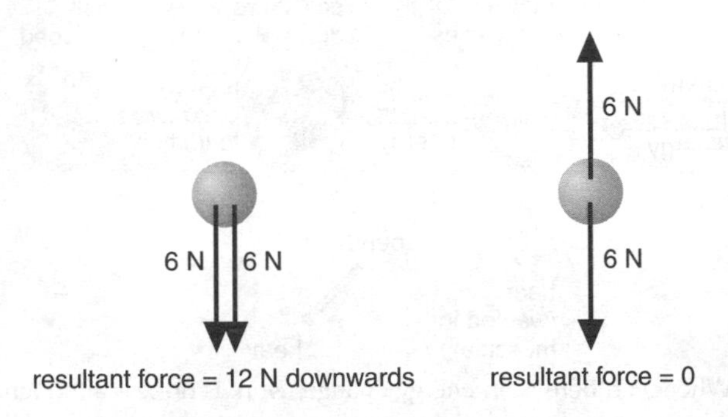

The more mass something has, the more force is needed to produce any given acceleration.

When balanced forces act on something, its acceleration is zero. This means that it is *either* stationary *or* moving at a steady velocity (steady speed in a straight line).

Weight and *g*

On Earth, everything feels the downward force of gravity. This gravitational force is called ***weight***. As for other forces, its SI unit is the newton (N).

Near the Earth's surface, the gravitational force on each kg is about 10 N: the ***gravitational field strength*** is 10 N kg^{-1}. This is represented by the symbol *g*.

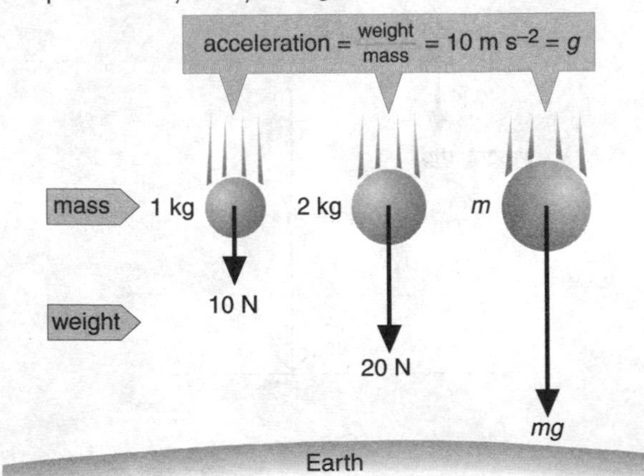

In the diagram above, all the masses are falling freely (gravity is the only force acting). From $F = ma$, it follows that all the masses have the same downward acceleration, *g*. This is the ***acceleration of free fall***.

You can think of *g*

either as a gravitational field strength of 10 N kg^{-1}

or as an acceleraton of free fall of 10 m s^{-2}.

In more accurate calculations, the value of *g* is normally taken to be 9.81, rather than 10.

Moments and balance

The turning effect of a force is called a ***moment***:

moment of force about a point = force × perpendicular distance* from point

* measured from the line of action of the force.

The dumb-bell below balances at point O because the two moments about O are equal but opposite.

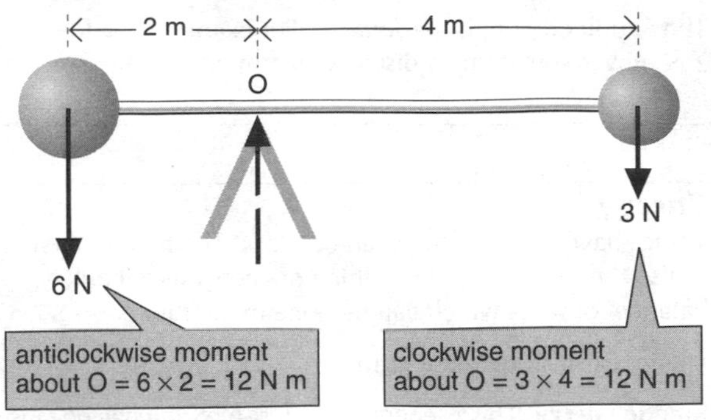

The dumb-bell is made up of smaller parts, each with its own weight. Together, these are equivalent to a single force, the total weight, acting through O. O is the ***centre of gravity*** of the dumb-bell.

Density

The density of an object is calculated like this:

$$\text{density} = \frac{\text{mass}}{\text{volume}}$$

The SI unit of density is the kilogram/cubic metre (kg m^{-3}).

For example, 2000 kg of water occupies a volume of 2 m^3. So the density of water is 1000 kg m^{-3}.

Density values, in kg m^{-3}

alcohol	800	iron	7 900
aluminium	2 700	lead	11 300

Pressure

Pressure is calculated like this:

$$\text{pressure} = \frac{\text{force}}{\text{area}}$$

The SI unit of pressure is the newton/square metre, also called the ***pascal*** (**Pa**). For example, if a force of 12 N acts over an area of 3 m^2, the pressure is 4 Pa.

Liquids and gases are called ***fluids***.

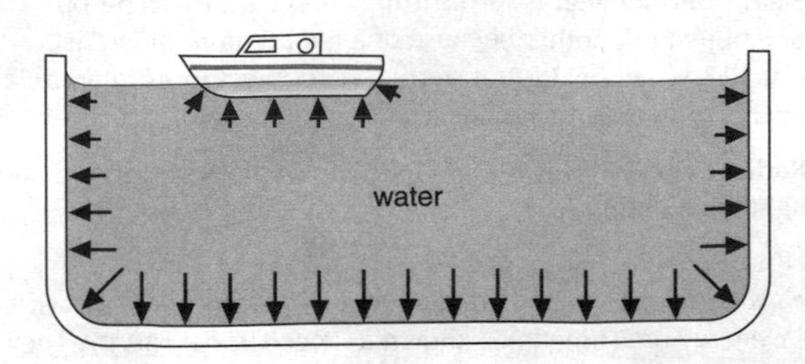

In a fluid:

- Pressure acts in all directions. The force produced is always at right-angles to the surface under pressure.
- Pressure increases with depth.

B2 Work, energy, and power

Work

Work is done whenever a force makes something move.
It is calculated like this:

work done = force × distance moved in direction of force

The SI unit of work is the ***joule*** (J). For example, if a force of 2 N moves something a distance of 3 m, then the work done is 6 J.

Energy

Things have energy if they can do work. The SI unit of energy is also the joule (J). You can think of energy as a 'bank balance' of work which can be done in the future.

Energy exists in different forms:

Kinetic energy This is energy which something has because it is moving.

Potential energy This is energy which something has because of its position, shape, or state. A stone about to fall from a cliff has ***gravitational*** potential energy. A stretched spring has ***elastic*** potential energy. Foods and fuels have ***chemical*** potential energy. Charge from a battery has ***electrical*** potential energy. Particles from the nucleus (centre) of an atom have ***nuclear*** potential energy.

Internal energy Matter is made up of tiny particles (e.g. atoms or molecules) which are in random motion. They have kinetic energy because of their motion, and potential energy because of the forces of attraction trying to pull them together. An object's internal energy is the total kinetic and potential energy of its particles.

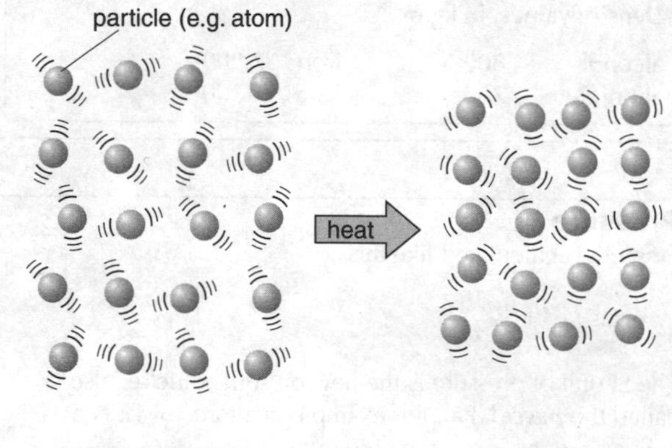

Heat (thermal energy) This is the energy transferred from one object to another because of a temperature difference. Usually, when heat is transferred, one object loses internal energy, and the other gains it.

Radiant energy This is often in the form of waves. Sound and light are examples.

Note:

- Kinetic energy, and gravitational and elastic potential energy are sometimes known as ***mechanical energy***. They are the forms of energy most associated with machines and motion.
- Gravitational potential energy is sometimes just called potential energy (or PE), even though there are other forms of potential energy as described above.

Energy changes

According to the ***law of conservation of energy***,

energy cannot be made or destroyed, but it can be changed from one form to another.

The diagram below shows the sequence of energy changes which occur when a ball is kicked along the ground. At every stage, energy is lost as heat. Even the sound waves heat the air as they die away. As in other energy chains, all the energy eventually becomes internal energy.

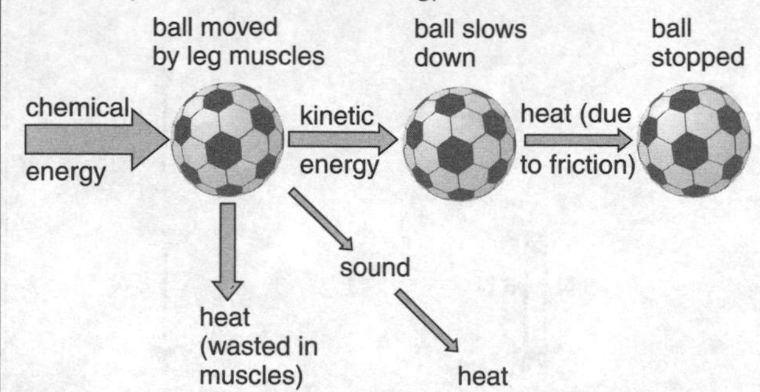

Whenever there is an energy change, work is done – although this may not always be obvious. For example, when a car's brakes are applied, the car slows down and the brakes heat up, so kinetic energy is being changed into internal energy. Work is done because tiny forces are making the particles of the brake materials move faster.

An energy change is sometimes called an energy transformation. Whenever it takes place,

work done = energy transformed

So, for each 1 J of energy transformed, 1 J of work is done.

Calculating potential energy (PE)

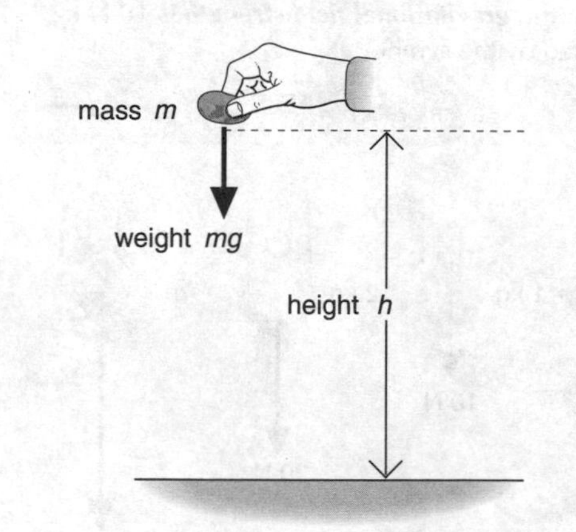

The stone above has potential energy. This is equal to the work done in lifting it to a height *h* above the ground.

The stone, mass *m*, has a weight of *mg*.
So the force needed to overcome gravity and lift it is *mg*.

As the stone is lifted through a height *h*,

work done = force × distance moved = $mg \times h$

So potential energy = mgh

For example, if a 2 kg stone is 5 m above the ground, and *g* is 10 N kg^{-1}, then the stone's PE = 2 × 10 × 5 = 100 J.

Calculating kinetic energy (KE)

The stone on the right has kinetic energy. This is equal to the work done in increasing the velocity from zero to v. B7 shows you how to calculate this. The result is

$$\text{kinetic energy} = \frac{1}{2}mv^2$$

For example, if a 2 kg stone has a speed of 10 m s^{-1}, its KE $= \frac{1}{2} \times 2 \times 10^2 = 100$ J

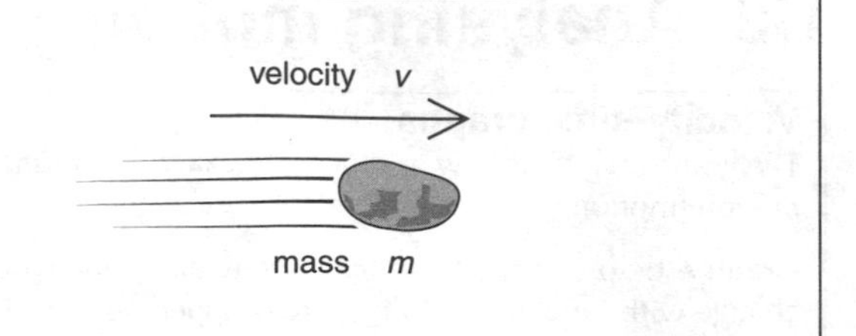

PE to KE

The diagram on the right shows how PE is changed into KE when something falls. The stone in this example starts with 100 J of PE. Air resistance is assumed to be zero, so no energy is lost to the air as the stone falls.

By the time the stone is about to hit the ground (with velocity v), all of its potential energy has been changed into kinetic energy. So

$$\frac{1}{2}mv^2 = mgh$$

Dividing both sides by m and rearranging gives

$$v = \sqrt{2gh}$$

In this example, $v = \sqrt{2 \times 10 \times 5} = 10$ m s^{-1}.

Note that v does not depend on m. A heavy stone hits the ground at exactly the same speed as a light one.

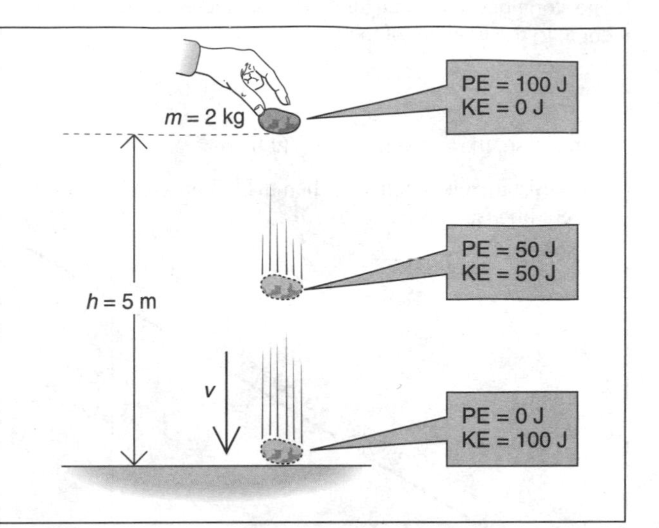

Vectors, scalars, and energy

Vectors have magnitude and direction. When adding vectors, you must allow for their direction. In B1, for example, there are diagrams showing two 6 N forces being added. In one, the resultant is 12 N. In the other, it is zero.

Scalars are quantities which have magnitude but no direction. Examples include mass, volume, energy, and work. Scalar addition is simple. If 6 kg of mass is added to 6 kg of mass, the result is always 12 kg. Similarly, if an object has 6 J of PE and 6 J of KE, the total energy is 12 J.

As energy is a scalar, PE and KE can be added without allowing for direction. The stone on the right has the same total PE + KE throughout its motion. As it starts with the same PE as the stone in the previous diagram, it has the same KE (and speed) when it is about to hit the ground.

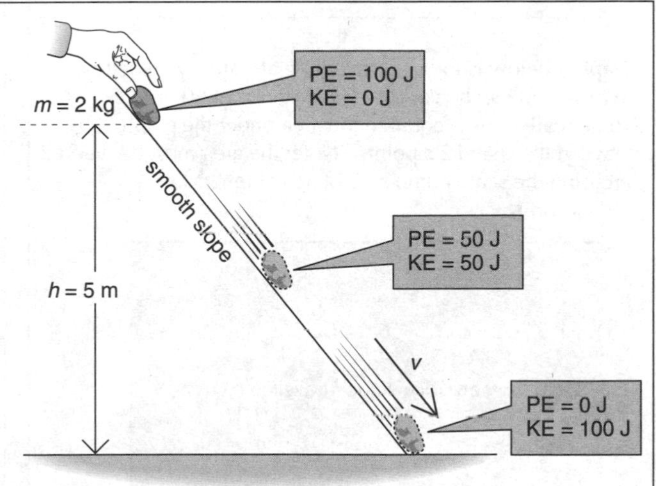

Power

Power is calculated like this:

$$\text{power} = \frac{\text{energy transferred}}{\text{time taken}} \quad \text{or} \quad \text{power} = \frac{\text{work done}}{\text{time taken}}$$

The SI unit of power is the **watt** (W). A power of 1 W means that energy is being transformed at the rate of 1 joule/second (J s^{-1}), so work is being done at the rate of 1 J s^{-1}.

Below, you can see how to calculate the power output of an electric motor which raises a mass of 2 kg through a height of 12 m in 3 s:

$$\text{PE gained} = mgh = 2 \times 10 \times 12 = 240 \text{ J}$$

$$\text{power} = \frac{\text{energy transferred}}{\text{time taken}} = \frac{240}{3} = 80 \text{ W}$$

motor

Efficiency

Energy changers such as motors waste some of the energy supplied to them. Their ***efficiency*** is calculated like this:

$$\text{efficiency} = \frac{\text{useful energy output}}{\text{energy input}} = \frac{\text{useful power output}}{\text{power input}}$$

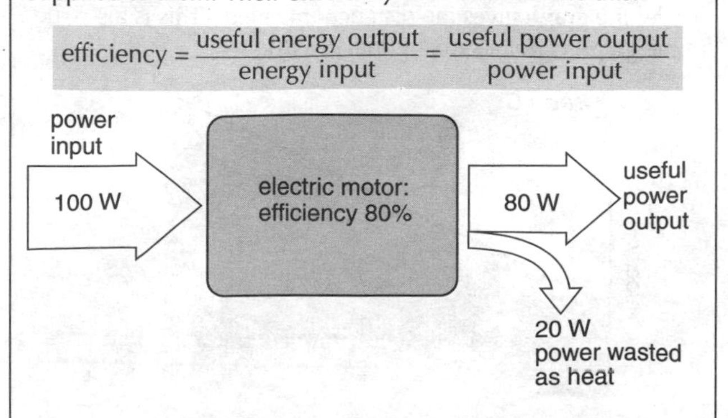

For example, if an electric motor's power input is 100 W, and its useful power output (mechanical) is 80 W, then its efficiency is 0.8. This can be expressed as 80%.

B3 Analysing motion

Velocity–time graphs

The graphs which follow are for three examples of ***linear*** motion (motion in a straight line).

Graph A below shows how the velocity of a stone would change with time, if the stone were dropped near the Earth's surface and there were no air resistance to slow it.
The stone has a *uniform* (unchanging) acceleration *a* which is equal to the gradient of the graph:

$$a = \frac{\Delta v}{\Delta t}$$

In this case, the acceleration is *g* ($9.81\ \text{m s}^{-2}$).

If air resistance is significant, then the graph is no longer a straight line (see B8).

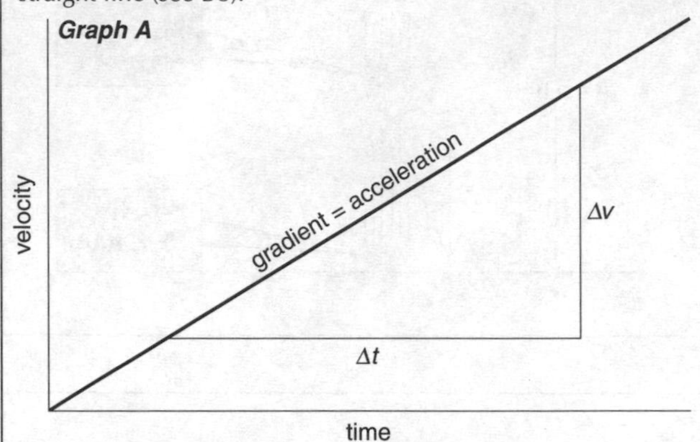

Graph B below is for a car travelling at a steady velocity of $30\ \text{m s}^{-1}$. In 2 s, the car travels a distance of 60 m. Numerically, this is equal to the area under the graph between the 0 and 2 s points. (Note: the area must be worked out using the scale numbers, not actual lengths.)

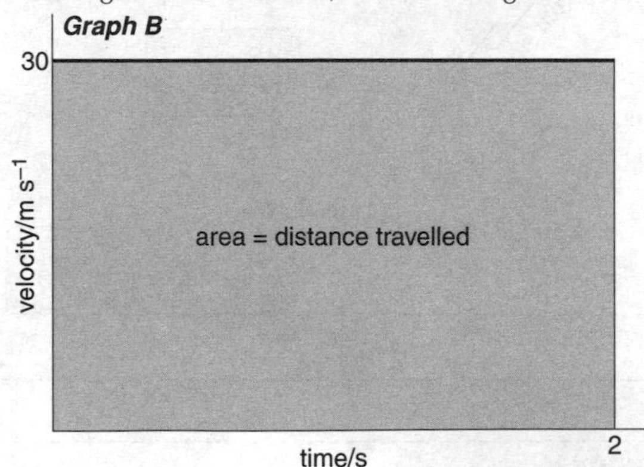

Graph C below is for a car with a changing velocity. However, the same principle applies as before: the area under the graph gives the distance travelled. (This is also true if the graph is not a straight line: see B8.)

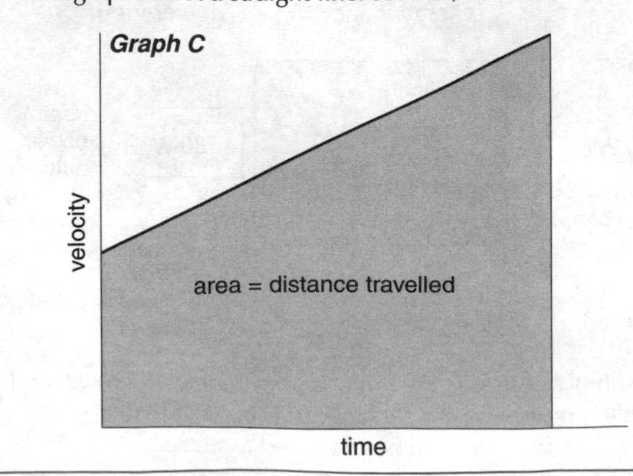

Equations of motion

The car below has uniform acceleration. In the following analysis, only motion between X and Y will be considered.

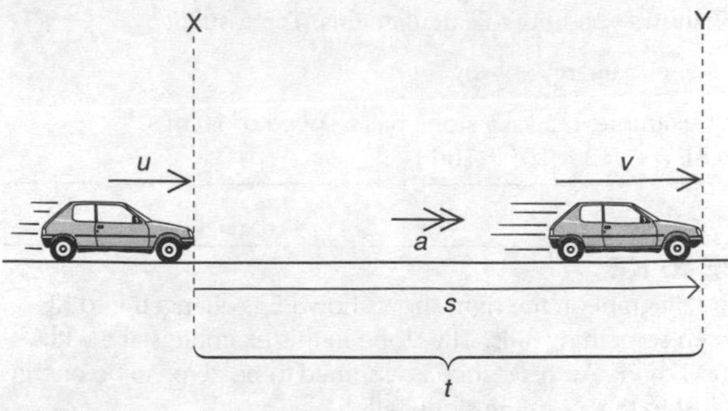

u = initial velocity (velocity on passing X)
v = final velocity (velocity on passing Y)
a = acceleration
s = displacement (in moving from X to Y)
t = time taken (to move from X to Y)

Here is a velocity–time graph for the car.

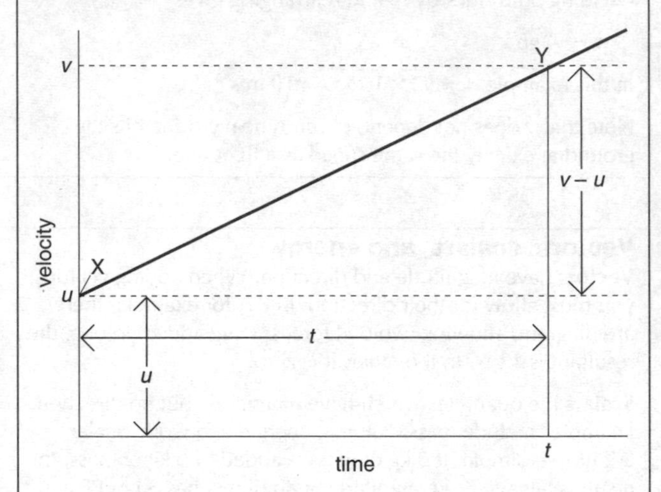

There are four equations (numbered 1–4 below) linking *u, v, a, s,* and *t*. They can be worked out as follows.

The acceleration is the gradient of the graph.
So $a = (v - u)/t$. This can be rearranged to give

$$v = u + at \qquad (1)$$

The distance travelled, *s* in this case, is the area under the graph. This is the area of one rectangle (height × base) plus the area of one triangle ($\frac{1}{2}$ × height × base). So it is $u \times t$ plus $\frac{1}{2} \times (v - u) \times t$. But $v - u = at$ from equation (1), so

$$s = ut + \tfrac{1}{2}at^2 \qquad (2)$$

As distance travelled = average velocity × time taken,

$$s = \tfrac{1}{2}(v + u)t \qquad (3)$$

If equations (1) and (3) are combined so that *t* is eliminated,

$$v^2 = u^2 + 2as \qquad (4)$$

Note:

- The equations are only valid for uniform acceleration.
- Each equation links a different combination of factors. You must decide which equation best suits the problem you are trying to solve.
- You must allow for vector directions. With horizontal motion, you might decide to call a vector to the right positive (+). With vertical motion, you might call a downward vector positive. So, for a stone thrown upwards at $30\ \text{m s}^{-1}$, $u = -30\ \text{m s}^{-1}$ and $g = +10\ \text{m s}^{-2}$.

Motion problems

Here are examples of how the equations of motion can be used to solve problems. For simplicity, units will not be shown in some equations. It will be assumed that air resistance is negligible and that g is 10 m s^{-2}.

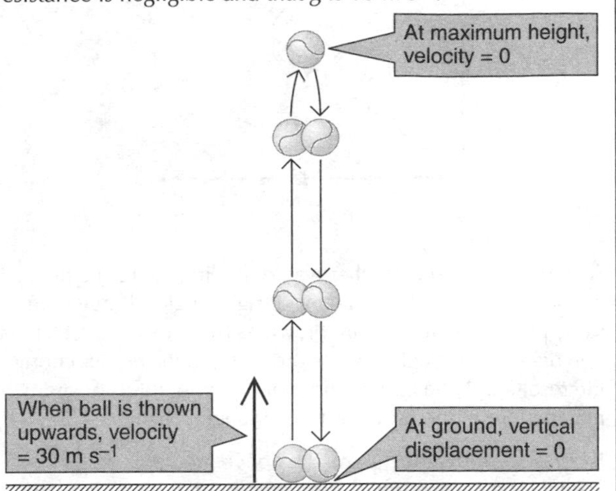

Example 1 *A ball is thrown upwards at 30 m s^{-1}. What time will it take to reach its highest point?*

The ball's motion only needs to be considered from when it is thrown to when it reaches its highest point. These are the 'initial' and 'final' states in any equation used.

When the ball is at it highest point, its velocity v will be zero. So, taking downward vectors as positive,

$u = -30$ m s^{-1} $\quad v = 0 \quad a = g = 10$ m s^{-2} $\quad t$ is to be found.

In this case, an equation linking u, v, a, and t is required. This is equation (1) on the opposite page:

$v = u + at$

So $0 = -30 + 10t$

Rearranged, this gives $t = 3.0$ s.

Example 2 *A ball is thrown upwards at 30 m s^{-1}. What is the maximum height reached?*

In this case,

$u = -30$ m s^{-1} $\quad v = 0 \quad a = g = 10$ m s^{-2} $\quad s$ is to be found.

This time, the equation required is (4) on the opposite page:

$v^2 = u^2 + 2as$

So $0 = (-30)^2 + (2 \times 10 \times s)$

This gives $s = -45$ m.

(Downwards is positive, so the negative value of s indicates an *upward* displacement.)

Example 3 *A ball is thrown upwards at 30 m s^{-1}. For what time is it in motion before it hits the ground?*

When the ball reaches the ground, it is back where it started, so its displacement s is zero. Therefore

$u = -30$ m s^{-1} $\quad s = 0 \quad a = g = 10$ m s^{-2} $\quad t$ is to be found.

This time, the equation required is (2) on the opposite page:

$s = ut + \frac{1}{2}at^2$

So $0 = (-30t) + (\frac{1}{2} \times 10 \times t^2)$

This gives $t = 6$ s.

(There is also a solution $t = 0$, indicating that the ball's displacement is also zero at the instant it is thrown.)

Measuring *g*

By measuring the time t it takes an object to fall through a measured height h, a value of g can be found (assuming that air resistance is negligible).

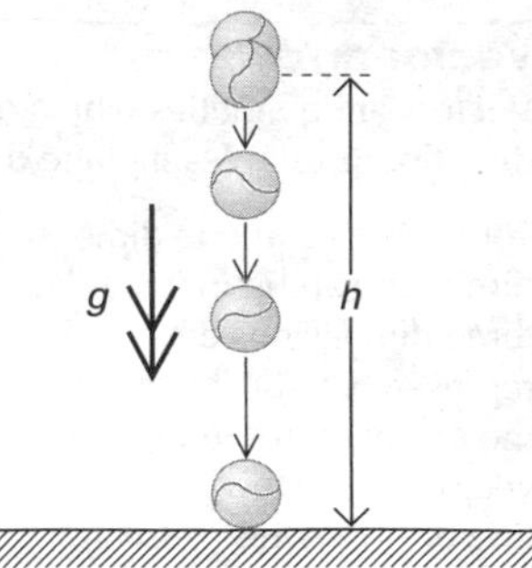

In the diagram on the right,
$u = 0 \quad a = g \quad s = h$

Applying equation (2) on the opposite page gives

$s = ut + \frac{1}{2}at^2$

So $h = 0t + \frac{1}{2}gt^2$

This gives $g = \dfrac{2h}{t^2}$.

Downwards and sideways

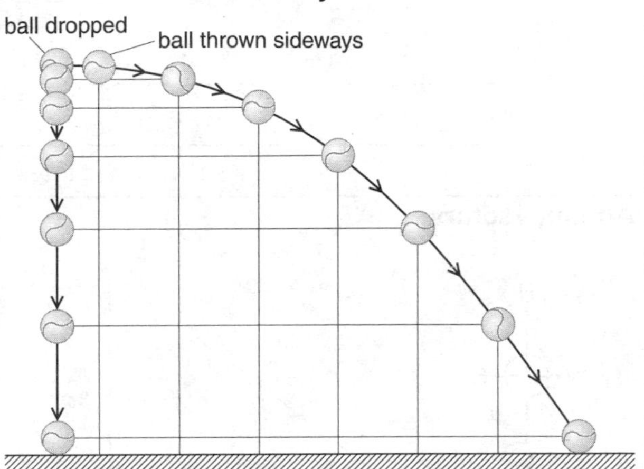

Above, one ball is dropped, while another is thrown sideways at the same time. There is no air resistance. The positions of the balls are shown at regular time intervals.

- Both balls hit the ground together. They have the same downward acceleration g.
- As it falls, the second ball moves sideways over the ground at a steady speed.

Results like this show that the vertical and horizontal motions are independent of each other.

Example *Below, a ball is thrown horizontally at 40 m s^{-1}. What horizontal distance does it travel before hitting the water? (Assume air resistance is negligible and g = 10 m s^{-2}).*

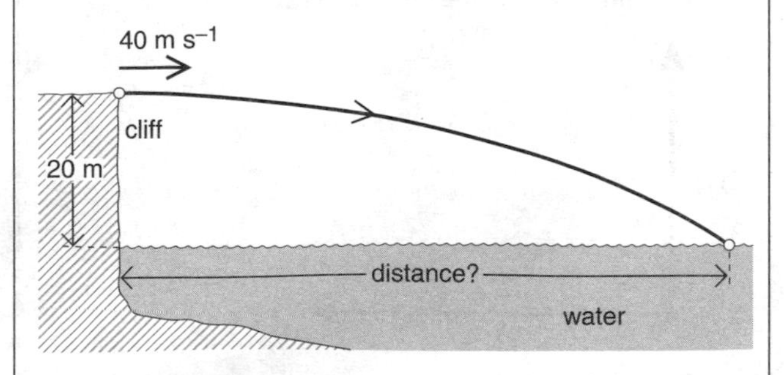

First, work out the time the ball would take to fall vertically to the sea. This can be done using the equation $s = ut + \frac{1}{2}at^2$, in which $u = 0$, $s = 20$ m, $a = g = 10$ m s^{-2}, and t is to be found. This gives $t = 2.0$ s.

Next, work out how far the ball will travel horizontally in this time (2 s) at a steady horizontal speed of 40 m s^{-1}.

As distance travelled = average speed × time, horizontal distance travelled = 40 × 2 = 80 m.

B4 Vectors

Vector arrows

Vectors are quantities which have both magnitude (size) and direction. Examples include displacement and force.

For problems in one dimension (e.g. vertical motion), vector direction can be indicated using + or –. But where two or three dimensions are involved, it is often more convenient to represent vectors by arrows, with the length and direction of the arrow representing the magnitude and direction of the vector. The arrowhead can either be drawn at the end of the line or somewhere else along it, as convenient. Here are two displacement vectors.

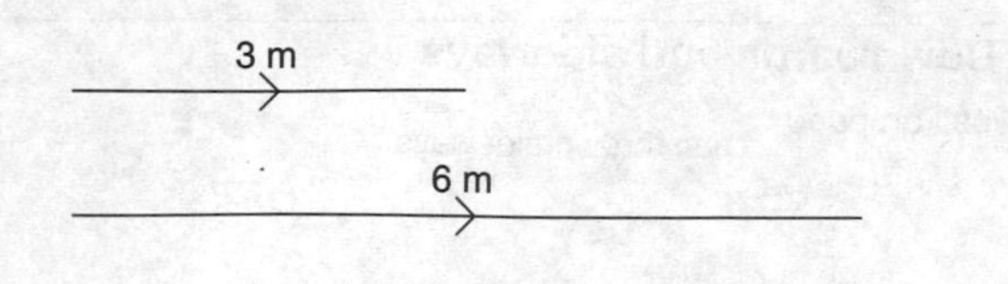

Adding vectors

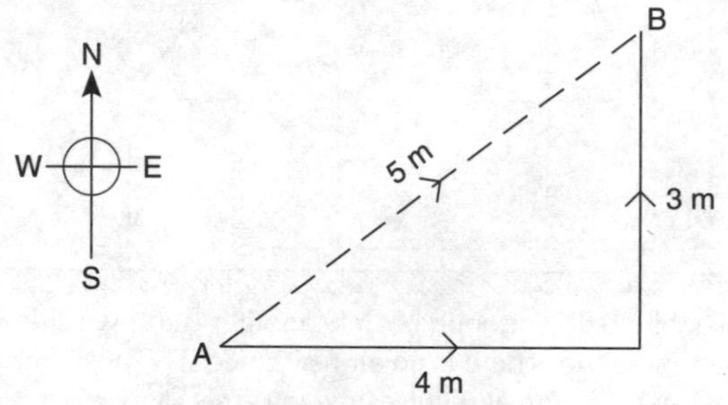

If someone starts at A, walks 4 m East and then 3 m North, they end up at B, as shown above. In this case, they are 5 m from where they started – a result which follows from Pythagoras' theorem. This is an example of vector addition. Two displacement vectors, of 3 m and 4 m, have been added to produce a ***resultant*** – a displacement vector of 5 m.

This principle works for any type of vector. Below, forces of 3 N and 4 N act at right-angles through the same point, O. The ***triangle of vectors*** gives their resultant. The vectors being added must be drawn head-to-tail. The resultant runs from the tail of the first arrow to the head of the second.

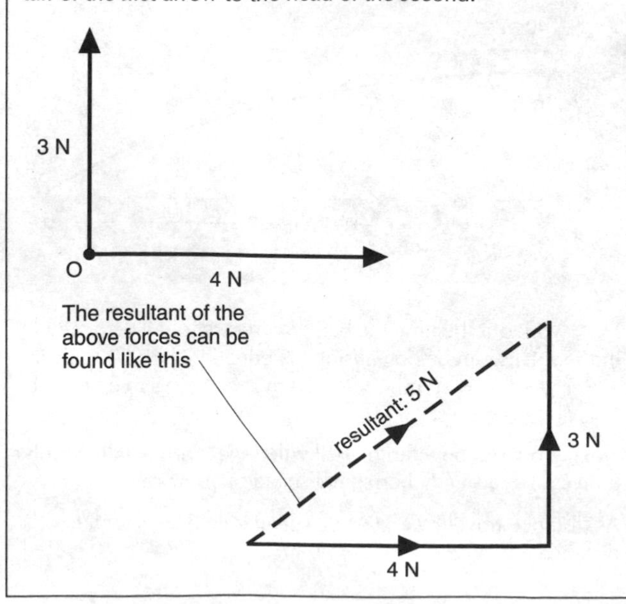

Parallelogram of vectors

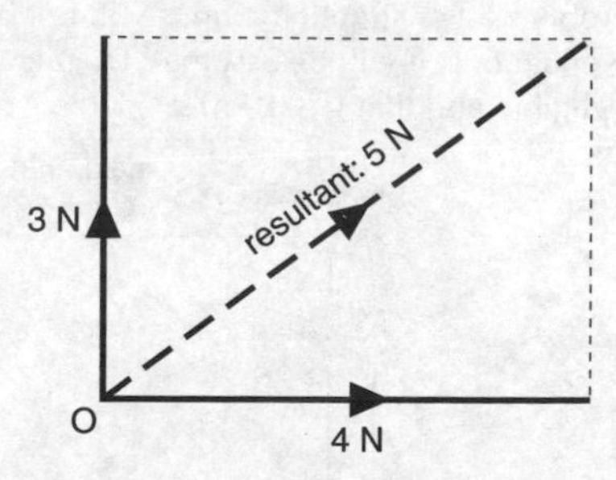

Above, you can see another way of finding the resultant of two forces, 3 N and 4 N, acting at right-angles through the same point. The vectors are drawn as two sides of a rectangle. The diagonal through O gives the magnitude and direction of the resultant. Note that the lines and angles in this diagram match those in the previous force triangle.

By drawing a parallelogram, the above method can also be used to add vectors which are not at right-angles. Here are two examples of a ***parallelogram of vectors***.

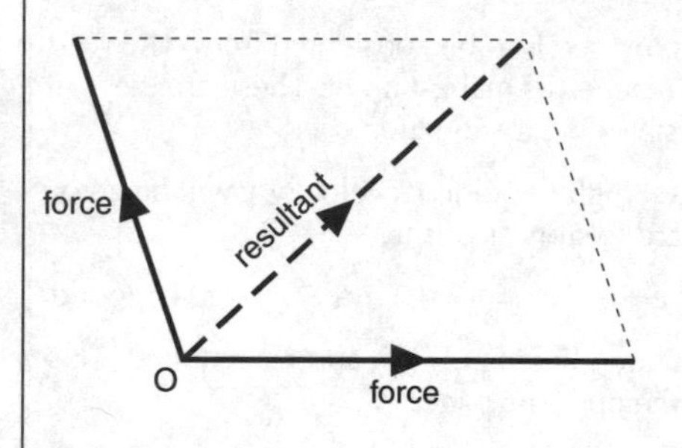

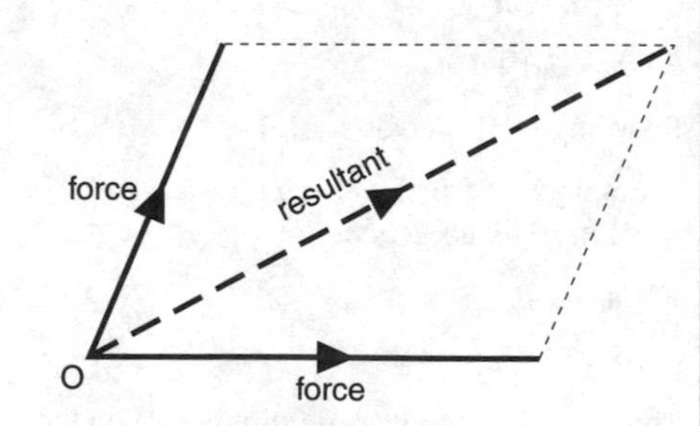

Note:

- The magnitude of the resultant depends on the relative directions of the vectors. For example, if forces of 3 N and 4 N are added, the resultant could be anything from 1 N (if the vectors are in opposite directions) to 7 N (if they are in the same direction).
- In the diagrams on this page, the resultant is always shown using a dashed arrow. This is to remind you that the resultant is a *replacement* for the other two vectors. There are *not* three vectors acting.

Multiplying vectors

When vectors are multiplied together, the product is not necessarily another vector. For example, work is the product of two vectors: force and displacement. But work is a scalar, not a vector. It has magnitude but no direction.

Methods of multiplying vectors have not been included in this book, other than for simple cases: for example, when force and displacement vectors are in the same direction.

Components

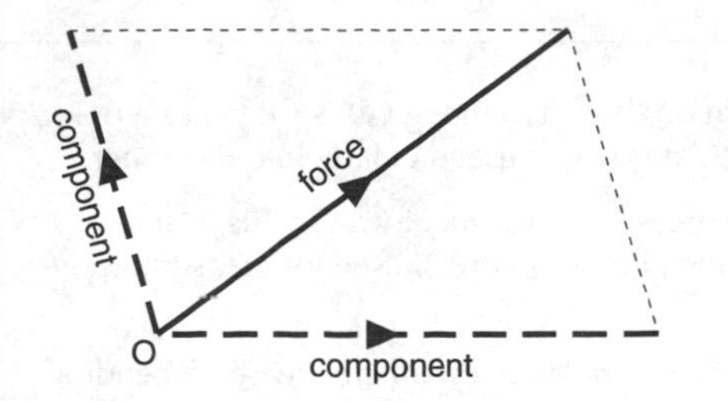

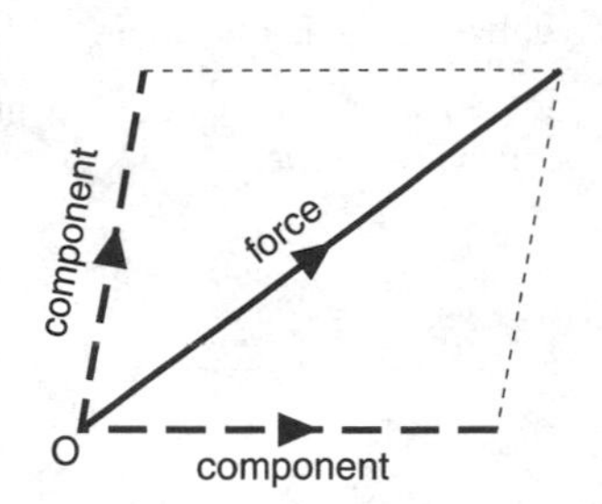

Two forces acting through a point can be replaced by a single force (the resultant) which has the same effect. Conversely, a single force can be replaced by two forces which have the same effect – a single force can be ***resolved*** into two ***components***. Two examples of the components of a force are shown above, though any number of other sets of components is possible.

Note:

- Any vector can be resolved into components.
- The components above are shown as dashed lines to remind you that they are a *replacement* for a single force. There are *not* three forces acting.

In working out the effects of a force (or other vector), the most useful components to consider are those at right-angles, as in the following example.

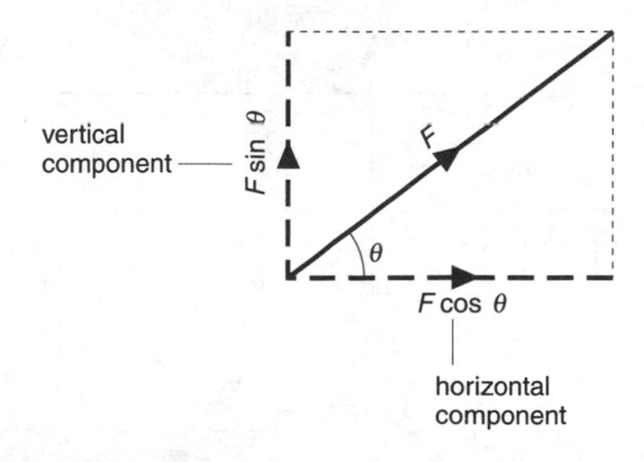

Below, you can see why the horizontal and vertical components have magnitudes of $F\cos\theta$ and $F\sin\theta$.

$$\cos\theta = \frac{F_x}{F}$$

So $\quad F_x = F\cos\theta$

$$\sin\theta = \frac{F_y}{F}$$

So $\quad F_y = F\sin\theta$

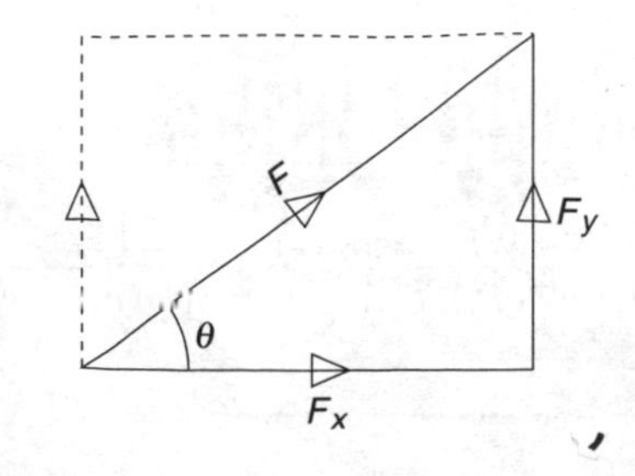

Equilibrium

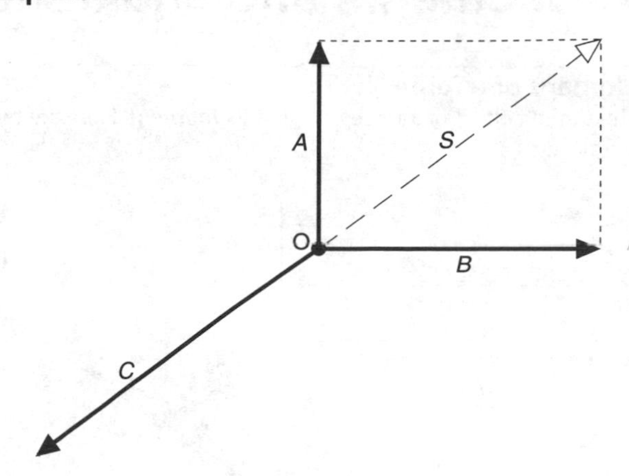

The particle O above has three forces acting on it – A, B, and C. Forces A and B can be replaced by a single force S. As force C is equal and opposite to S, the resultant of A, B, and C, is zero. This means that the three forces are in balance – the system is in ***equilibrium***.

If three forces are in equilibrium, they can be represented by the three sides of a triangle, as shown below. Note that the sides and angles match those in the previous force diagram. The forces can be drawn in any order, provided that the head of each arrow joins with the tail of another.

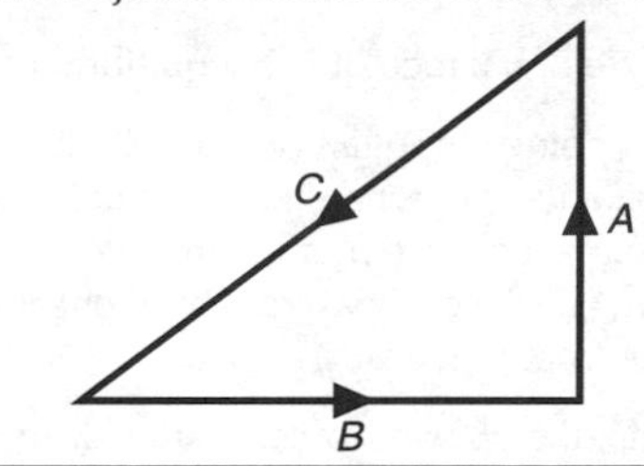

Resolving problem

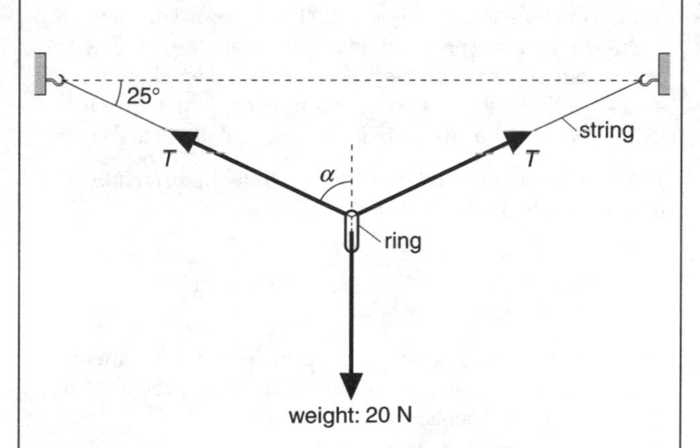

Example *Above, a ring is suspended from the middle of a piece of string. What is the tension in the string?*

Force T is the tension. It is present in both halves of the string. As angle α is 65°, this force has a component (upwards) of $T\cos 65°$. So

total of upward components on ring = $2T\cos 65°$

As the system is in equilibrium, the total of upward components must equal the downward force on the ring.

So $\quad 2T\cos 65° = 20$

This gives $\quad T = 24$ N

B5 Moments and equilibrium

Moment of a force

The turning effect of a force is called its ***moment***. Here are two examples:

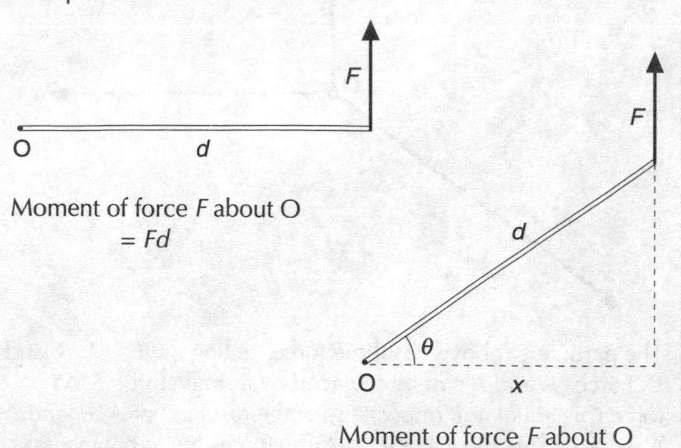

Moment of force F about O
$= Fd$

Moment of force F about O
$= Fx = Fd \cos \theta$

Note:

- In the diagram on the left, although O is shown as a point, it is really an *axis* going perpendicularly into the paper.
- Moments are measured in N m. However this is not the same unit as the N m, or J (joule), used for measuring energy.
- A moment can be *clockwise* or *anticlockwise*, depending on its *sense* (direction of turning). This can be indicated with a + or –. For example,

anticlockwise moment of 2 N m = +2 N m
clockwise moment of 2 N m = –2 N m

Principle of moments

The beam in the diagram on the right has weights on it. (The beam itself is of negligible weight.) The total weight is supported by an upward force R from the fulcrum.

The beam is in a state of balance. It is in equilibrium.

As the beam is not tipping to the left or right, the turning effects on it must balance. So, when moments are taken about O, as shown, the total clockwise moment must equal the total anticlockwise moment. (Note: R has zero moment about O because its distance from O is zero.)

As the beam is static, the upward force on it must equal the total downward force. So $R = 10 + 8 + 4 = 22$ N.

The beam is not turning about O. But it is not turning about any other axis either. So you would expect the moments about *any* axis to balance. This is exactly the case, as you can see in the next diagram. The beam and weights are the same as before, but this time, moments have been taken about point P instead of O. (Note: R does have a moment about P, so the value of R must be known before the calculation can be done.)

The examples shown on the right illustrate the ***principle of moments***, which can be stated as follows:

> If an object is in equilibrium, the sum of the clockwise moment about any axis is equal to the sum of the anticlockwise moments.

Here is another way of stating the principle. In it, moments are regarded as + or –, and the ***resultant moment*** is the algebraic sum of all the moments:

> If a rigid object is in equilibrium, the resultant moment about any axis is zero.

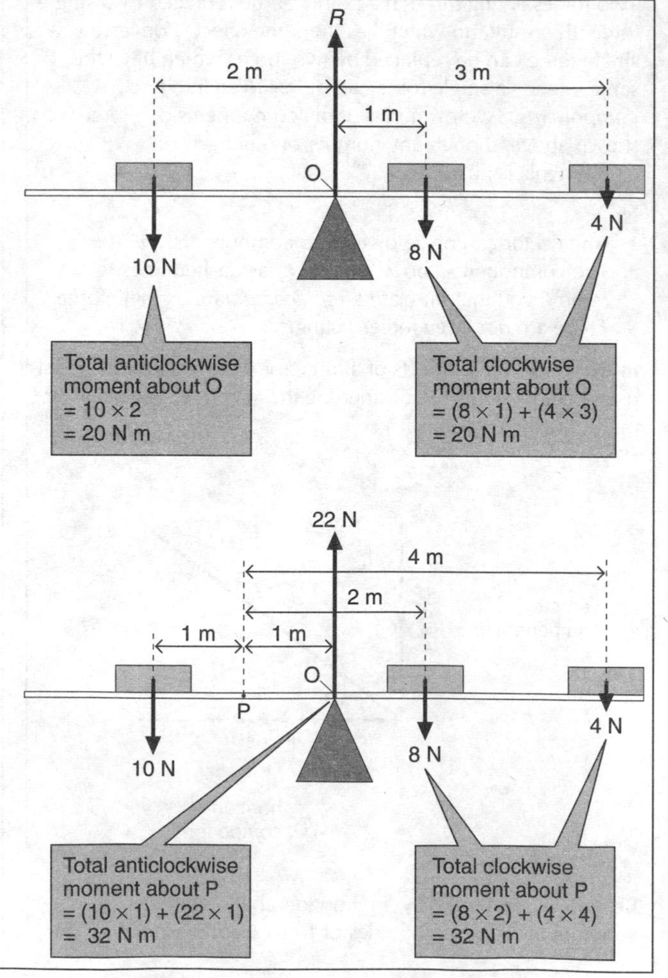

Centre of gravity

All the particles in an object have weight. The weight of the whole object is the resultant of all these tiny, downward gravitational forces. It appears to act through a single point called the ***centre of gravity***.

In the case of a rectangular beam with an even weight distribution, the centre of gravity is in the middle. Unless negligible, the weight must be included when analysing the forces and moments acting on the beam.

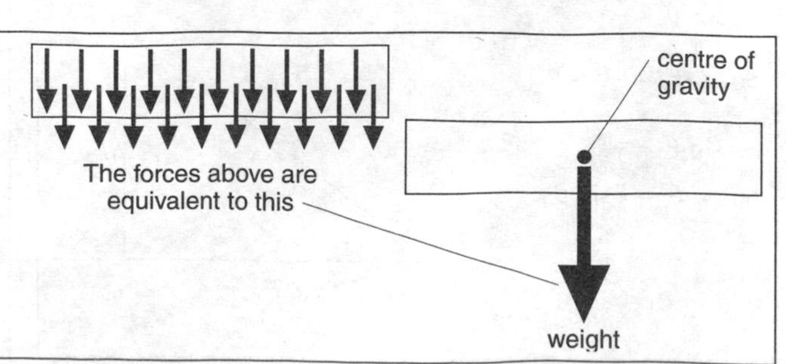

Conditions for equilibrium

There are two types of motion: ***translational*** (from one place to another) and ***rotational*** (turning). If a static, rigid object is in equilibrium, then

- the forces on it must balance, otherwise they would cause translational motion,
- the moments must balance, otherwise they would cause rotational motion.

The balanced beam on the opposite page is a simple system in which the forces are all in the same plane. A ***coplanar*** system like this is in equilibrium if

- the vertical components of all the forces balance,
- the horizontal components of all the forces balance,
- the moments about any axis balance.

To check for equilibrium, components can be taken in any two directions. However, vertical and horizontal components are often the simplest to consider. The balanced beam is especially simple because there are no horizontal forces.

Equilibrium problem

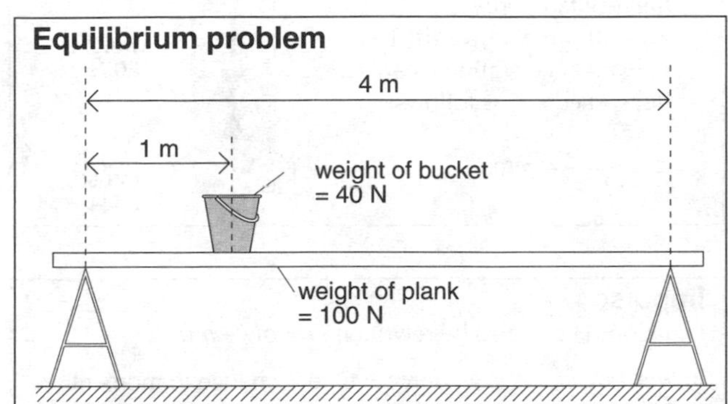

Example *A plank with a bucket on it is supported by two trestles. What force does each trestle exert on the plank?*

The first stage is to draw a ***free-body diagram*** showing just the rigid body (the plank) and the forces acting on it:

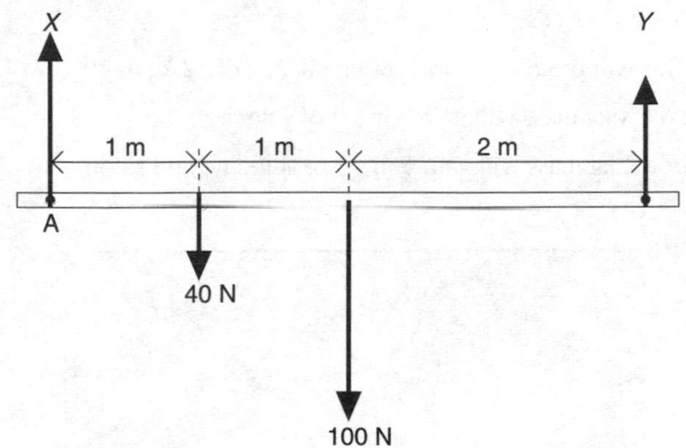

The body is in equilibrium, so the moments must balance, and the forces also. X and Y are the unknown forces.

Taking moments about A:

total clockwise moment = total anticlockwise moment

$$(40 \times 1) + (100 \times 2) = (Y \times 4)$$

This gives $Y = 60$ N.

Note the advantage of taking moments about A: X has a zero moment, so it does not feature in the equation.

Comparing the vertical forces:

total upward force = total downward force

$$Y + X = 40 + 100$$

As Y is 60 N, this gives $X = 80$ N.

Couples and torque

A pair of equal but opposite forces, as below, is called a ***couple***. It has a turning effect but no resultant force.

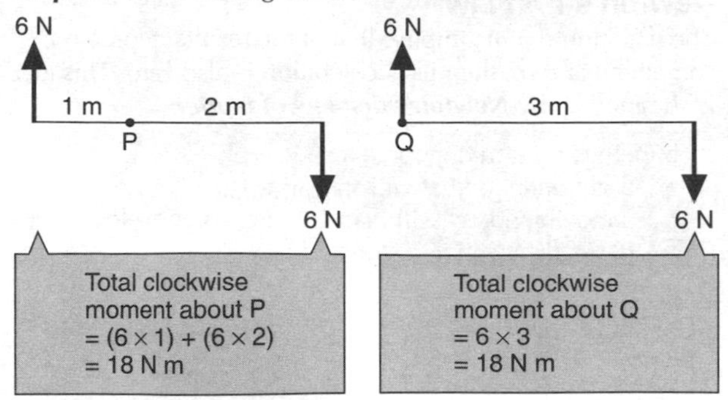

To find the total moment of a couple, you could choose any axis, work out the two moments and add them up. Whichever axis you choose, the answer is the same, so the simplest way of calculating the total moment is like this:

moment of couple = one force × perpendicular distance between forces

Note:

- The total moment of a couple is called a ***torque***.
- Strictly speaking, a couple is any system of forces which has a turning effect only i.e. one which produces rotational motion without translational (linear) motion.

Stability

For a static object, there are three types of equilibrium, as shown below. Whether the equilibrium is ***stable***, ***unstable***, or ***neutral*** depends on the couple formed by the weight and the reaction when the object is displaced.

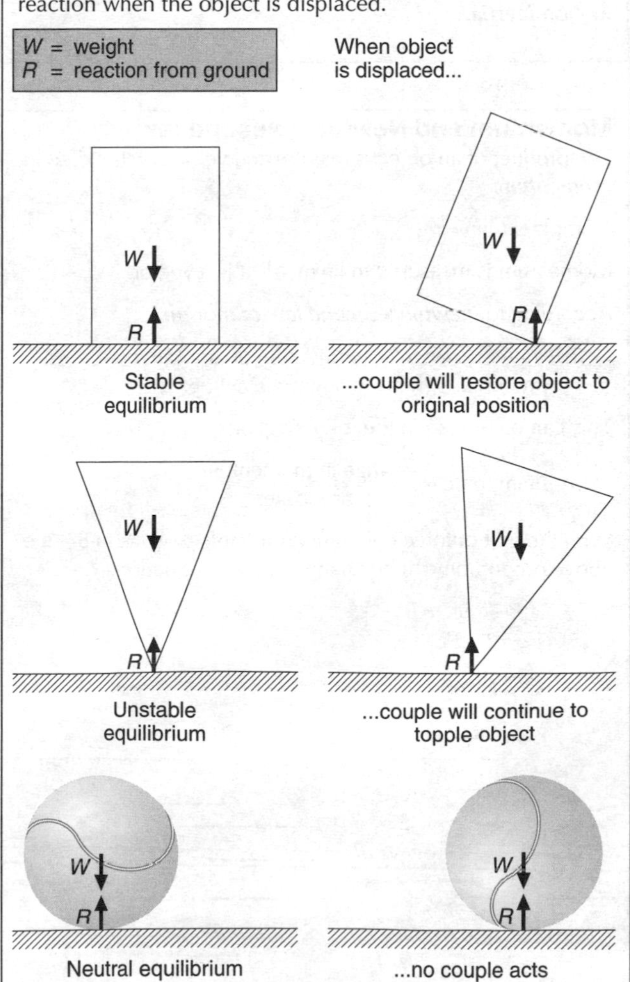

B6 Motion and momentum

Newton's first law

The equation $F = ma$ implies that, if the resultant force on something is zero, then its acceleration is also zero. This idea is summed up by ***Newton's first law of motion***:

> If there is no resultant force acting,
> - a stationary object will stay at rest,
> - a moving object will maintain a constant velocity (a steady speed in a straight line).

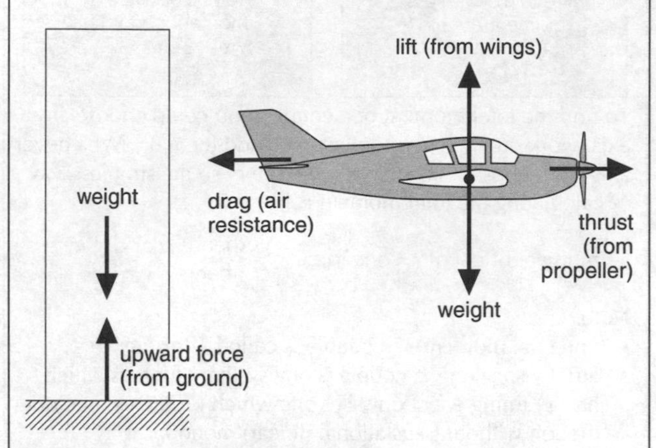

From Newton's first law, it follows that if an object is at rest or moving at constant velocity, then the forces on it must be balanced, as in the examples above.

The more mass an object has, the more it resists any change in motion (because more force is needed for any given acceleration). Newton called this resistance to change in motion ***inertia***.

Momentum and Newton's second law

The product of an object's mass m and velocity v is called its ***momentum***:

> momentum = mv

Momentum is measured in kg m s^{-1}. It is a vector.

According to ***Newton's second law of motion***:

> The rate of change of momentum of an object is proportional to the resultant force acting.

This can be written in the following form:

$$\text{resultant force} \propto \frac{\text{change in momentum}}{\text{time taken}}$$

With the unit of force defined in a suitable way (as in SI), the above proportion can be changed into an equation:

$$F = \frac{mv - mu}{t} \qquad (1)$$

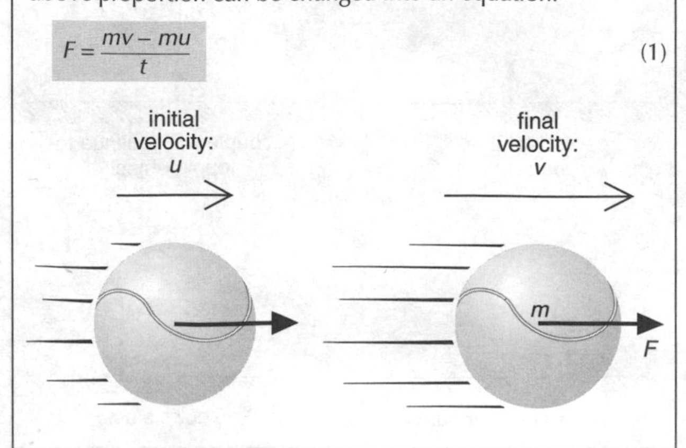

Linked equations

Equation (1) can be rewritten $F = \dfrac{m(v - u)}{t}$

But acceleration $a = \dfrac{(v - u)}{t}$. So $F = ma$ (2)

Equations (1) and (2) are therefore different versions of the same principle.

Note:

- In arriving at the equation $F = ma$ above, the mass m is assumed to be constant. But according to Einstein (see H12), mass increases with velocity (though insignificantly for velocities much below that of light). This means that $F = ma$ is really only an approximation, though an acceptable one for most practical purposes.
- When using equations (1) and (2), remember that F is the resultant force acting. For example, on the right, the resultant force is 26 – 20 = 6 N upwards. The upward acceleration a can be worked out as follows:

$$a = \frac{F}{m} = \frac{6}{2} = 3 \text{ m s}^{-2}$$

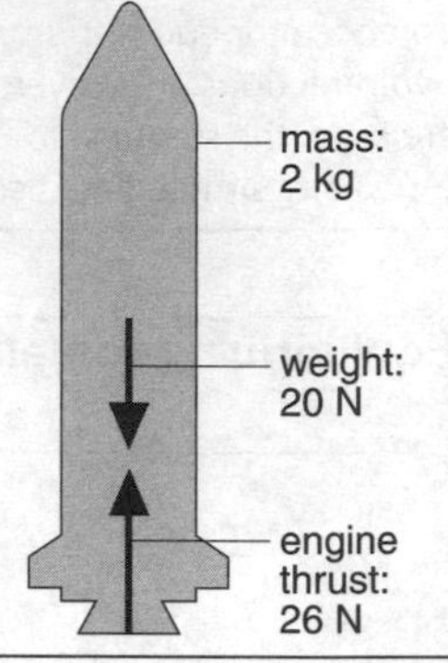

Impulse

Equation (1) can also be rewritten $Ft = mv - mu$

In words force × time = change in momentum

The quantity 'force × time' is called an ***impulse***.

A given impulse always produces the same change in momentum, irrespective of the mass. For example, if a resultant force of 6 N acts for 2 s, the impulse delivered is 6 × 2 = 12 N s.

This will produce a momentum change of 12 kg m s^{-1}

So a 4 kg mass will gain 3 m s^{-1} of velocity

or a 2 kg mass will gain 6 m s^{-1} of velocity, and so on.

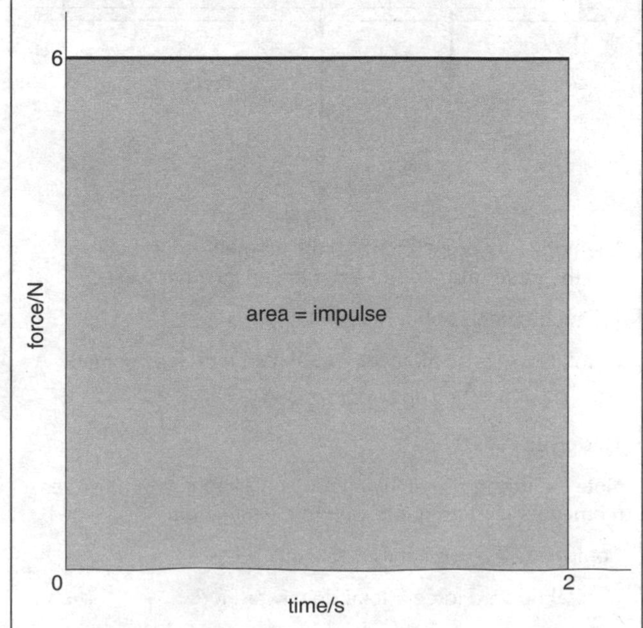

The graph above is for a uniform force of 6 N. In 2 s, the impulse delivered is 12 N s. Numerically, this is equal to the area of the graph between the 0 and 2 s points.

Newton's third law

A single force cannot exist by itself. Forces are always pushes or pulls between *two* objects, so they always occur in pairs. One force acts on one object; its equal but opposite partner acts on the other. This idea is summed up by ***Newton's third law of motion***:

> If A is exerting a force on B, then B is exerting an equal but opposite force on A.

The law is sometimes expressed as follows:

> To every action, there is an equal but opposite reaction.

Examples of action–reaction pairs are given below.

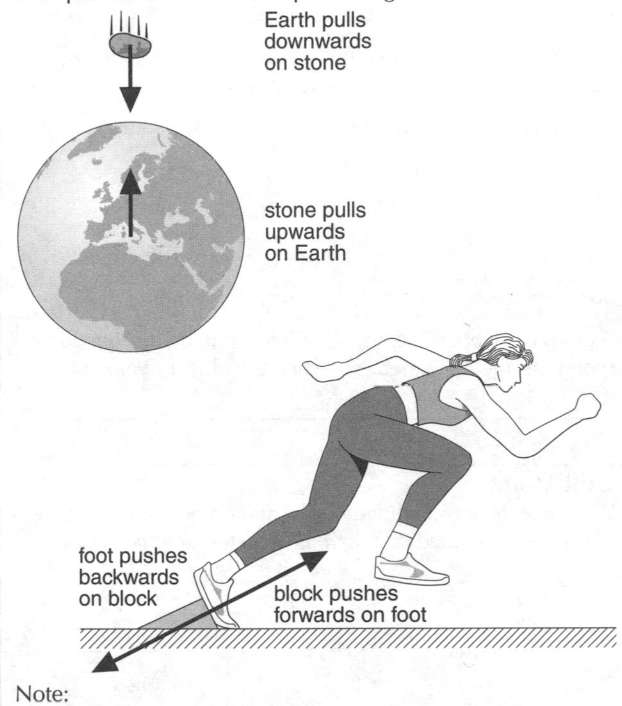

Note:

- It does not matter which force you call the action and which the reaction. One cannot exist without the other.
- The action and reaction do not cancel each other out because they are acting on *different* objects.

Momentum problem

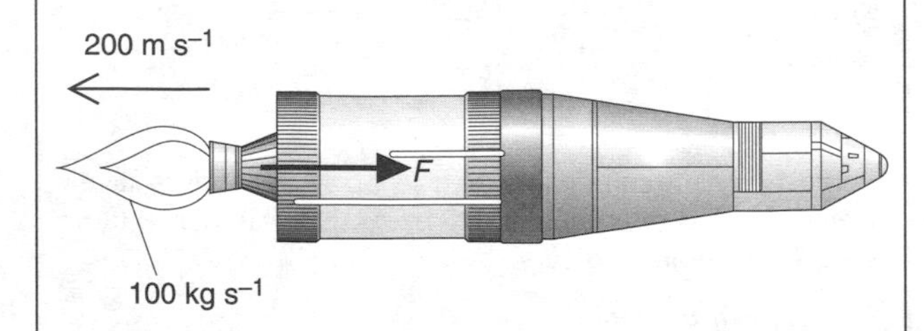

Example *A rocket engine ejects 100 kg of exhaust gas per second at a velocity (relative to the rocket) of 200 m s⁻¹. What is the forward thrust (force) on the rocket?*

By Newton's third law, the forward force on the rocket is equal to the backward force pushing out the exhaust gas. By Newton's second law, this force F is equal to the momentum gained per second by the gas, so it can be calculated using equation (1) with the following values:

$m = 100$ kg $\quad t = 1$ s $\quad u = 0 \quad v = 200$ m s^{-1}

So $F = \dfrac{mv - mu}{t} = \dfrac{(100 \times 200) - (100 \times 0)}{1} = 20\,000$ N

Conservation of momentum

Trolleys A and B below are initially at rest. When a spring between them is released, they are pushed apart.

By Newton's third law, the force exerted by A on B is equal (but opposite) to the force exerted by B on A. These equal forces also act for the same time, so they deliver equal (but opposite) impulses. As a result, A gains the same momentum to the left as B gains to the right.

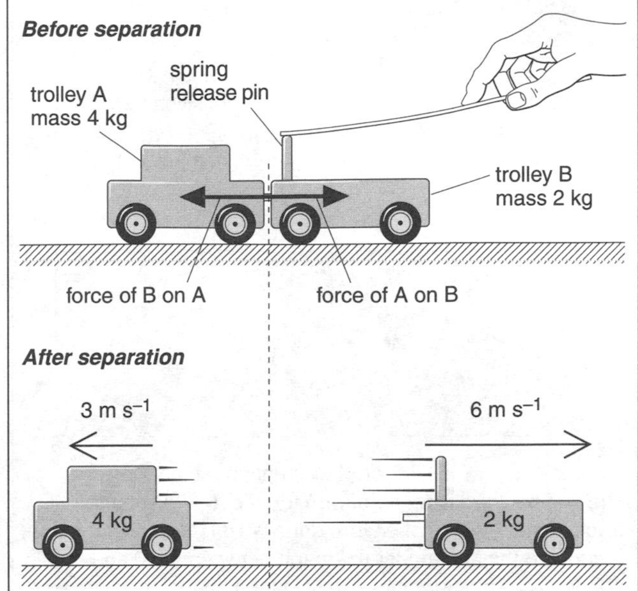

Momentum is a vector, so its direction can be indicated using + or –. If vectors to the right are taken as +,

before the trolleys separate
total momentum = 0

after the trolleys separate
momentum of A = 4 × (–3) = –12 kg m s^{-1}
momentum of B = 2 × (+6) = +12 kg m s^{-1}
so total momentum = 0 kg m s^{-1}

Together, trolleys A and B make up a ***system***. The total momentum of this system is the same (zero) before the trolleys push on each other as it is afterwards. This illustrates the ***law of conservation of momentum***:

> When the objects in a system interact, their total momentum remains constant, provided that there is no external force on the system.

Below, the separating trolleys are shown with velocities of v_1 and v_2 instead of actual values. In cases like this, it is always best to choose the same direction as positive for all vectors. It does not matter that A is really moving to the left. If A's velocity is 3 m s^{-1} to the left, then $v_1 = -3$ m s^{-1}.

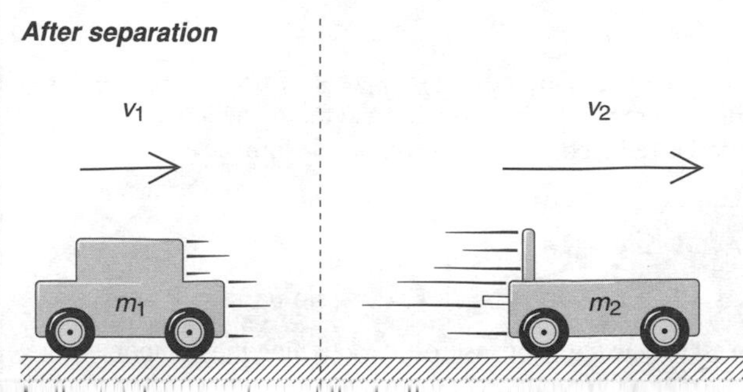

As the total momentum of the trolleys is zero,

$m_1v_1 + m_2v_2 = 0$

So, if v_2 is positive, v_1 must be negative.

B7 Work, energy, and momentum

Work

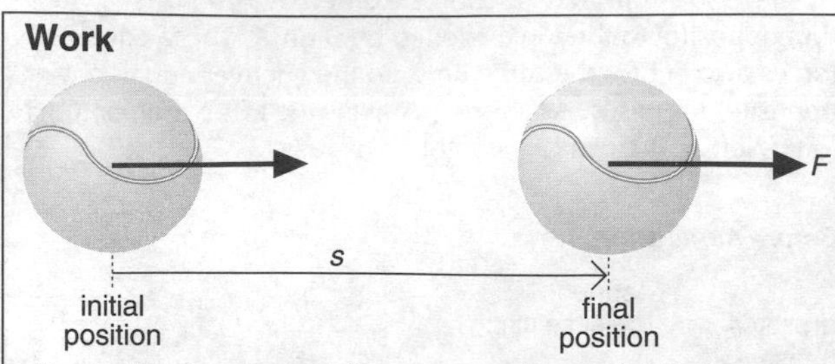

Above, F is the resultant force on an object. If W is the work done when the force has caused a displacement s, then

$$W = Fs$$

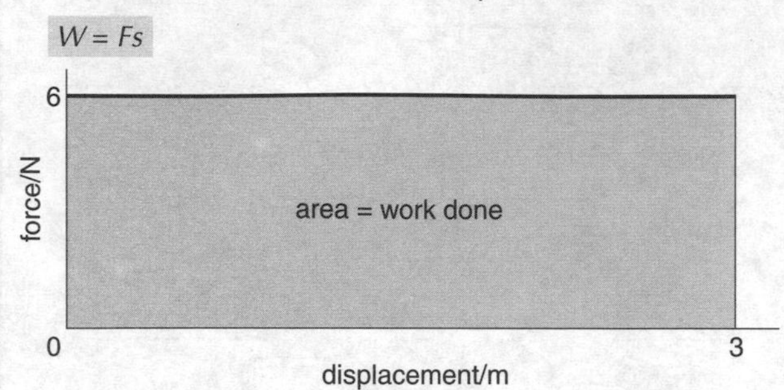

The graph above is for a uniform force of 6 N. When the displacement is 3 m, the work done is 18 J. Numerically, this is equal to the area under the graph between the 0 and 3 m points. (The same principle applies for a changing force: see B8.)

Using a ramp Below, a load is raised, first by lifting it vertically, and then by pulling it up a frictionless ramp. The force needed in each case is shown, but not the balancing force. (F_1 must balance the weight, so $F_1 = mg$.)

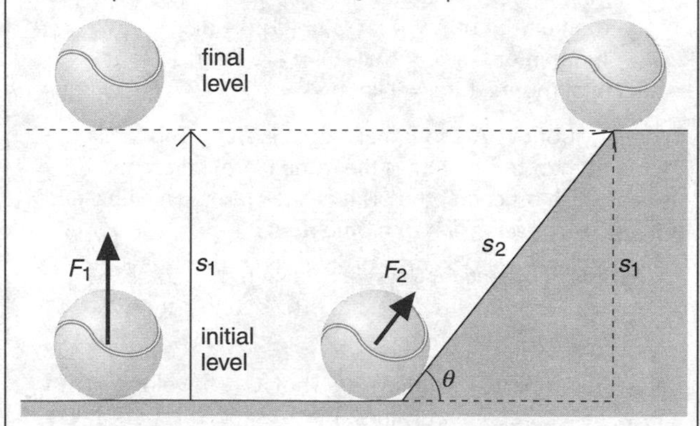

The gain in potential energy is the same in both cases. So, by the law of conservation of energy, the work done must also be the same. Therefore

$$F_2 s_2 = F_1 s_1 \qquad (1)$$

As $s_2 > s_1$, it follows that $F_2 < F_1$. So, by using the ramp, the displacement is increased, but the force needed to raise the load is reduced. The ramp is a simple form of machine.

Equation (1) leads to two further results:

As $s_1 = s_2 \sin\theta$, $\quad F_2 = F_1 \sin\theta$

As $F_1 = mg$, $\quad F_2 = mg \sin\theta$

You can also get the last result by finding the component of mg down the ramp. F_2 is the force needed to balance it.

The frictionless ramp wastes no energy. But this is not true of most machines. Where there is friction, the work done *by* a machine is less than the work done *on* it.

Finding an equation for kinetic energy (KE)

Below, an object of mass m is accelerated from velocity u to v by a resultant force F. While gaining this velocity, its displacement is s and its acceleration is a.

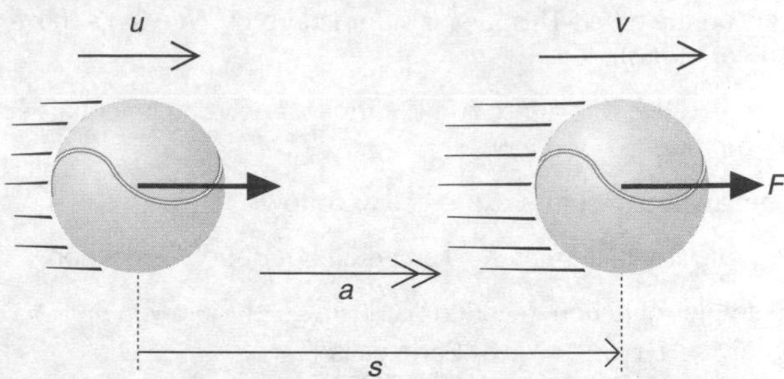

From the law of conservation of energy, the KE gained by the object is equal to the work done on it, Fs.

From equation (4) in B3, $v^2 - u^2 = 2as$

$\therefore \quad as = \frac{1}{2}v^2 - \frac{1}{2}u^2$

So $\quad mas = \frac{1}{2}mv^2 - \frac{1}{2}mu^2$

But $\quad mas = Fs \quad$ (because $F = ma$)

So $\quad Fs = \frac{1}{2}mv^2 - \frac{1}{2}mu^2$.

As Fs is the work done, the right-hand side of the equation represents the KE gained. So, when the object's velocity is v, its KE $= \frac{1}{2}mv^2$.

Collisions

Whenever objects collide, their total momentum is conserved, provided that there is no external force acting.

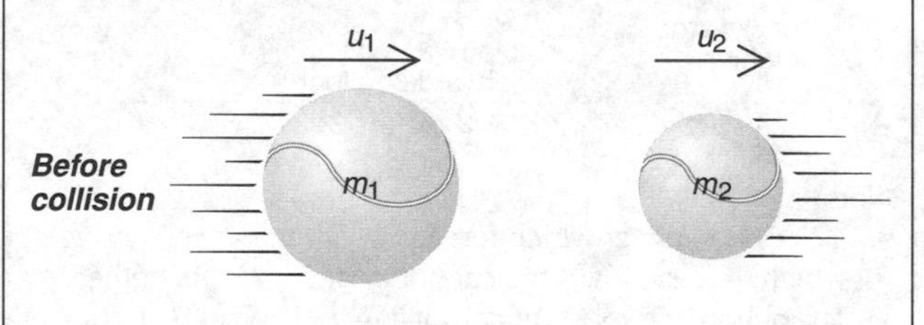

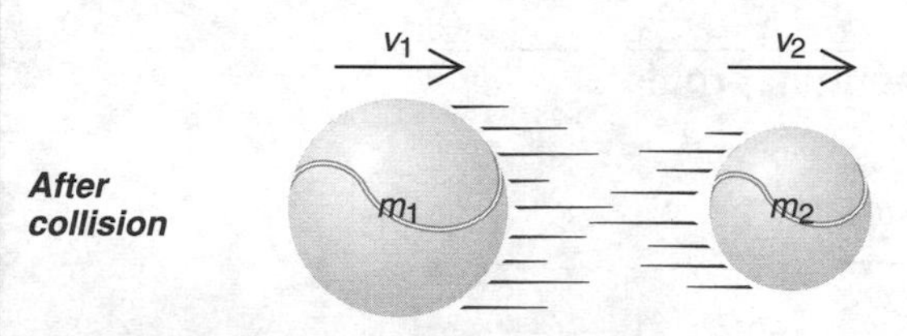

Above, two balls collide and then separate. All vectors have been defined as positive to the right. As the total momentum is the same before and after,

$$m_1u_1 + m_2u_2 = m_1v_1 + m_2v_2$$

Elastic collision An elastic collision is one in which the total kinetic energy of the colliding objects remains constant. In other words, no energy is converted into heat (or other forms). If the above collision is elastic,

$$\tfrac{1}{2}m_1u_1^2 + \tfrac{1}{2}m_2u_2^2 = \tfrac{1}{2}m_1v_1^2 + \tfrac{1}{2}m_2v_2^2$$

One consequence of the above is that the speed of separation of A and B is the same after the collision as before:

$$u_1 - u_2 = -(v_1 - v_2)$$

Inelastic collision In an inelastic collision, kinetic energy is converted into heat. The total amount of *energy* is conserved, but the total amount of *kinetic energy* is not.

Collision problems

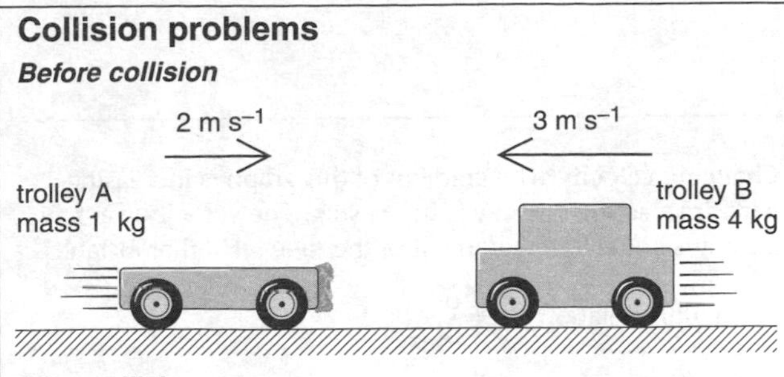

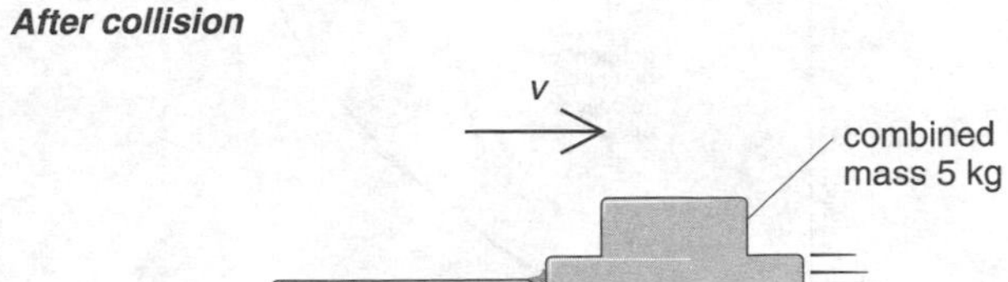

Example 1 *The trolleys above collide and stick together. What is their velocity after the collision? (Assume no friction.)*

All vectors to the right will be taken as positive.
The unknown velocity is v (to the right).

momentum = mass × velocity

before the collision

momentum of A = $1 \times 2 = 2$ kg m s^{-1}

momentum of B = $4 \times (-3) = -12$ kg m s^{-1}

∴ total momentum = -10 kg m s^{-1} (2)

After the collision

A and B have a combined mass of 5 kg, and a combined velocity of v. So total momentum = $5 \times v$.

As the total momentum is the same before and after,

$5v = -10$ which gives $v = -2$ m s^{-1}

So the trolleys have a velocity of 2 m s^{-1} to the *left*.

Example 2 *When the trolleys collide, how much of their total kinetic energy is lost (converted into other forms)?*

$KE = \frac{1}{2}mv^2$

before the collision

KE of A = $\frac{1}{2} \times 1 \times 2^2 = 2$ J

KE of B = $\frac{1}{2} \times 4 \times (-3)^2 = 18$ J

∴ total KE = 20 J (3)

after the collision

total KE = $\frac{1}{2} \times 5 \times (-2)^2 = 10$ J

Comparing the total KEs before and after, 10 J of KE is lost.

Example 3 *If the collision had been elastic, what would the velocities of the trolleys have been after separation?*

Let v_1 be the final velocity of A and v_2 be the final velocity of B (both defined as positive to the right).

As both total momentum and total KE are conserved,

total momentum after collision = -10 kg m s^{-1} (from 2)
total KE after collision = 20 J (from 3)

So $(1 \times v_1) + (4 \times v_2) = -10$
And $(\frac{1}{2} \times 1 \times v_1^2) + (\frac{1}{2} \times 4 \times v_2^2) = 20$

Solving these equations for v_1 and v_2 gives

$v_1 = -6$ m s^{-1} and $v_2 = -1$ m s^{-1}

Note:

- There is an alternative solution which gives the velocities before the collision: 2 m s^{-1} and -3 m s^{-1}.

Recoiling particles

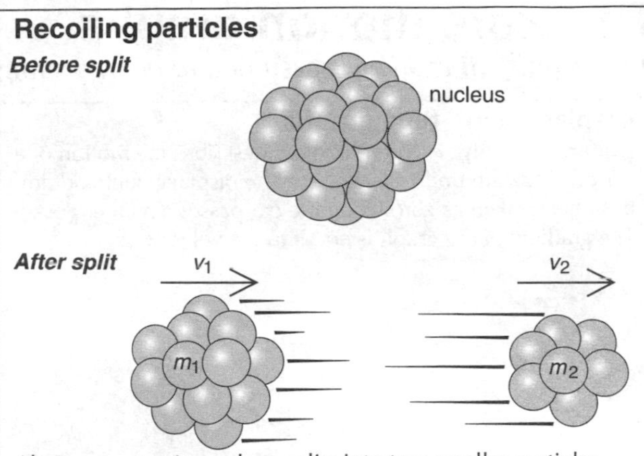

Above, an atomic nucleus splits into two smaller particles with a loss of nuclear energy (see also G3). The particles share the energy released (as kinetic energy) and shoot apart. All vectors have been defined as positive to the right.

As the total momentum is conserved, $m_1v_1 + m_2v_2 = 0$ (4)

Also KE of A = $\frac{1}{2}m_1v_1^2$ (5)

and KE of B = $\frac{1}{2}m_2v_2^2$ (6)

From (4), (5), and (6), the following can be obtained:

$$\frac{\text{KE of A}}{\text{KE of B}} = \frac{m_2}{m_1}$$

This means, for example, that if A has 9 times the mass of B, then B will shoot out with 9 times the KE of A. In other words, it will have 90% of the available energy. The energy is only shared equally if A and B have the same mass.

Power and velocity

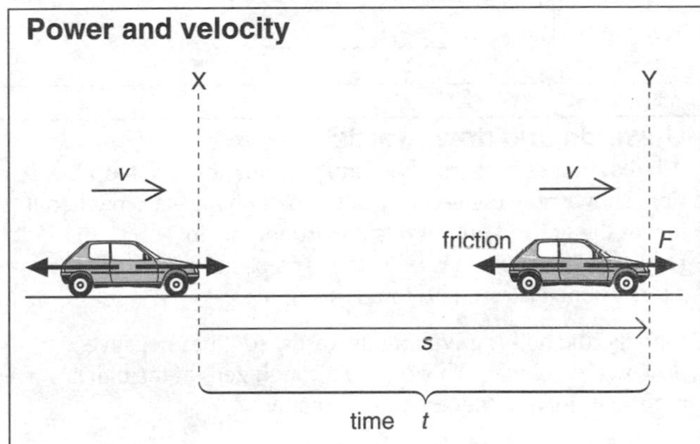

Above, the car's engine provides (via the driven wheels) a forward force F which balances the total frictional force (mainly air resistance) on the car. As a result, the car maintains a steady velocity v. The displacement of the car is s in time t. P is the power being delivered to the wheels.

In moving from X to Y, work done (by F) = Fs.

$$\text{power} = P = \frac{\text{work done}}{\text{time taken}} = \frac{Fs}{t}$$

But $v = \frac{s}{t}$ so $P = Fv$

i.e. power delivered = force × velocity

For example, if a force of 200 N is needed to maintain a steady velocity of 5 m s^{-1} against frictional forces,

power delivered = 200 × 5 = 1000 W

All of this power is wasted as heat in overcoming friction. Without friction, no forward force would be needed to maintain a steady velocity, so no work would be done.

B8 More motion graphs

In this unit, all motion is assumed to be in a straight line.

Displacement–time graphs

Uniform velocity The graph below describes the motion of a car moving with uniform velocity. The displacement and time have been taken as zero when the car passes a marker post. The gradient of the graph is equal to the velocity v:

$$v = \frac{\Delta s}{\Delta t}$$

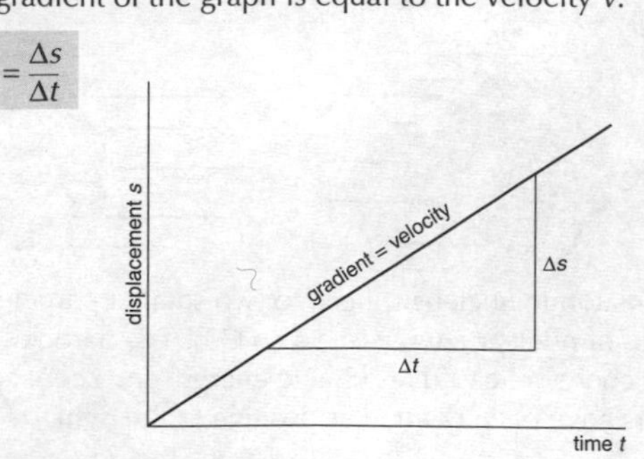

Changing velocity The gradient of this graph is increasing with time, so the velocity is increasing. The velocity v at any instant is equal to the gradient of the *tangent* at that instant.

In calculus notation $v = \frac{ds}{dt}$

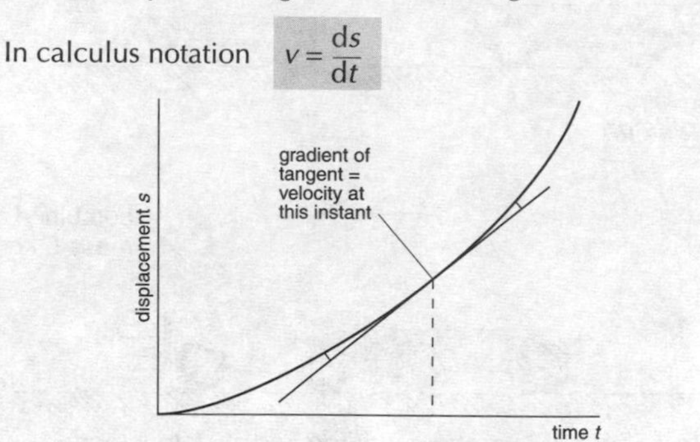

Velocity–time graphs

Uniform acceleration The graph below describes the motion of a car gaining velocity at a steady rate. The time has been taken as zero when the car is stationary. The gradient of the graph is equal to the acceleration.

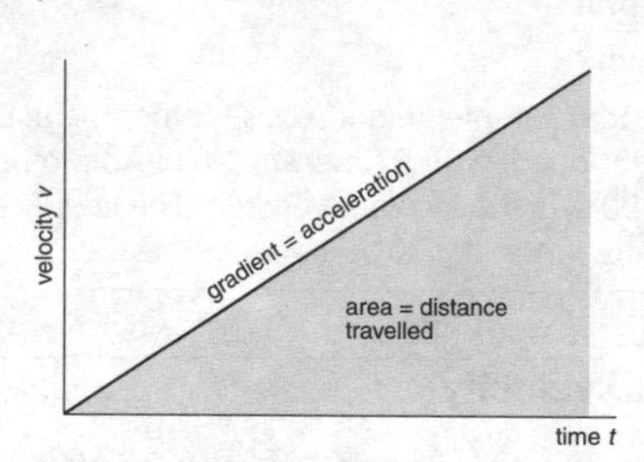

Changing acceleration The acceleration a at any instant is equal to the gradient of the *tangent* at that instant.

In calculus notation $a = \frac{dv}{dt}$

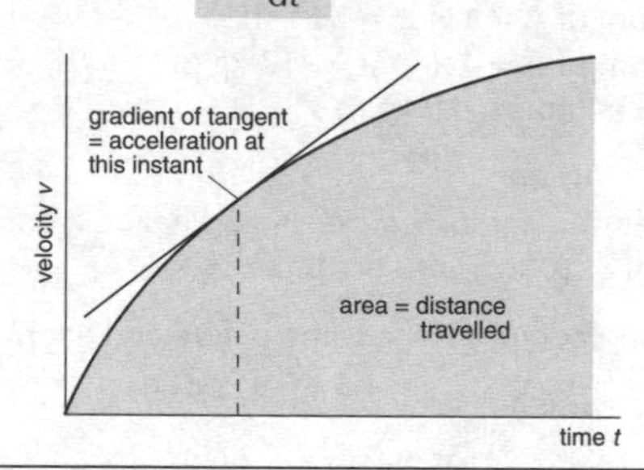

Upwards and downwards

A ball bounces upwards from the ground. The graph on the right shows how the velocity of the ball changes from when it leaves the ground until it hits the ground again. Downward velocity has been taken as positive. Air resistance is assumed to be negligible.

Initially, the ball is travelling upwards, so it has negative downward velocity. This passes through zero at the ball's highest point and then becomes positive.

The gradient of the graph is constant and equal to g.

Note:

- The ball has downward acceleration g, even when it is travelling upwards. (Algebraically, losing upward velocity is the same as gaining downward velocity.)
- The ball has downward acceleration g, even when its velocity is zero (at its highest point).

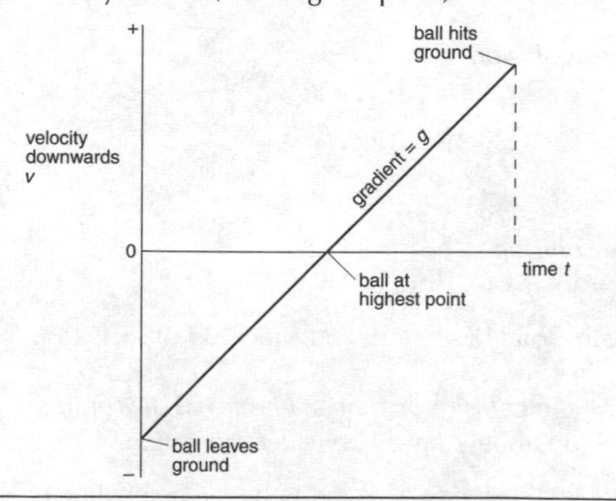

Terminal velocity

Air resistance on a falling object can be significant. As the velocity increases, the air resistance increases, until it eventually balances the weight. The resultant force is then zero, so there is no further gain in velocity. The object has reached its ***terminal velocity***.

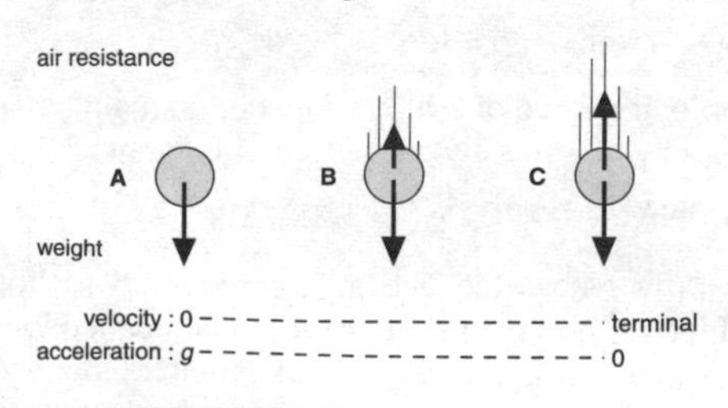

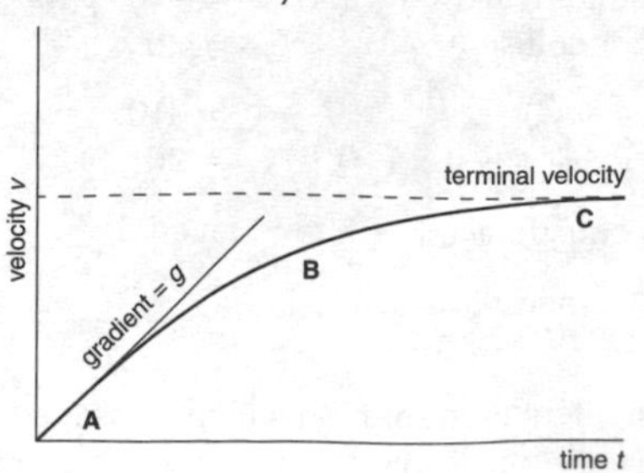

Force, impulse, and work

The area under a force–time graph is equal to the impulse delivered by the force.

The area under a force–displacement graph is equal to the work done by the force.

The graphs on the right are for a non-uniform force: for example, the force used to stretch a spring.

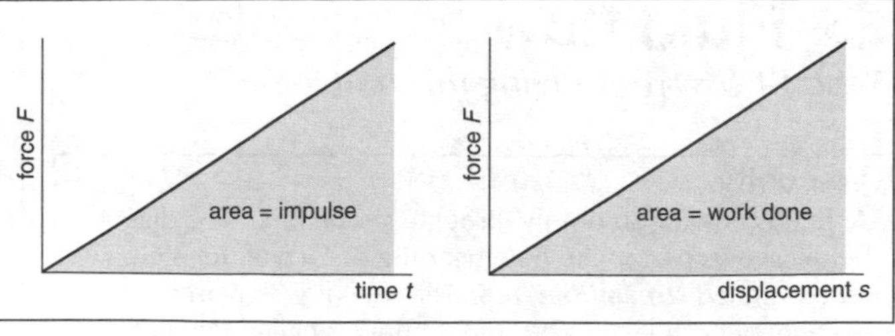

Aircraft principles

Aircraft propulsion

Read B6 first.

To move forward, an aircraft pushes a mass of gas backwards so that, by Newton's third law, there is an equal forward force on the aircraft. Here are two ways of producing a backward flow of gas:

Jet engine Air is drawn in at the front by a large fan, and pushed out at the back. Exhaust gases are also ejected, at a higher speed.

Propeller This is driven by the shaft of a jet engine or piston engine. Its blades are angled so that air is pushed backwards as it rotates.

Note:

- *Momentum problem* in B6 shows how to calculate the thrust (force) of a rocket engine. The same principles can be applied to a jet engine or propeller.

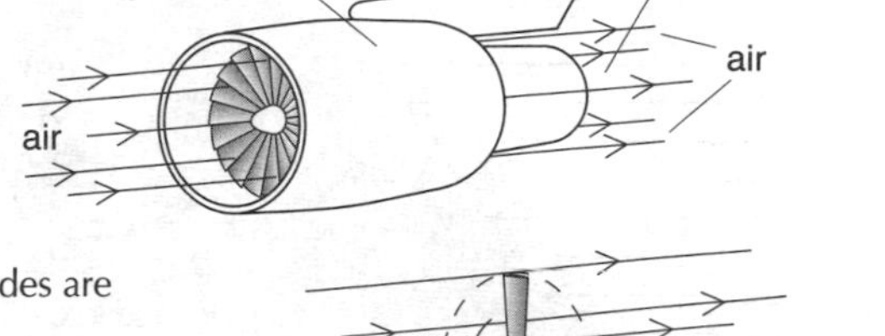

Lift and drag

Read B9 first.

A wing is an ***aerofoil*** – a shape which produces more lift than drag. For a wing of horizontal area S moving at velocity v through air of density ρ, the lift F_L is given by

$$F_L = \tfrac{1}{2} S C_L \rho v^2 \qquad (1)$$

where C_L is the lift *coefficient* of the aerofoil section.

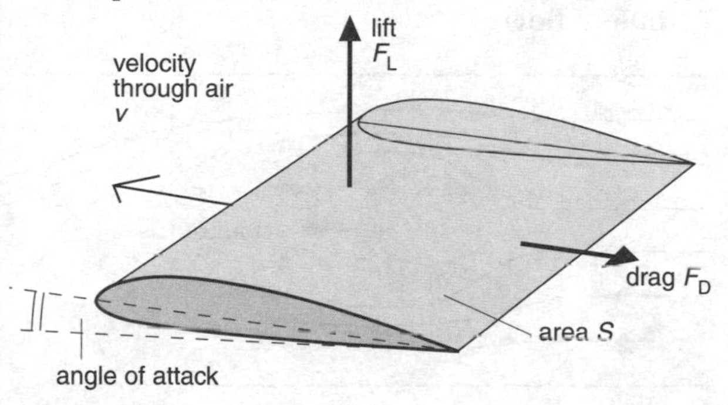

Angle of attack The value of C_L depends on this angle (shown above). Up to a certain limit, increasing the angle of attack increases C_L and, therefore, increases the lift.

For level flight, the lift must balance the aircraft's weight (see B6). If the speed decreases, then according to the above equation, the lift would also decrease if there were no change in C_L. To maintain lift, the pilot must pull the nose of the aircraft up slightly to increase the angle of attack.

Stalling If the angle of attack becomes too high, the airflow behind the wing becomes very turbulent and there is a sudden loss of lift. The wing is ***stalled***:

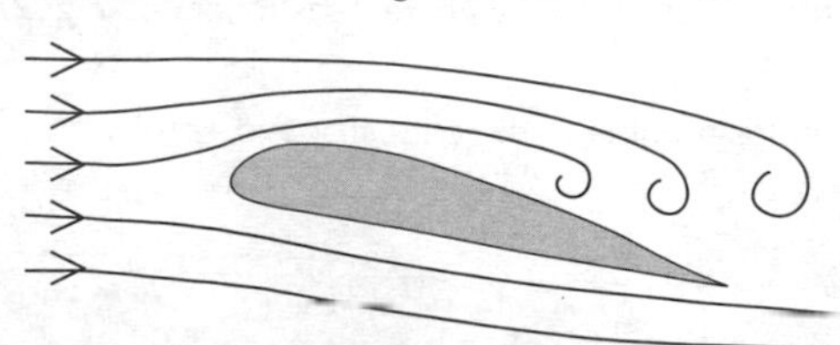

Linking lift and drag Equation (1) is similar in form to that for drag in B9: $F_D = \tfrac{1}{2} A C_D \rho v^2$. Lift and drag are related. If the lift on an aerofoil increases, so does the drag.

Helicopters

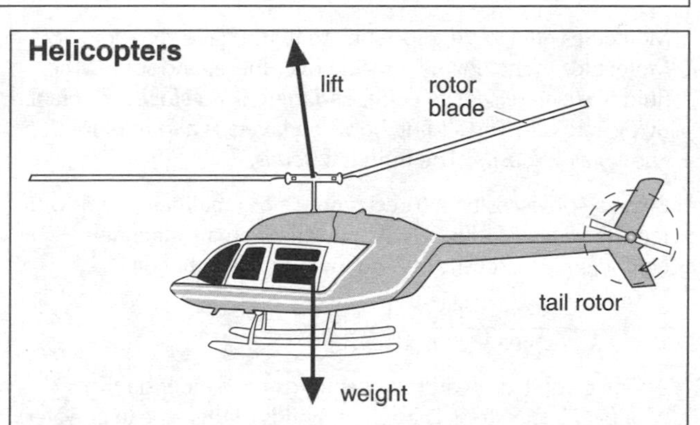

A helicopter's rotor blades are aerofoils. Their motion creates the airflow needed for lift. Each blade is hinged at the rotor hub so that it can move up and down, and there is a lever mechanism for varying its angle of attack. By making each blade rise and fall as it goes round, the plane of the rotor can be tilted to give the horizontal component of force needed for forwards, backwards, or sideways motion.

As the engine exerts a torque on the rotor, there is an equal but opposite torque on the engine. The tail rotor balances this torque and stops the helicopter spinning round.

Hovercraft

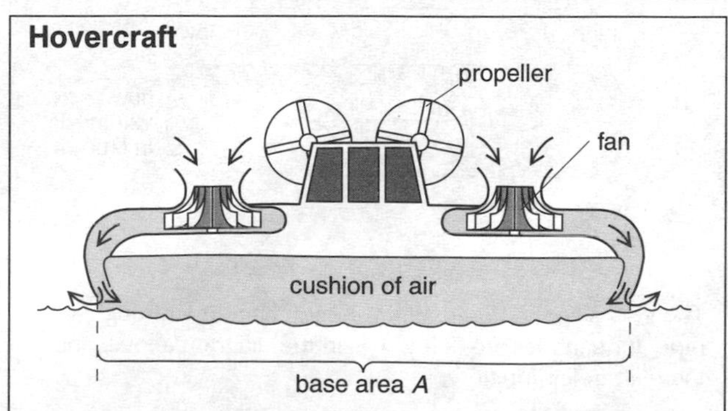

A hovercraft is supported by a 'cushion' of air. If its base area is A, and the trapped air has an excess pressure Δp above atmospheric pressure, then the upward force on the hovercraft is $\Delta p\, A$. This balances the weight. Air is constantly leaking from under the hovercraft. The fans maintain excess pressure by replacing the lost air.

B9 Fluid flow

Read F1 before studying this unit.

Viscosity

A fluid is flowing smoothly through a wide pipe. The diagram below shows part of the flow near the surface of the pipe. The arrows called ***streamlines***, represent the direction and velocity of each layer. The smooth flow is called ***laminar*** (layered) or ***streamline*** flow.

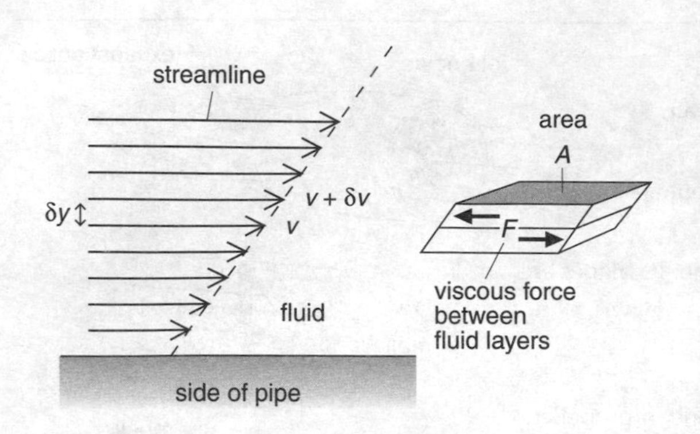

Molecules next to the pipe stick to it and have zero velocity. Molecules in the next layer slide over these, and so on. The fluid is *sheared* (see also H5), and there is a ***velocity gradient*** $\delta v/\delta y$ across it. The sliding between layers is a form of friction known as ***viscosity***. The fluid is ***viscous***.

Because of viscosity, a force is needed to maintain the flow. If F is the viscous force between layers of area A, then the coefficient of viscosity η is defined by this equation:

$$\eta = \frac{\text{shear stress}}{\text{velocity gradient}} = \frac{F/A}{\delta v/\delta y}$$

At any given temperature, most fluids have a constant η, whatever shear stress is applied. Fluids of this type (e.g. water) are called ***Newtonian fluids***. However, some liquids are ***thixotropic***: when the shear stress is increased, η decreases. Some paints and glues are like this. They are very viscous (i.e. semi-solid) until stirred.

Liquid flow through a pipe

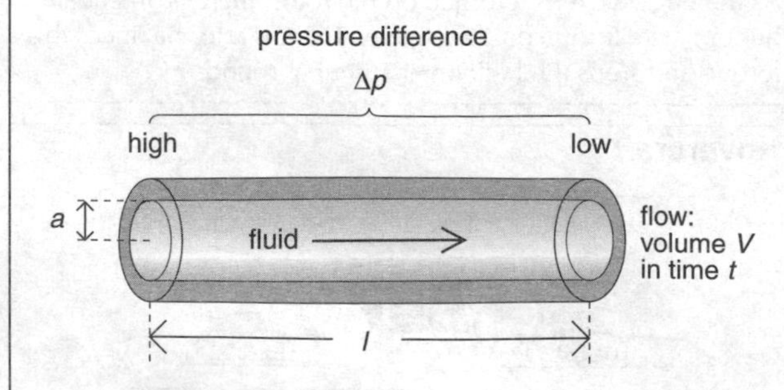

The viscosity of a liquid affects how it can flow through a pipe. If quantities are defined as in the diagram above, and there is streamline flow,

$$\frac{V}{t} = \frac{\pi a^4 \Delta p}{8\eta l}$$

This is called ***Poiseuille's equation***.

- As $V/t \propto a^4$, halving the radius of a pipe reduces the rate of flow to 1/16th for the same pressure difference.

Stokes' law

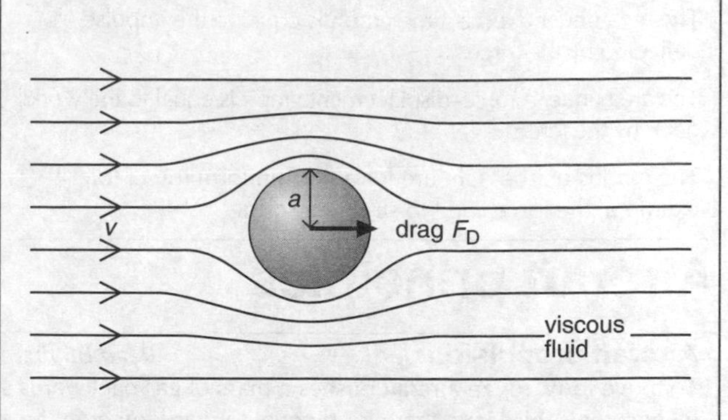

Above, a sphere is moving through a fluid at a speed v (for simplicity, in this and later diagrams, the air is shown moving, rather than the object). If the flow is streamline, as shown, then the ***drag*** F_D (resisting force from the fluid) is given by this equation, called ***Stokes' law***:

$$F_D = 6\pi\eta a v$$

Note:

- In this case, drag ∝ speed

A falling sphere will reach its ***terminal velocity*** when the forces on it balance (see right, and also B8), i.e.

weight = drag + upthrust

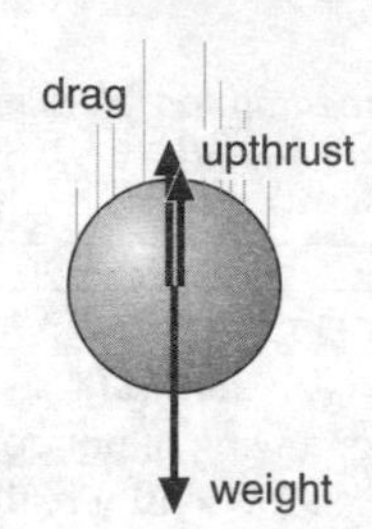

Turbulent flow

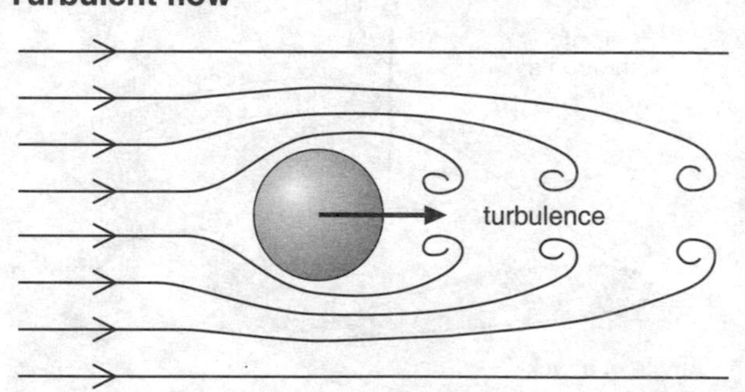

When a sphere (or other object) moves through a fluid, or a fluid flows through a pipe, the flow is only streamline beneath a certain ***critical speed***. Beyond this speed, it becomes *turbulent*, as shown above. Turbulence arises in most practical situations involving fluid flow.

Note:

- Poiseuille's equation and Stokes' law only apply at low speeds, where the flow is non-turbulent.

Reynolds' number (Re) This is defined as $\rho v l/\eta$, where ρ is the density of the fluid, v its speed (or that of an object moving through it), and l is a characteristic length (e.g. $2r$ for a sphere).

If the Reynolds number at the critical speed v_c is $(Re)_c$,

$$v_c = \frac{(Re)_c \eta}{\rho l}$$

Note:

- A dimensions check on this equation shows that Reynolds' number is dimensionless, i.e. it has no units.
- For a fluid flowing through a pipe, the flow usually becomes turbulent if the Reynolds number > 2500.

Drag from turbulent flow

Above the critical velocity, when the flow is turbulent, the drag F_D becomes dependent on the momentum changes in the fluid, rather than on the viscosity. It therefore depends on the density ρ of the fluid. For a sphere of radius r,

$$F_D = Br^2\rho v^2$$

where B is a number related to the Reynolds number.

Note:

- In this case, drag ∝ (speed)^2.
- Vehicles and aircraft are 'streamlined' in order to increase the critical speed and reduce drag (see B8, B10).

Drag coefficient The drag F_D on the moving vehicle (e.g. a car) can be worked out using its drag coefficient C_D. This is defined by the following equation:

$$F_D = \tfrac{1}{2}AC_D\rho v^2$$

where A is the cross-sectional (i.e. frontal) area of the vehicle, v its speed, and ρ is the density of the air.

Car designers try to achieve as low a drag coefficient as possible (see B10). A low value for C_D would be 0.30.

The equation of continuity

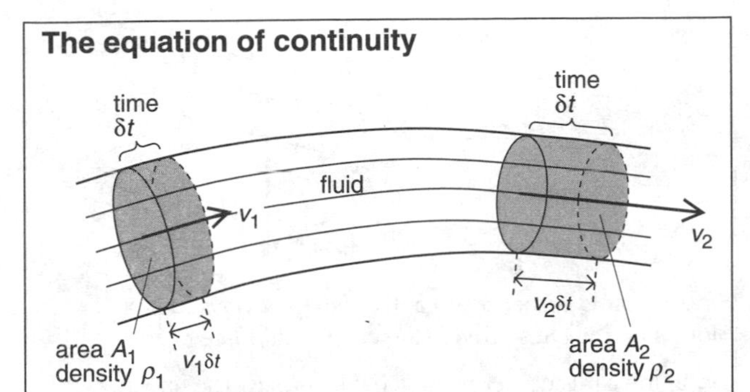

Above, in time δt, the same mass of fluid must pass through A_2 as through A_1, otherwise mass would not be conserved. But mass = density × volume. So

$$\rho_1 A_1 v_1 \delta t = \rho_2 A_2 v_2 \delta t$$

So $\rho_1 A_1 v_1 = \rho_2 A_2 v_2$

This is called the ***equation of continuity***.

If the fluid is *incompressible* (e.g. a liquid), then $\rho_1 = \rho_2$.

So $A_1 v_1 = A_2 v_2$ (1)

Bernoulli's equation

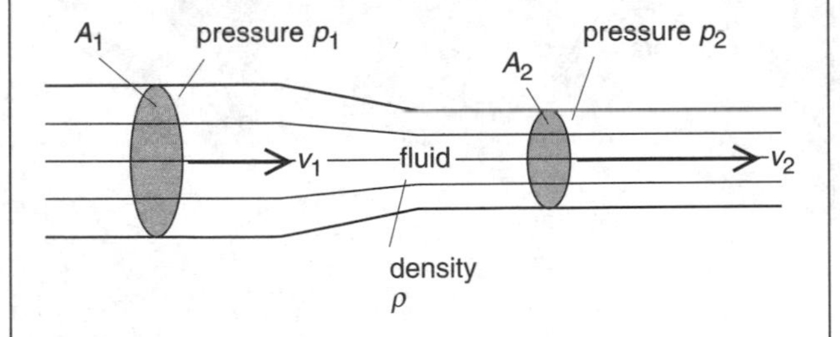

The fluid in the pipe above is incompressible. It is also non-viscous, i.e. in pushing the fluid through the pipe, there are no viscous forces to overcome, so no energy losses.

As $A_2 < A_1$, it follows from equation (1), that $v_2 > v_1$, i.e. the narrowing of the pipe makes the fluid speed up. As the fluid gains speed, work must be done to give it extra KE. And as this requires a resultant force, p_2 must be less than p_1. So the *increase* in velocity is accompanied by a *reduction* in pressure. This is called the ***Bernoulli effect***.

Working through the above steps mathematically, and assuming that no energy is wasted, gives the following:

$$p_1 + \tfrac{1}{2}\rho v_1^2 = p_2 + \tfrac{1}{2}\rho v_2^2 \qquad (2)$$

This is one form of ***Bernoulli's equation***.

Note:

- Bernoulli's equation follows directly from the law of conservation of energy.

Using the Bernoulli effect

Equation (2) is only valid for a non-viscous, incompressible fluid in a horizontal pipe. However the Bernoulli effect applies in situations where the fluid is both viscous and compressible. Examples include the following.

Aerofoil (wing) This is shaped so that the airflow speeds up across its top surface, causing a pressure drop above the wing and, therefore, a pressure difference across it. The result is an upward force which contributes to the total ***lift***. (Most of the lift is due to the ***angle of attack***: see B8.)

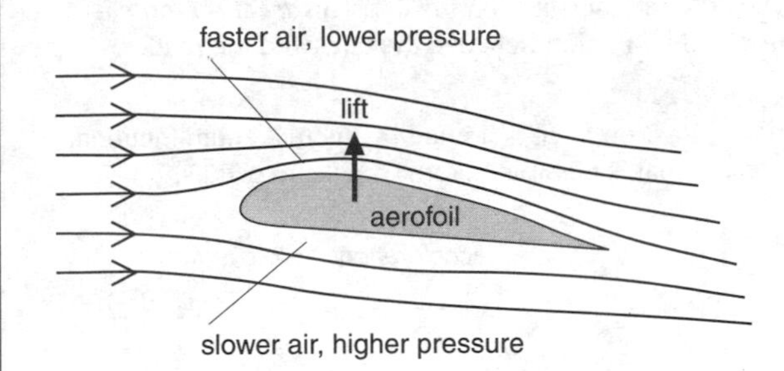

Venturi meter This can be used to find the rate of flow of a fluid. The faster the flow, the greater the pressure difference between X and Y (see right) and, therefore, the greater the height difference h on the manometer.

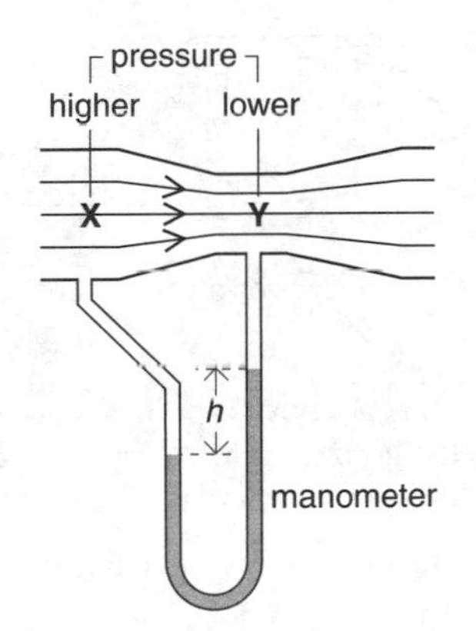

Spinning ball In some sports, 'spin' is used to make the ball 'swing'. Below, a spinning ball is moving through the air. Being viscous, air is dragged around by the surface of the ball, so the airflow is speeded up on side X and slowed down on side Y. This causes a pressure difference which produces a force.

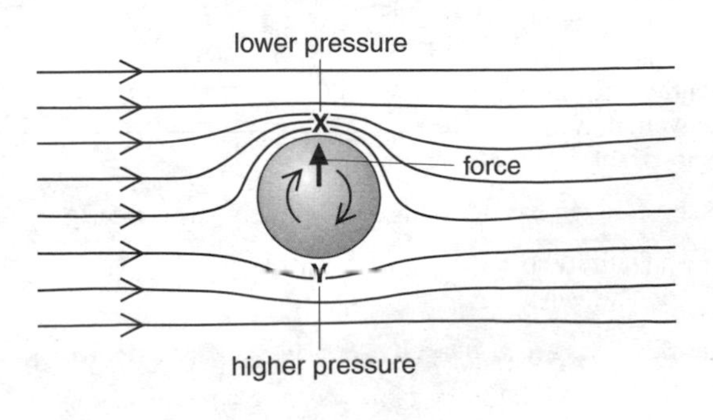

B10 Cars in motion

Traction forces

Read B5 and B6 first.

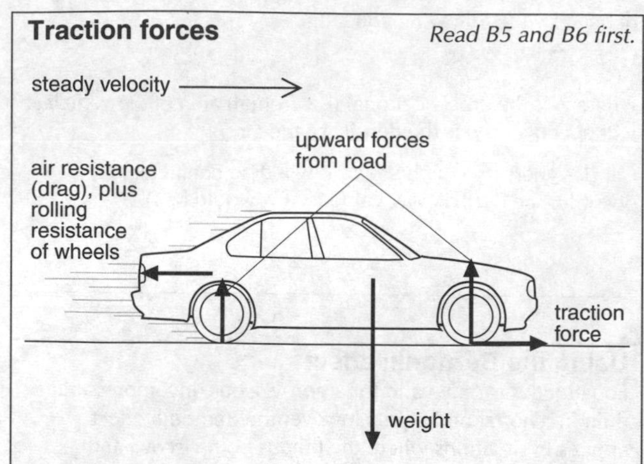

The car above is maintaining a steady velocity, so the forces on it must be balanced (i.e. in equilibrium). Also, the car has no rotational motion, so the moments of the forces about any point must be balanced.

The wheels driven by the engine exert a rearward force on the road, so the road exerts an equal forward force on the wheels – and therefore on the car. This ***traction force*** is provided by friction between the tyres and the road.

Note:

- The traction force is limited by the maximum frictional force that is possible before wheel slip occurs.

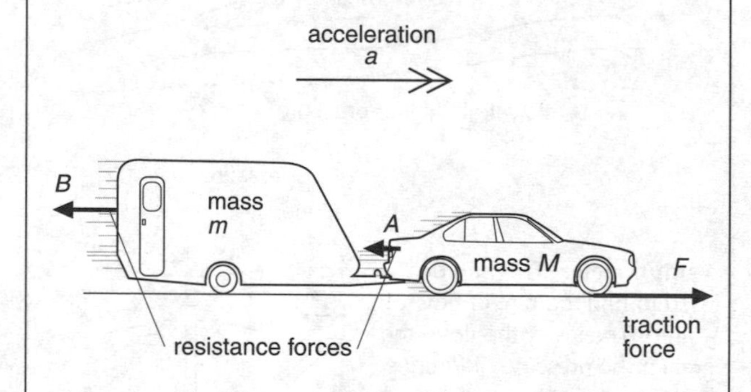

The car above is pulling a caravan. It is accelerating because the horizontal forces on it are unbalanced. (*For simplicity, the balanced vertical forces have not been shown.*)

Treating the car and caravan as a single object,

$$\text{acceleration} = a = \frac{\text{resultant force}}{\text{total mass}} = \frac{F - (A + B)}{M + m}$$

The traction force F' on the caravan comes from the car's tow bar. To calculate this force, it is best to start by drawing another diagram for the caravan alone, as on the right.

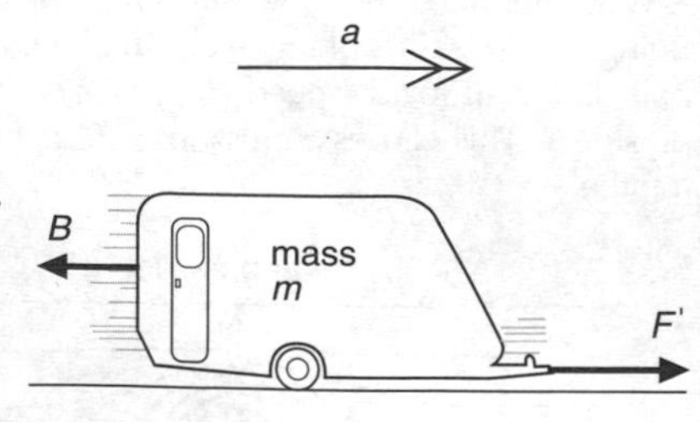

As the caravan has the same acceleration a as the car:

$$\text{resultant force on caravan} = F' - B = \text{mass of caravan} \times \text{acceleration} = ma$$

So $$F' = ma + B$$

Cornering

Read B11 first.

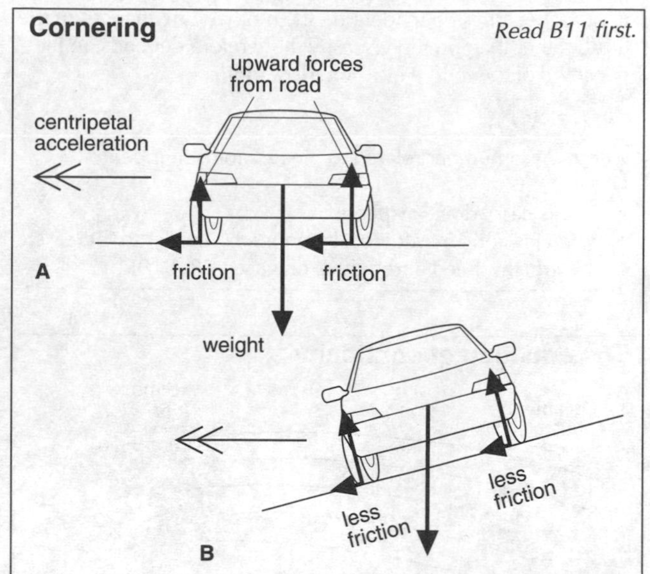

In A, a car is cornering on a flat road. Sideways frictional forces on the tyres provide the centripetal force needed.

In B, the car is cornering on a banked road. The upward forces from the road now have horizontal components. These provide some of the centripetal force required, so less sideways friction on the tyres is needed.

Braking

Car brakes are operated hydraulically (see F1).

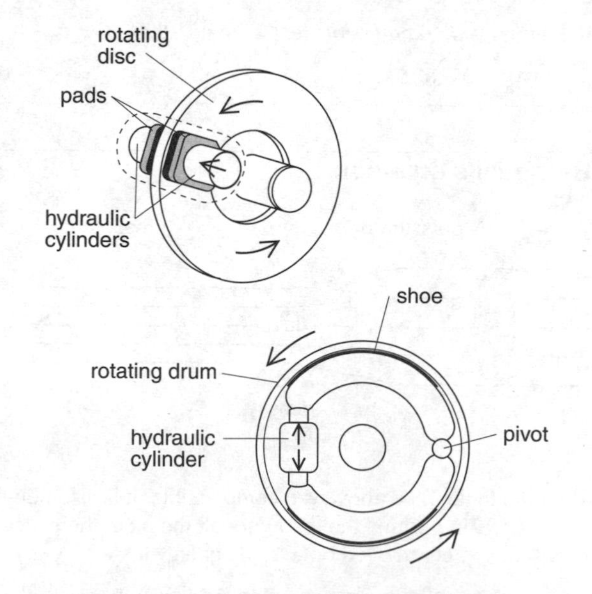

In a ***disc brake***, two friction pads are pushed against a steel disc which rotates with the wheel. In a ***drum brake***, two curved friction strips, called ***shoes***, are pushed against the inside of a steel drum which rotates with the wheel.

When the brakes are applied, the wheels exert a forward force on the road, so the road exerts an equal backward force on the wheels – and therefore on the vehicle. The braking force is limited by the maximum frictional force that is possible before skidding occurs.

Energy dissipation During braking, the car's kinetic energy is transferred into internal energy in the brakes. So the brakes heat up. The energy transferred Q and the temperature rise ΔT are linked by $Q = mc\Delta T$ (see F3).

Drag coefficient

The drag F_D on a moving vehicle such as a car can be worked out using its drag coefficient C_D. This is defined by the following equation:

$$F_D = \frac{1}{2} A C_D \rho v^2$$

where A is the cross-sectional (i.e. frontal) area of the vehicle, v is its speed, and ρ is the density of the air.

Car designers try to achieve as low a drag coefficient as possible. A low value for C_D would be 0.30.

This **drag force** is what is commonly called air resistance.

Maximum speed of a car

Provided that the maximum frictional force between the tyres and the road is not exceeded the maximum traction force F between the road and a car is fixed by the size of the engine.

When the traction force is equal to the sum of all the resistive forces acting the maximum speed is reached.

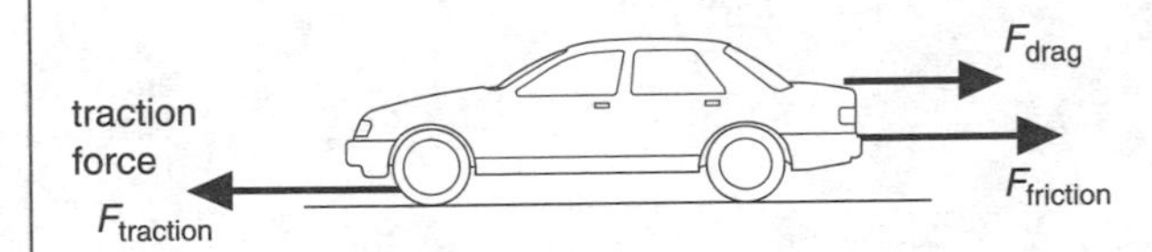

The frictional force is approximately constant when motion has started. Initially the drag force will be small and the car is able to accelerate. As speed builds up eventually a speed will be reached when

$$F_{traction} = F_{friction} + F_{drag}$$

$$F_{traction} = F_{friction} + \frac{1}{2} A C_D \rho v^2$$

This equation shows that higher maximum speeds can be achieved by

- increasing the traction force that the engine can provide
- reducing the drag coefficient by streamlining the car by reducing the frontal area of the vehicle and drag coefficient.

Maximum speed and power of an engine

At the maximum speed the output power of the engine

$$P_{max} = \text{traction force} \times \text{velocity}$$

The power used to overcome friction is much smaller than that used to overcome the drag force, so the drag force is the major limiting factor. Hence

$$P_{max} = \frac{1}{2} A C_D \rho v^2 \times v$$

$$P_{max} \propto v^3 \text{ (approximately)}$$

This means that to increase the speed by about 25% the engine power output has to be doubled.

Note: The above analysis applies to any object (from people to aircraft) moving against air resistance.

Minimum stopping distances

When a driver notices a hazard that requires an emergency stop, he or she has to react to the emergency and apply the brakes. The brakes then have to bring the car to rest.
The minimum stopping distance d_m is given by

$$d_m = vt + \frac{v^2}{2a}$$

where v is the speed of the vehicle, t is the reaction time of the driver, and a is the deceleration of the vehicle.

(The first term is the speed of travel multiplied by the reaction time or 'thinking distance' and the second term is derived from $v^2 = u^2 + 2as$.)

A typical reaction time is 0.7 s and a typical deceleration is 7.5 m s^{-2}.

The graph shows the stopping distances for this data.

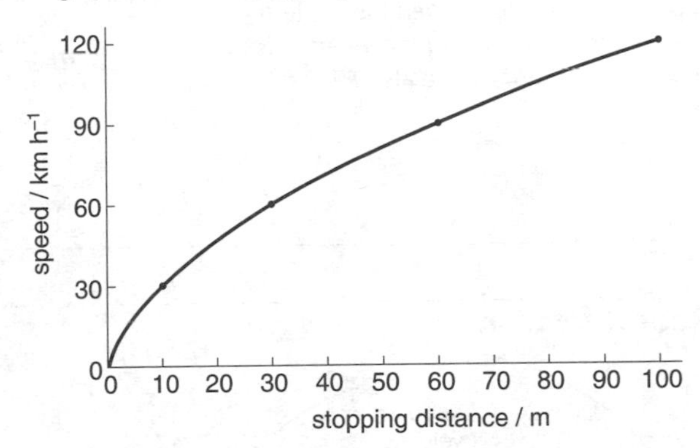

Stopping distances will increase if

- the reaction time is increased, owing to driver fatigue for example
- a car is heavily loaded
- condition of the car's brakes is poor
- road conditions are poor, owing to rain or ice for example.

Car safety

In the event of an accident occurring cars and equipment in them are designed to protect passengers.

Seat belts and air bags

Car seat belts and air bags work on the principle that the force on a passenger is reduced by

- losing momentum over a longer period of time (recall that $F = \Delta(mv)/t$)
- losing kinetic energy over a longer distance (recall that $F = \Delta(E_K)/s$)

The design is such that the KE is reduced to zero in as long a time as possible but before the passenger hits a rigid part of the car.

The energy of the motion is converted into thermal energy and elastic energy.

Crumple zones

Cars will deform on impact and this could crush the passengers. Cars are therefore designed with weaker sections at the front so that these parts deform easily. This deformation transforms most of the energy involved in the collision so that the passenger compartment remains intact.

B11 Circular motion

Angular displacement

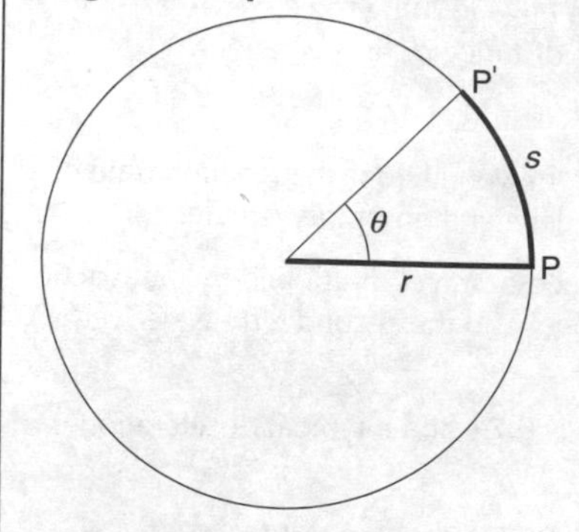

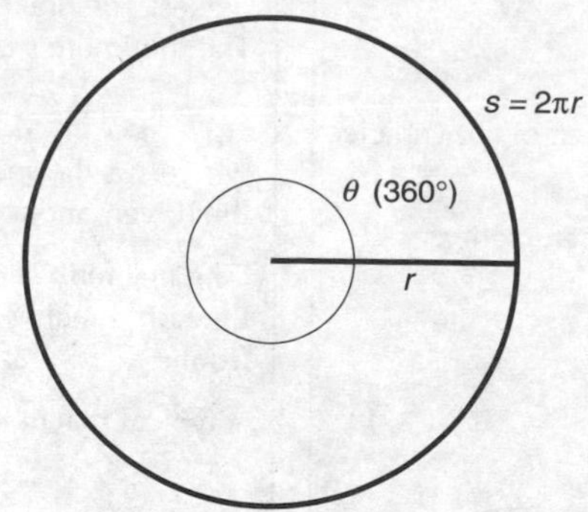

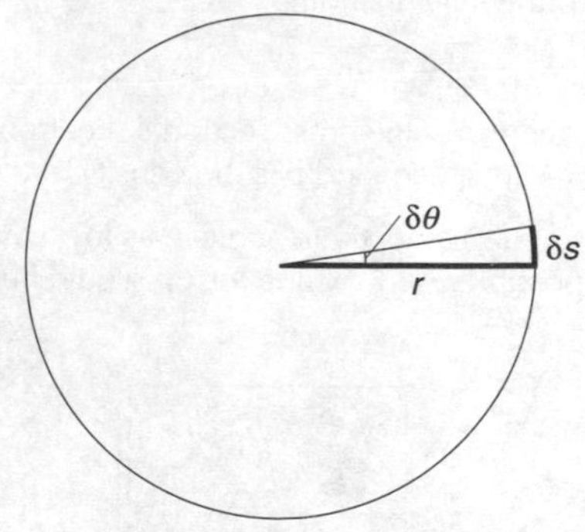

If point P moves to P′, then the angle θ is called the ***angular displacement***. It is measured in ***radians*** (see also A1):

$$\theta = \frac{s}{r}$$

If s is the full circumference of the circle,

$$\theta = \frac{2\pi r}{r} = 2\pi$$

So $\quad 2\pi$ radians = 360°

$\therefore$ 1 radian = $\frac{360}{2\pi}$ = 57.3°

Above, $\delta\theta$ is a very small angle. ($\delta\theta$ counts as one symbol.) δs is so small that it can either be the arc of a circle or the side of a triangle. So

$$\sin \delta\theta = \frac{s}{r} = \delta\theta$$

i.e. for *small* angles $\sin \delta\theta = \delta\theta$.

Rate of rotation

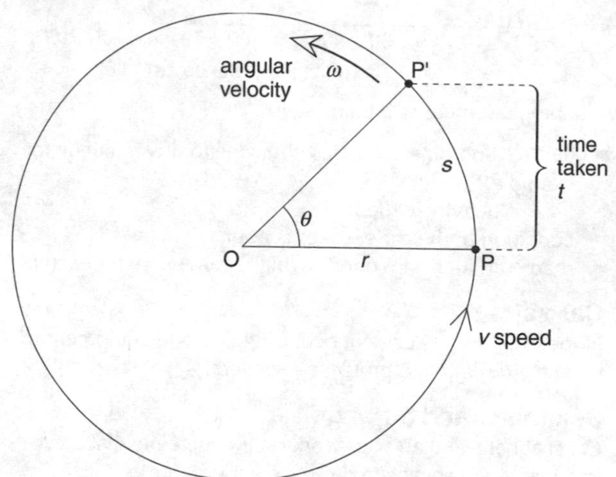

P is a point on a wheel which is turning at a steady rate. In time t, it moves to P′. The rate of rotation can be measured either as an ***angular velocity*** or as a ***frequency***.

Angular velocity ω

$$\text{angular velocity} = \frac{\text{angular displacement}}{\text{time taken}}$$

In symbols $\quad \omega = \frac{\theta}{t}$

For example, if a wheel turns through 10 radians in 2 seconds, then ω = 5 rad s^{-1}.

Frequency f

$$\text{frequency} = \frac{\text{number of rotations}}{\text{time taken}}$$

Frequency is measured in hertz (Hz). For example, if a wheel completes 12 rotations in 4 seconds, then f = 3 Hz.

Period T This is time taken for one rotation. If a wheel makes 3 complete rotations per second (f = 3 Hz), then the time taken for one rotation is $\frac{1}{3}$ second. So

$$T = \frac{1}{f}$$

Linking *v*, *ω*, and *r*

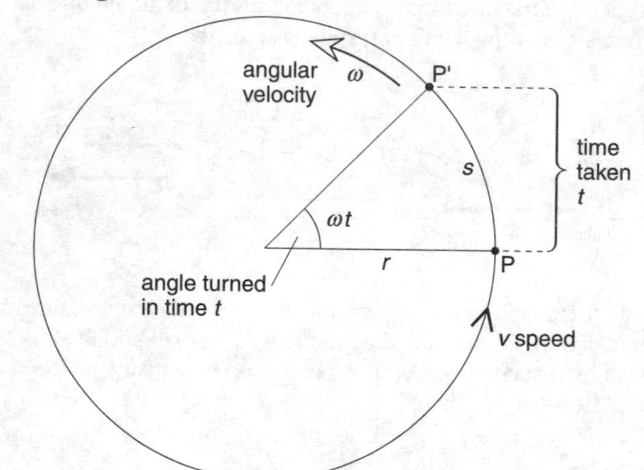

Above, a particle is moving in circle with a steady speed v. (It is not a steady velocity because the direction of the velocity vector is changing.) The particle moves a distance s in time t, so

$$v = \frac{s}{t}$$

As the angular velocity is ω, the angle turned in time t is ωt (from the equation on the left). But $\omega t = s/r$. So $s = \omega t r$. Substituting this in the above equation gives

$$v = \omega r$$

Linking ω, f, and T As there are 2π radians in one full rotation (360°),

$$\omega = 2\pi f$$

For example, a wheel turning at 3 rotations per second (f = 3 Hz) has an angular velocity of 6π radians per second.

As $T = 1/f$, it follows from the previous equation that

$$T = \frac{2\pi}{\omega}$$

Centripetal acceleration

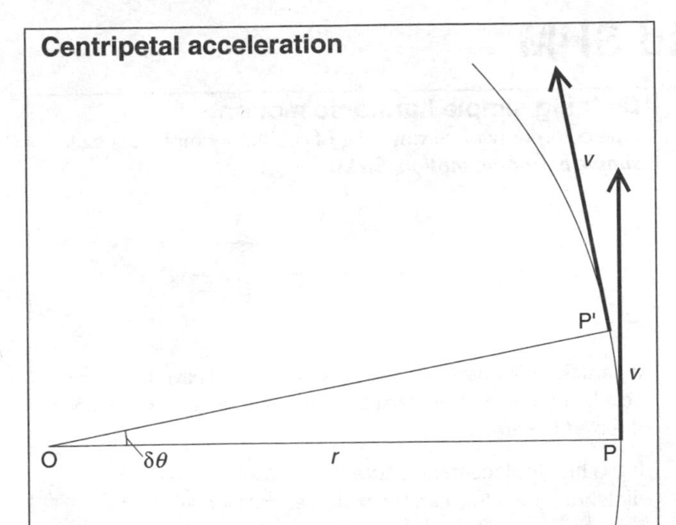

Above, a particle is moving in a circle with a steady speed v. The diagram shows how the velocity vector changes direction as the particle moves from P to P′ in time δt.

Below, the velocity vectors from the previous diagram have been used in a triangle of vectors (see B4).

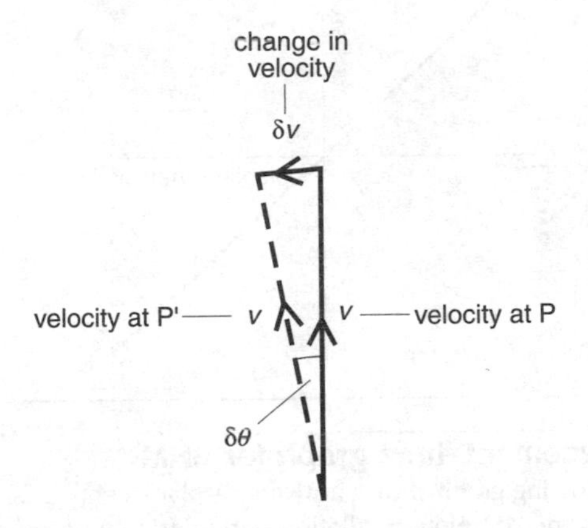

The δv vector represents the *change* in velocity because it is the velocity vector which must be *added* to the velocity at P to produce the new velocity (the resultant) at P′. Note that the change in velocity is towards O. In other words, the particle has an *acceleration* towards the centre of the circle. This is called ***centripetal acceleration***.

If a is the centripetal acceleration, $a = \dfrac{\delta v}{\delta t}$

But, from the triangle above, $\delta\theta = \dfrac{\delta v}{v}$. So $\delta v = v\delta\theta$

Substituting this in the previous equation, $a = \dfrac{v\delta\theta}{\delta t}$

But $\delta\theta = \omega\delta t$. So $a = v\omega$

Using $v = \omega r$, two more versions of the above equation can be obtained. So

$a = v\omega$ $\qquad a = \dfrac{v^2}{r}$ $\qquad a = \omega^2 r$

For example, if a particle is moving at a steady speed of $3\ \text{m s}^{-1}$ in a circle of radius 2 m, its centripetal acceleration a is found using the middle equation: $a = 3^2/2 = 4.5\ \text{m s}^{-2}$.

Note:

- When something accelerates, its velocity changes. As velocity is a vector, this can mean a change in *speed* or *direction* (or both). Centripetal acceleration is produced by a change in direction, not speed.

Centripetal force

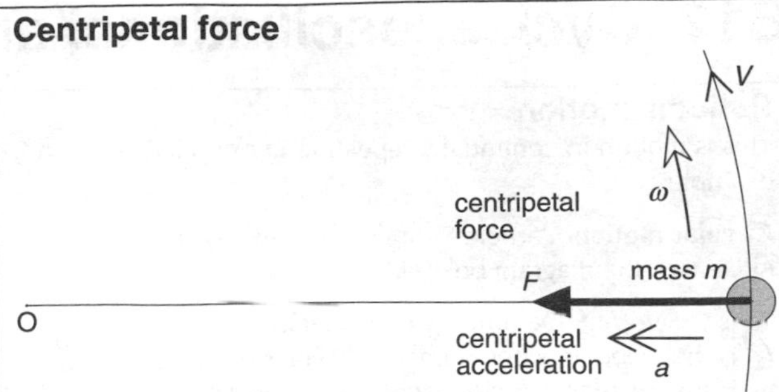

To produce centripetal acceleration, a ***centripetal force*** is needed. It must act towards the centre of the circle. The centripetal force F, mass m, and centripetal acceleration a are linked by the equation $F = ma$. So, using the equations for a in the previous column,

$F = mv\omega$ $\qquad F = \dfrac{mv^2}{r}$ $\qquad F = m\omega^2 r$

Note:

- Centripetal force is *not* produced by circular motion. It is the force *needed* for circular motion. Without it, the object would travel in a straight line.

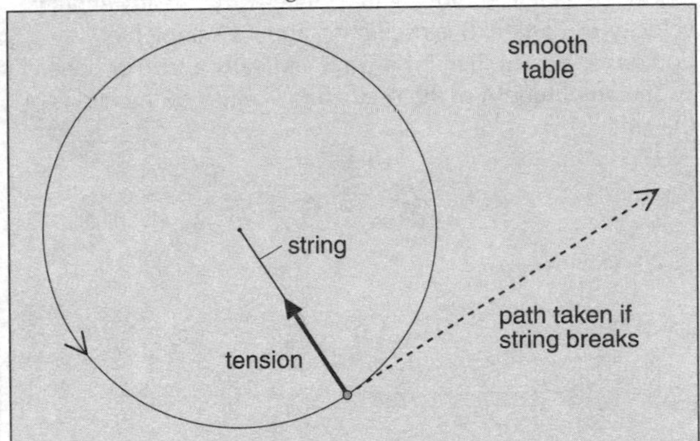

Above, a mass moves in a circle on a smooth table. The tension in the string provides the centripetal force needed. There is no outward 'centrifugal force' on the mass. If the string breaks, the mass travels along a tangent.

Angle of bank An aircraft must bank to turn. This is so that the lift L (from the wings) and the weight mg can produce a resultant to provide the centripetal force F, where

$$F = \frac{mv^2}{r}$$

In the triangle of vectors (below right):

$L\cos\theta = mg$ and $L\sin\theta = \dfrac{mv^2}{r}$

Dividing the second equation by the first gives $\tan\theta = v^2/r$, where θ is the angle of bank required for the turn.

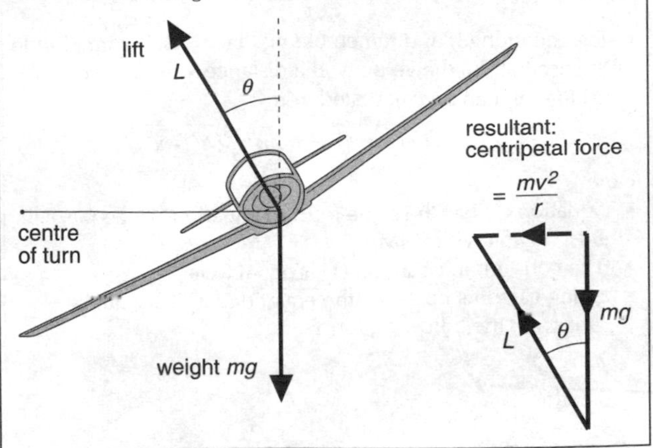

B12 Cycles, oscillations, and SHM

Periodic motion

This is motion in continually repeating ***cycles***. Here are two examples:

Circular motion Particle P moving at a steady speed in a circle (see the diagram below).

T is the ***period*** (the time for one cycle).
f is the ***frequency*** (the number of cycles per second).
ω is the ***angular velocity*** (measured in rad s^{-1}).

T, f, and ω are linked by the equations below:

$$T = \frac{1}{f} \qquad \omega = 2\pi f \qquad T = \frac{2\pi}{\omega}$$

(see B11)

Oscillatory motion (e.g. a swinging pendulum)
T, f, and ω are also used when describing oscillatory motion, although ω has no direct physical meaning. They are linked by the same equations as for circular motion.

Linking circular motion and SHM

Below, particle P is moving in a circle with a steady angular velocity ω. Particle B is oscillating about O along the horizontal axis so that it is always vertically above or beneath P. The amplitude A of the oscillation is equal to the radius of the circle, r.

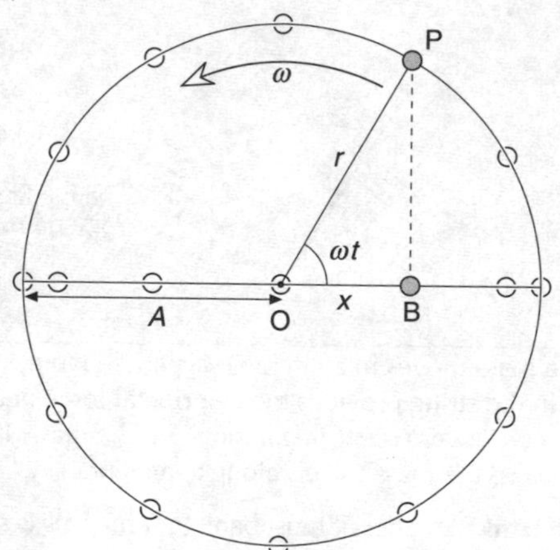

For particle B, $\quad x = r\cos\omega t \qquad (2)$

Using calculus, B's velocity v and acceleration a can be found from the above equation. These are the results:

$$v = -r\omega\sin\omega t \qquad (3)$$

$$a = -r\omega^2\cos\omega t \qquad (4)$$

From equations (4) and (2), it follows that

$$a = -\omega^2 x \qquad (5)$$

From the diagram, $\sin\omega t = \frac{\sqrt{r^2 - x^2}}{r}$

Using equation (3) and remembering that $r = A$, the amplitude of the oscillation, the velocity at a distance x from the centre of oscillation can be calculated from

$$v = \omega\sqrt{A^2 - x^2} = 2\pi f\sqrt{(A^2 - x^2)} \qquad (6)$$

Note:

- Equation (5) has the same form as equation (1). So particle B is moving with SHM.
- The constant in equation (1) is equal to ω^2.
- Using calculus notation, the equation for SHM can be written in the following form:

$$\frac{d^2x}{dt^2} = -\omega^2 x$$

Defining simple harmonic motion

One commonly occurring type of oscillatory motion is called ***simple harmonic motion*** **(SHM)**.

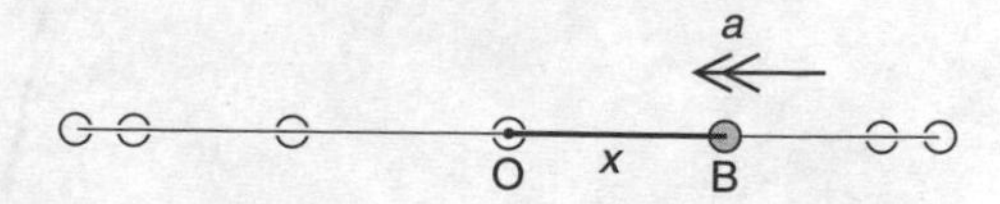

If particle B (above) oscillates about O with SHM, its acceleration is proportional to its displacement from O, and directed towards O.

If x is the displacement, and a is the acceleration (in the x direction), then this can be expressed mathematically:

$$a = -\text{(positive constant)}\,x \qquad (1)$$

The minus sign indicates that a is always in the opposite direction to x.

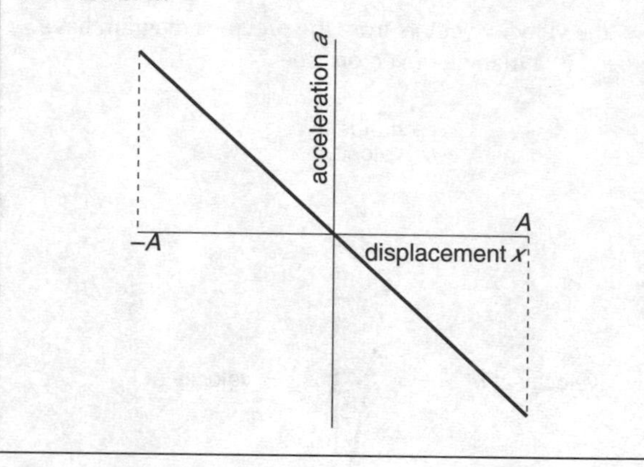

Displacement–time graph for SHM

The following graph shows how the displacement varies with time for one complete oscillation starting from the centre of oscillation.

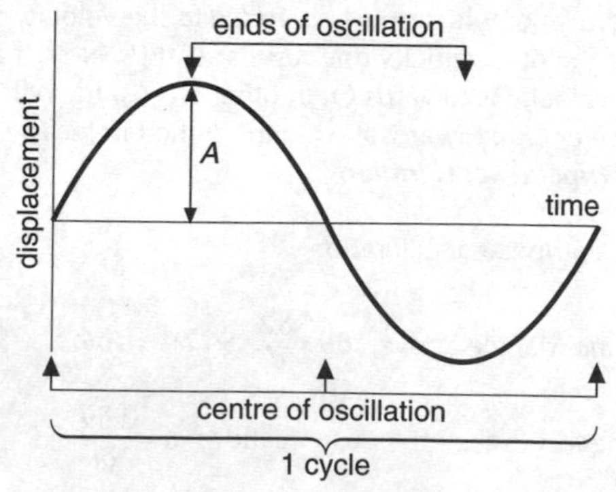

A is the amplitude of the oscillation.

Phase of the oscillation

An oscillation that has the same period but reaches its peak at a different time to that shown above is said to ***oscillate out of phase.*** The phase difference between oscillators is quoted as an angle not a time. Two important examples are shown below:

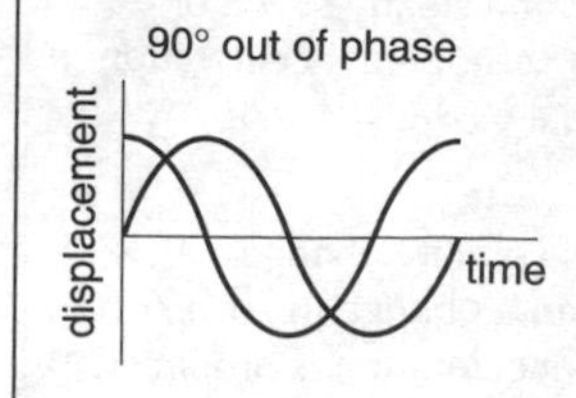

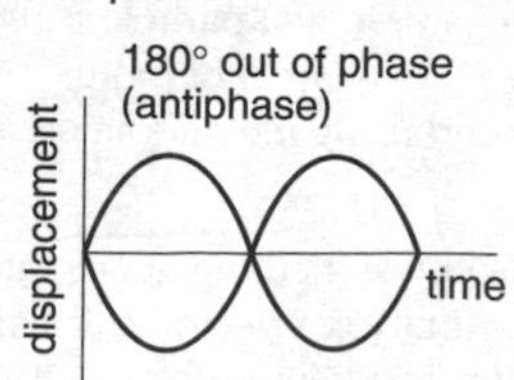

SHM and a mass on a spring

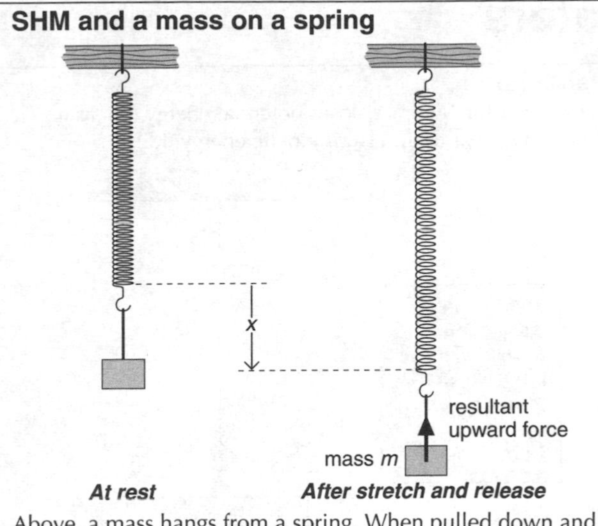

Above, a mass hangs from a spring. When pulled down and released, the mass makes small, vertical oscillations.

Springs obey Hooke's law. This means that extension x of a spring is directly proportional to the applied force (or load) F.

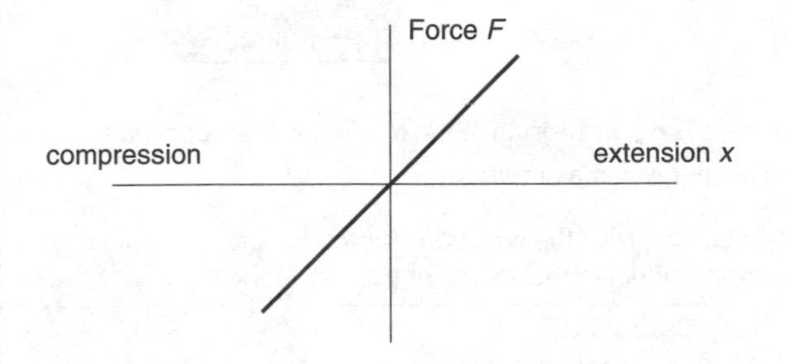

The graph of applied force against extension is a straight line through the origin.

The gradient of this graph is equal to the ***spring constant*** or ***stiffness of the spring***, k:

$$k = \frac{F}{x}$$

So, if the mass m is pulled down by x and then released,

resultant upward force on mass = kx

But force = mass × acceleration

So acceleration (upwards) = $\frac{kx}{m}$

So acceleration (in x direction) $a = -\frac{kx}{m}$

Comparing this with equation (5) shows that the motion is SHM and that

$$\frac{k}{m} = \omega^2$$

As $T = \frac{2\pi}{\omega}$ $\quad T = 2\pi\sqrt{\frac{m}{k}}$

In any oscillating system to which Hooke's law applies, the motion is SHM.

SHM and the simple pendulum

Provided its swings are small, and air resistance is neglible, a simple pendulum moves with SHM. The following analysis shows why.

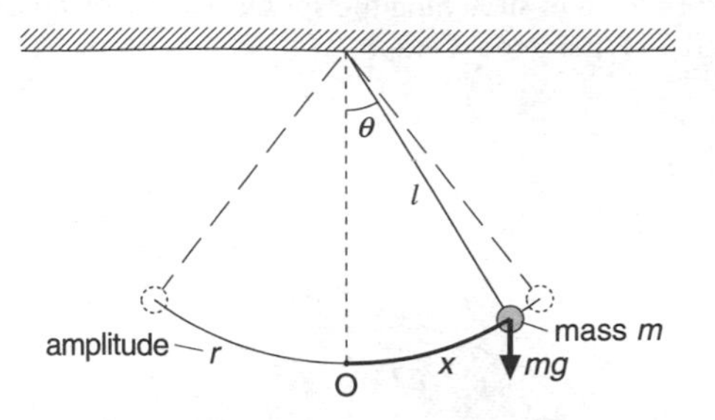

The mass m (above) has been displaced by x. It is being pulled towards O by a component of its weight:

force (towards O) = $mg \sin\theta$

But for very small angles $\sin\theta = \frac{x}{l}$ (see B11)

So force (towards O) = $\frac{mgx}{l}$

But force = mass × acceleration

So acceleration (towards O) = $\frac{gx}{l}$

So acceleration (in x direction) $a = \frac{-gx}{l}$

Comparing this with equation (5) shows that the motion is SHM and that

$$\frac{g}{l} = \omega^2$$

As $T = \frac{2\pi}{\omega}$ $\quad T = 2\pi\sqrt{\frac{l}{g}}$

Note:

- T does not depend on the amplitude (for smaller swings, the period stays the same). This is true for *all* SHM.

The graphs on the right are for an object moving with SHM: for example, a pendulum making small swings.
At the ends of each oscillation, the velocity is zero. The displacement and acceleration have their peak values, but when one is positive, the other is negative, and vice versa.
At the centre, the velocity has its peak positive or negative value, but the displacement and acceleration are both zero.

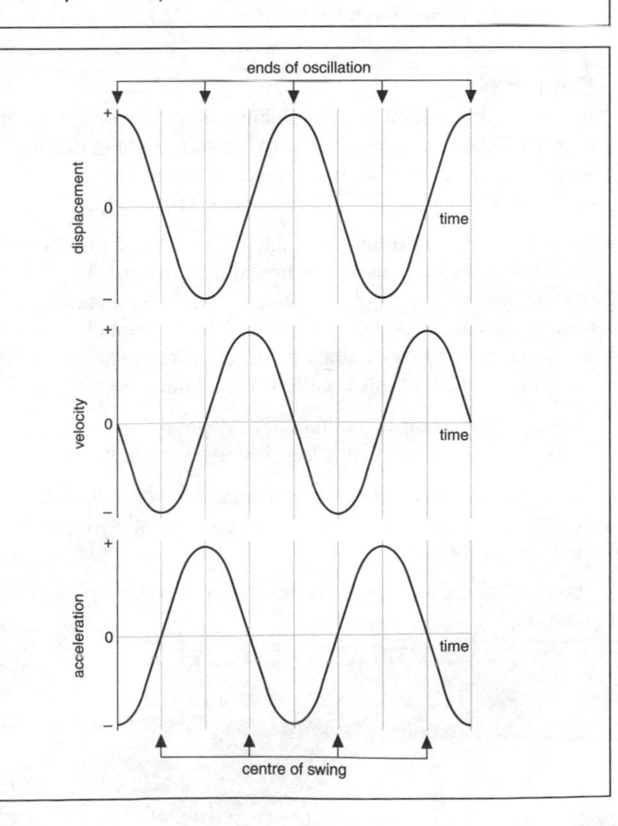

B13 Energy changes in oscillators

Mass–spring system

The elastic (or strain energy) stored in a mass–spring system is the work done in stretching the spring. This is the area under the force–displacement graph.

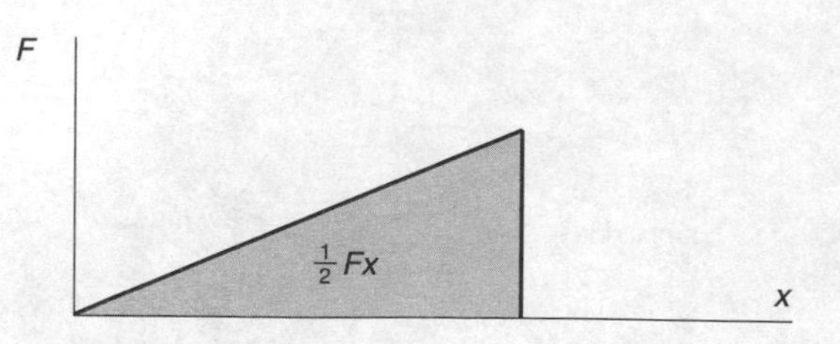

$$\text{elastic energy stored} = \tfrac{1}{2}Fx$$

Since $F = kx$, another useful equation for stored energy is

$$\text{elastic energy stored} = \tfrac{1}{2}kx^2$$

As the spring moves toward the equilibrium position it loses elastic stored energy and gains kinetic energy. But in the absence of any damping the total energy remains constant.

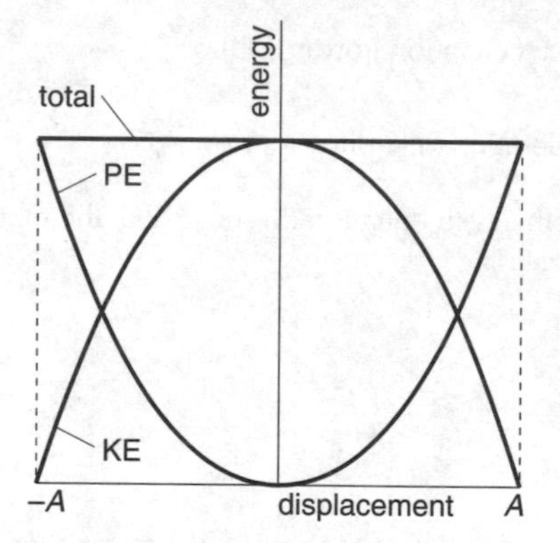

$$\text{maximum elastic stored energy} = \tfrac{1}{2}kA^2$$
$$\text{maximum kinetic energy} = \tfrac{1}{2}m(A\omega)^2$$

Pendulum

For a pendulum the mass loses potential energy (PE) as it swings downwards and gains kinetic energy (KE).

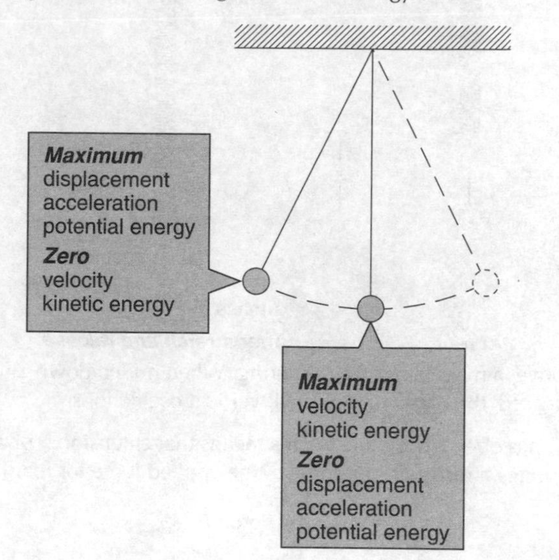

If there is no air resistance the total PE + KE is constant.

total energy = maximum KE = $\frac{1}{2}m(A\omega)^2$

Notice that in all the equations for total energy the total energy is proportional to the *square* of the amplitude (A^2).

Damping

A mass–spring system or a pendulum will not go on swinging for ever. Energy is gradually lost to the surroundings due to air resistance or some other resistive force and the oscillations die away. This effect is called **damping**.

In road vehicles, dampers (wrongly called 'shock absorbers') are fitted to the suspension springs so that unwanted oscillations die away quickly. Some systems (for example moving-coil ammeters and voltmeters) have so much damping that no real oscillations occur. The minimum damping needed for this is called **critical damping**.

The rate at which the amplitude falls depends on the fraction of the existing energy that is lost during each oscillation.

In a **lightly damped** system only a small fraction is lost so that the amplitude of one oscillation is only slightly lower than the one before.

The graphs here are for oscillations with different degrees of damping.

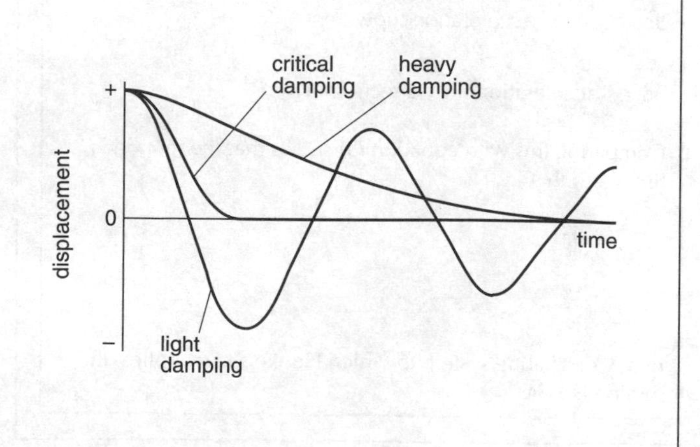

Forced oscillations and resonance

Natural frequency
This is the frequency of the oscillation that occurs when the mass of an oscillator (such as a pendulum bob or mass of a mass–spring system) is displaced and then released. The only forces acting are the internal forces of the oscillating system.

Forced oscillations
Forced oscillations occur when an external periodic force acts on an object that is free to oscillate.

Examples include:
- engine vibrations making bus windows oscillate
- the spinning drum causing vibrations in a washing machine
- the body of a guitar vibrating when a string is plucked.

The body that is forced to oscillate vibrates at the same frequency as that of the external source that is providing the energy.

The amplitude of the oscillations produced depends on
- how close the external frequency is to the natural frequency of the oscillator
- the degree of damping of the oscillating system.

Resonance
Resonance occurs when the frequency of the external source that is driving the oscillation is equal to the natural frequency of the oscillator that is being driven into oscillations.

When resonance occurs the amplitude of the resulting oscillations is a maximum.

The following graph shows how the amplitude of an oscillator with natural frequency f_0 varies with the frequency of the frequency of the source that is driving the oscillations.

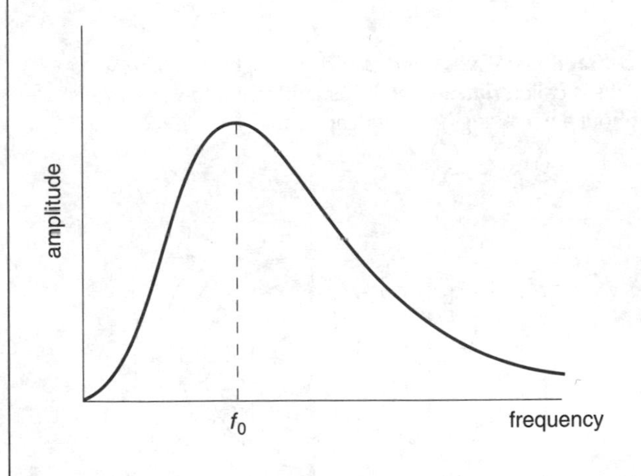

Graphs such as these are called ***frequency response graphs***.

Effect of damping on resonance
When a system is lightly damped, it loses very little energy during an oscillation. If it is being forced to oscillate it will retain most of the energy put into it so that the energy stored builds up and the amplitude becomes very large.

When a system is heavily damped, energy is lost quickly so that the amplitude is lower.

The graphs shows the frequency response for lightly and heavily damped oscillators that have the same natural frequency.

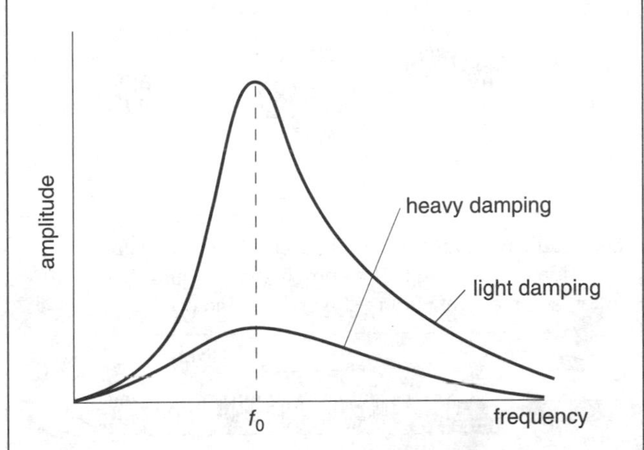

Note:
The amplitude of an oscillation stops increasing when the energy put in each cycle is equal to that lost during the cycle.

Electrical resonance
Electrical oscillators made from combinations of capacitors and inductors (coils) can also be forced into oscillations or be made to resonate.

Resonance in an *RCL* circuit

See also D4.

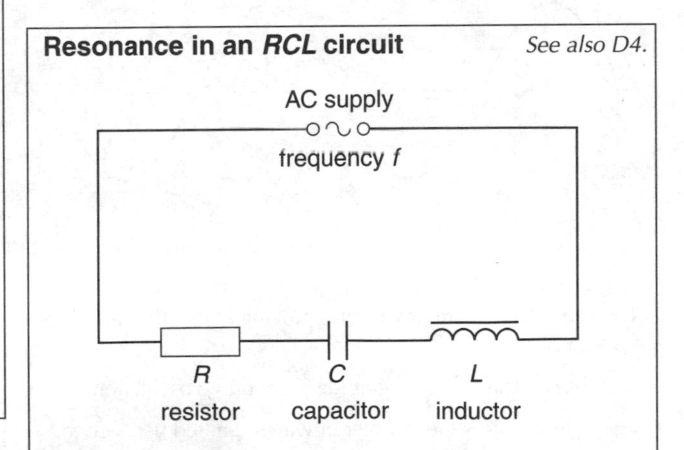

In the circuit shown the current in the circuit reaches a maximum at the resonant frequency. A circuit such as this is used to pick out the required transmission in a radio receiver. Only the frequencies in a small range around the frequency of the radio channel are selected.

The natural frequency of an electrical oscillator depends on the capacitance and inductance of the coil used. By varying the capacitance you can tune in to different channels.

The range of frequencies selected depends on the damping which in turn depends on the resistance in the circuit.

C1 Waves and rays

Types of wave motion

Waves transfer energy from one place to another. Where ever there is wave motion, there must be:

- a source of oscillation,
- a material or field which can transmit oscillations.

Wave motion can be demonstrated using a 'slinky' spring, as shown below. The moving waves are called ***progressive waves***. There are two main types.

Transverse waves The oscillations are at right-angles to the direction of travel:

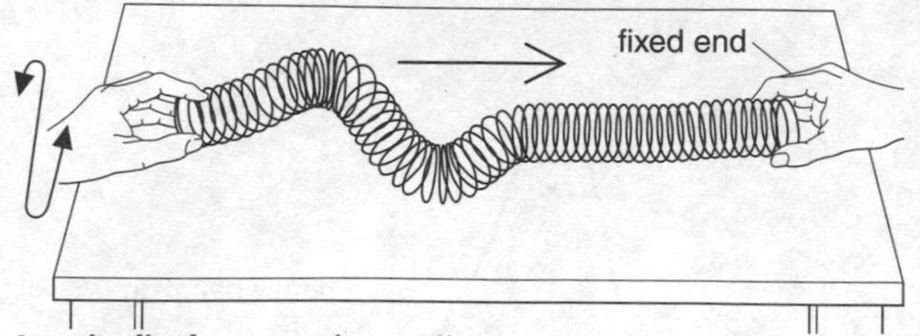

Longitudinal waves The oscillations are in line with the direction of travel, so that a compression ('squash') is followed by a rarefaction ('stretch'), and so on.

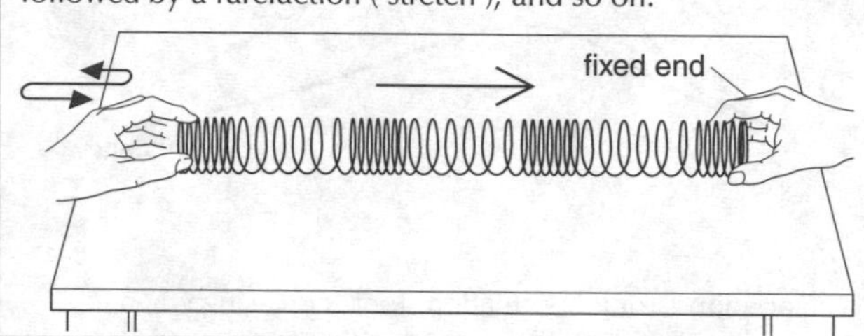

Wave features

wavelength

amplitude

wavelength

The waves above are tiny ripples moving across the surface of some water.

Amplitude This is the magnitude (size) of the oscillation.

Frequency This is the number of waves emitted per second. The SI unit is the ***hertz*** (**Hz**). A frequency of, say, 5 Hz means that 5 waves are being emitted per second.

Wavelength In the example above, this is the distance between one wave crest and the next.

Speed The speed, frequency, and wavelength of a wave are linked by this equation:

speed = frequency × wavelength

For example, if the frequency is 5 Hz and the wavelength is 2 m, then the wave speed is 10 m s^{-1}. You could predict this result without the equation. If there are 5 waves per second and each occupies a length of 2 m, then each wave will travel 2 m in $\frac{1}{5}$ s, or 10 m in a second.

Waves in a ripple tank

Wave effects can be investigated using a **ripple tank** in which ripples travel across the surface of shallow water.

Reflection Waves striking an obstacle are reflected. The angle of incidence is equal to the angle of reflection.

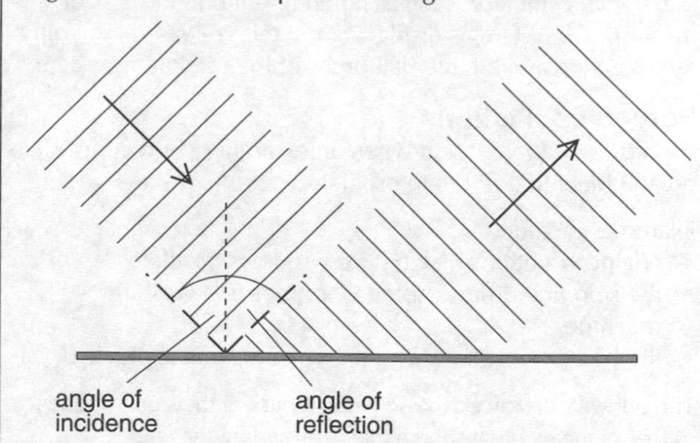

Refraction When waves are slowed down, they are ***refracted*** (bent), provided the angle of incidence is not zero. In a ripple tank, the waves can be slowed by using a flat piece of plastic to make the water shallower.

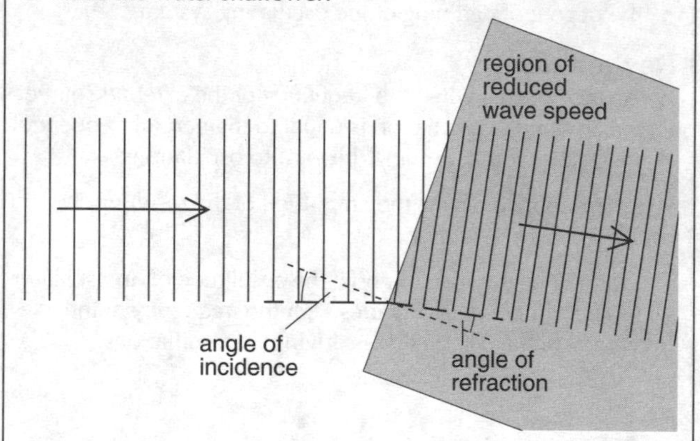

Diffraction Waves bend round the edges of a narrow gap. This is called ***diffraction***. It is significant if the gap size is about a wavelength. Wider gaps cause less diffraction.

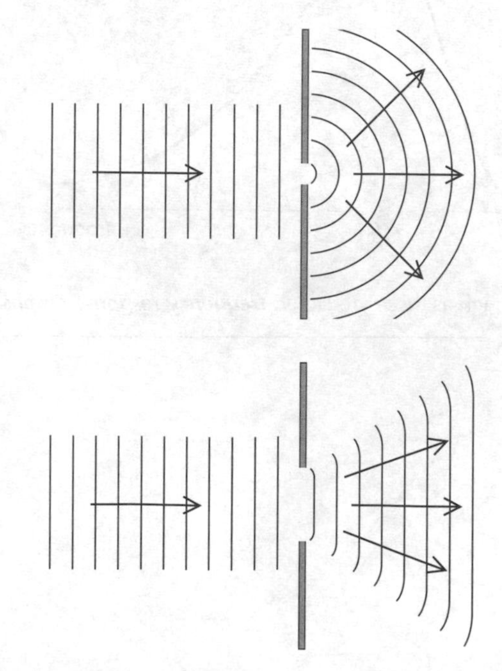

Interference If two identical sets of waves overlap, they may either reinforce or cancel each other, depending on whether they are in phase ('in step') or out of phase.

Light waves and rays

Light waves are transverse electric and magnetic field ripples. They can travel through empty space (a vacuum).

Light waves come from atoms. A burst of wave energy is given off whenever an electron loses energy by dropping to a lower orbit. Sometimes, this burst of wave energy acts like a particle, a **photon**.

The speed of light in empty space is 300 000 km s^{-1}. It is less in transparent materials such as air and water, which is why these materials refract light.

Our eyes experience different wavelengths as different colours. These range from red (0.000 7 mm) down to violet (0.000 4 mm). As light waves are so short, they are not noticeably diffracted by everyday objects.

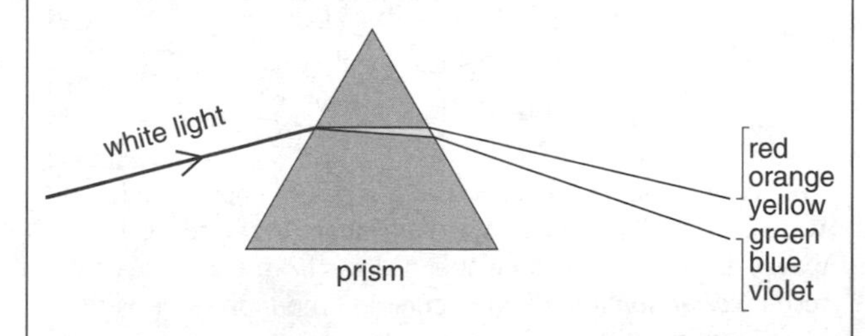

White light is not a single colour, but a mixture of all the colours of the rainbow. When white light enters a prism, the different wavelengths (i.e. colours) are slowed by different amounts, so they are refracted by different amounts. As a result, the white light splits into a range of colours called a ***spectrum***. The spreading effect is known as ***dispersion***. Violet light has the lowest speed in glass, so it is refracted most.

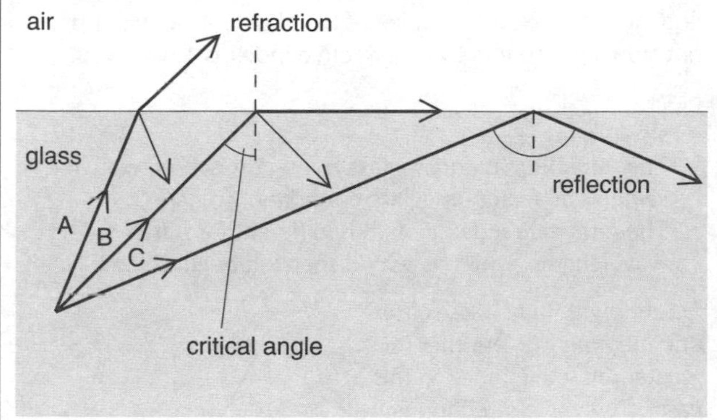

In diagrams, a ***ray*** is a line showing the direction in which light is travelling. In the diagram above:

- Ray A is mainly refracted when it passes from glass to air, although some light is also reflected.
- Ray B is also refracted, but only just. Beyond the ***critical angle***, no refraction can occur.
- Ray C strikes the surface at too great an angle for any refraction to occur. So all the light is reflected. This is called ***total internal reflection***.

In an optical fibre, light entering one end of the glass or plastic fibre is totally internally reflected until it comes out of the other.

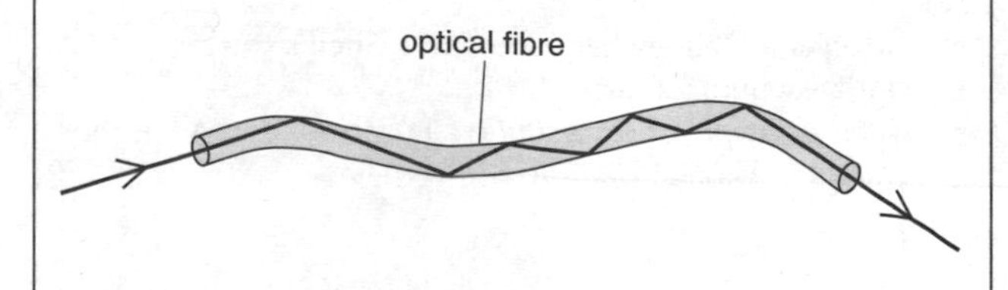

The electromagnetic spectrum

Light is one member of a whole family of transverse waves called the ***electromagnetic spectrum***. In empty space, these waves all travel at the same speed: 300 000 km s^{-1}.

Electromagnetic waves are emitted whenever electrons or other charged particles oscillate or lose energy. The greater the energy change, the lower the wavelength.

Wavelength in metres	Wave type	Typical sources, uses, and effects
10^{5}		
	radio waves: LW, MW, SW, VHF, UHF	From electrons oscillating in aerial. Used for communication.
3×10^{-2}		
	microwaves	From electrons oscillating in magnetron. Used for radar, communication, cooking.
10^{-3}		
	infrared	From hot objects. Used for heating.
7×10^{-7}		
	light	From very hot objects. Only form of radiation visible to human eye. Used for communication.
4×10^{-7}		
	ultraviolet	From very hot objects. Ionizes atoms. Causes fluorescence (makes some chemicals glow). Kills germs. Causes suntan.
10^{-9}		
	X-rays	From electrons stopped rapidly in X-ray tube. Causes ionization and fluorescence. Used for X-ray photography.
	gamma rays	From radioactive materials. Uses and effects as for X-rays.

Sound waves

Sound waves are longitudinal. When, say, a loudspeaker cone vibrates, it sends compressions and rarefactions through the air which the ear can detect.

Speed The speed of sound waves in dry air at room temperature is about 340 m s^{-1}. This rises if the air is warmer or damper. Sound travels faster through liquids than through gases, and faster still through solids.

Loudness The greater the amplitude of the sound waves, the louder the sound.

Wavelength This can vary from about 15 mm to 15 m, so sound waves will diffract round everyday objects.

Pitch The higher the frequency, the higher the pitch:

pitch: low — high

range of human hearing | ultrasound

frequency: 20 Hz — 20 000 Hz

C2 Moving waves

Progressive wave motion

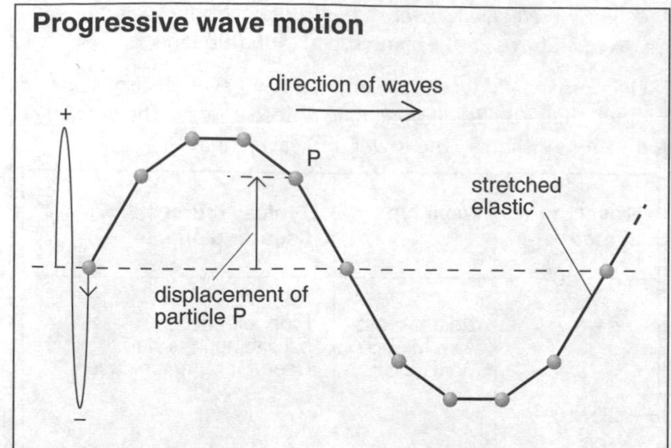

Above is one model of how waves travel. The first particle is oscillated up and down with SHM (see B12). This pulls on the next particle, making it oscillate up and down slightly later, and so on, along the line. As a result, ***progressive*** (moving) waves are seen travelling from left to right. The waves in this case are ***transverse*** (see C1) because the oscillations are at right-angles to the direction of travel.

The ***displacement*** of a particle at any instant is measured from the centre line, with a + or – to indicate an *upward* or *downward* direction.

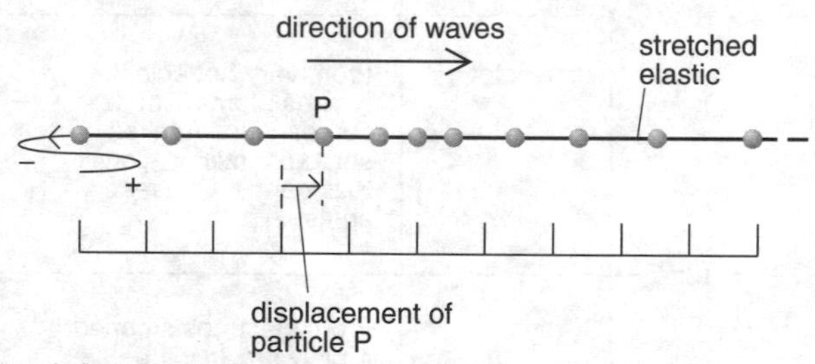

With ***longitudinal*** waves as above, the oscillations are in the direction of travel (see C1). So a particle can have a displacement to the *right* (+) or *left* (–).

The speed of waves depends on the properties of the ***medium*** (material) through which they are travelling. For example, if the particles above are lighter, or the elastic tighter, then each particle is affected more rapidly by the one before, so the wave speed is greater.

The speed of electromagnetic waves

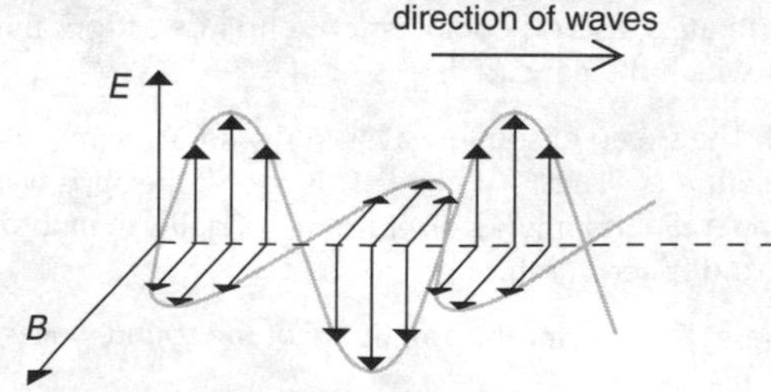

Electromagnetic waves such as light (see C1) are transverse waves. However, it is not particles which oscillate, but an electric field (E) coupled with a magnetic field (B), as shown above. The speed of the waves through space, called the ***speed of light***, depends on the permittivity ε_0 and permeability μ_0 of a vacuum (see E1 and E6):

$$\text{speed of light (in vacuum) } c = \frac{1}{\sqrt{\varepsilon_0 \mu_0}} = 3.0 \times 10^8 \text{ m s}^{-1}$$

Refraction of light waves

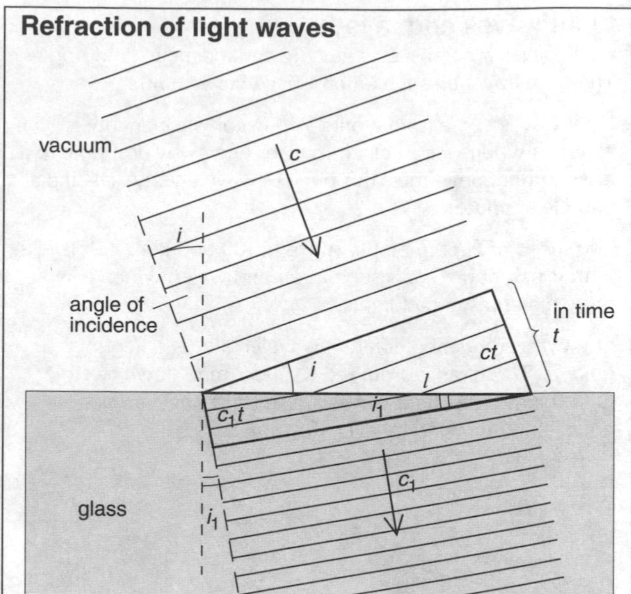

Above, light waves are shown as a series of lines called ***wavefronts***. All points on a wavefront are in phase. As the waves enter the glass, their speed slows from c to c_1. As a result, ***refraction*** (bending) occurs. In time t, one side of the beam travels a shorter distance in the glass (c_1t) than the other side does in the vacuum (ct). From the triangles in the diagram:

$$\sin i = \frac{ct}{l} \quad \text{and} \quad \sin i_1 = \frac{c_1 t}{l} \quad \therefore \quad \frac{c}{c_1} = \frac{\sin i}{\sin i_1} = n_1$$

n is a constant called the ***refractive index*** of the medium (glass).

When light enters a typical glass, its speed slows from 3.0×10^8 m s^{-1} to 2.0×10^8 m s^{-1}. So, from the above equation, the refractive index of the glass is 1.5. Water does not slow light so much. Its refractive index is 1.3.

Note:
- $\sin i_1 \propto \sin i$.
- The refraction at an air–glass boundary is effectively the same as at a vacuum–glass boundary.
- The refractive index is slightly different for different wavelengths, which is why dispersion occurs (see C1).

On the right, light passes from one medium into another (of greater refractive index in this case). The wave direction is indicated by a single ray. The following equation applies:

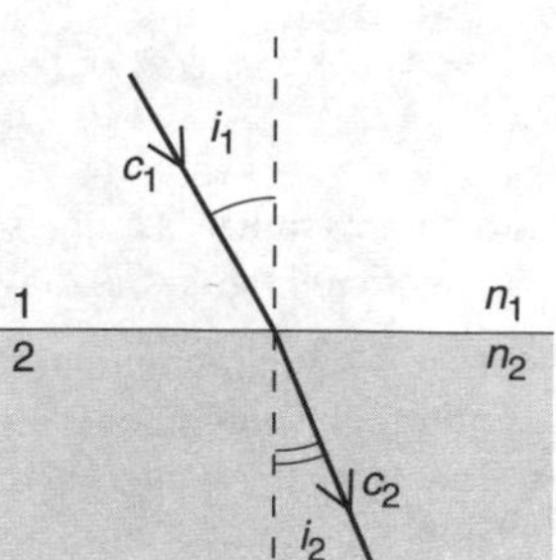

$$\frac{c_1}{c_2} = \frac{\sin i_1}{\sin i_2} = {}_1n_2 \qquad (1)$$

${}_1n_2$ is the ***relative refractive index*** for light passing from medium 1 to medium 2. It can be shown that

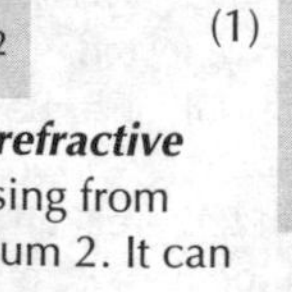

$${}_1n_2 = n_2/n_1$$

From this and equation (1), it follows that

$$n_1 \sin i_1 = n_2 \sin i_2 \quad \text{and} \quad n_1 c_1 = n_2 c_2$$

Note:
- The equation above left is known as **Snell's Law**.
- n for a vacuum (or air) = 1.
- By the ***principle of reversibility***, a ray from B to A has the same path as one from A to B.

So $\quad {}_2n_1 = \dfrac{1}{{}_1n_2}$

Speed, frequency, and wavelength

The waves on the right have a frequency f. So in time $1/f$, they move forward one wavelength, λ. As their speed c equals distance (λ) divided by time ($1/f$),

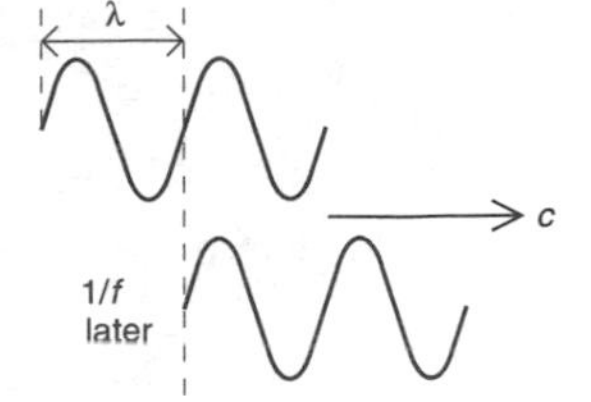

$$c = f\lambda$$

When refraction occurs, the frequency of the waves is unchanged. However, if the speed decreases, it follows from the above equation that the wavelength must also decrease.

Critical angle

On the right, light travels towards a boundary with a medium of lower refractive index. i_c is the ***critical angle*** (see C1). For angles greater than this, all the light is reflected by the surface and none is refracted. There is ***total internal reflection***.

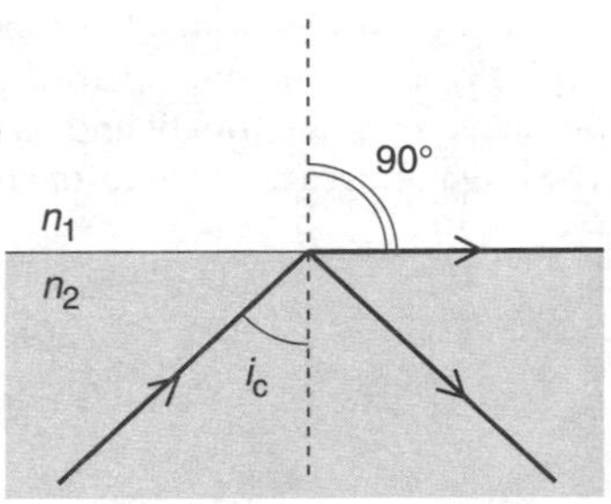

From equation (1)

$${}_1n_2 = \frac{\sin 90°}{\sin i_c} = \frac{1}{\sin i_c}$$

So $$\sin i_c = \frac{1}{{}_1n_2}$$

For an air–glass boundary, ${}_1n_2$ is effectively equal to n_2. If $n_2 = 1.5$, then $\sin i_c = 1/1.5$. So $i_c = 42°$ for the glass.

Optical fibres

Optical fibres (see also C1) can carry data, in the form of infrared pulses. They make use of total internal reflection.

Step-index multimode fibre This has a core surrounded by a cladding of lower refractive index. In the core, zig-zag paths (modes) of many different lengths are possible, so different pulses may overlap by the time they reach the end.

step-index multimode

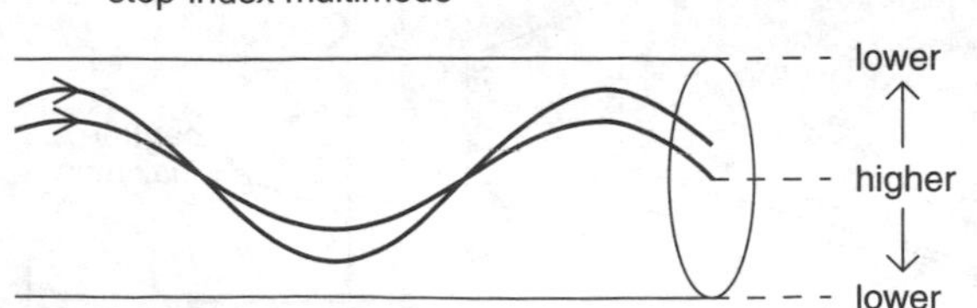

graded-index multimode

Graded-index multimode fibre The refractive index gradually reduces from the centre out. This means that the pulses take curved paths. But the longer paths are faster, so the travel times are about the same for all of them.

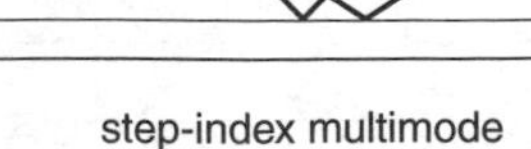

Graded-index monomode fibre This has a very narrow core, so that only one path is possible. As a result, the fibre can carry more pulses per second without overlap.

Polarization

In the top diagram on the opposite page, the particles oscillate in a vertical ***plane of vibration***. For a light wave, the plane of vibration is taken as that of the E vector. Light is usually a mixture of waves with different planes of vibration. It is ***unpolarized***. Polaroid transmits light in one plane of vibration only. Light like this is ***polarized***.

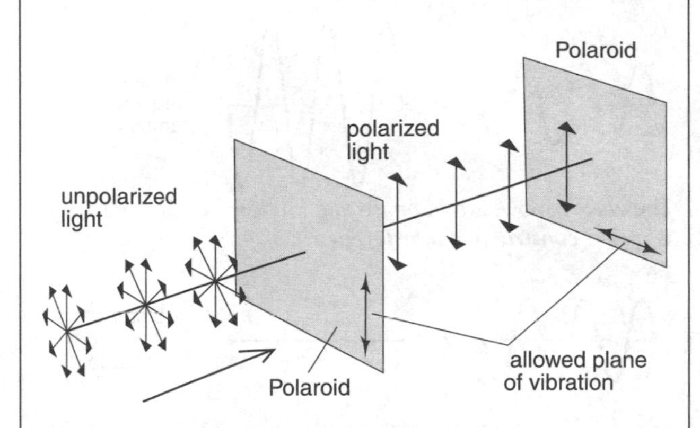

Above, polarized light from one Polaroid strikes a second. The light is blocked because its plane of vibration has no component in the allowed direction.

Only tranverse waves can be polarized. Experiments with Polaroid provide evidence that light waves are transverse.

Polarization by reflection When an unpolarized light ray strikes the surface of a transparent medium such as water, the refracted ray is partly polarized. At most angles, the reflected ray is also partly polarized.

But if the reflected ray is at 90° to the refracted ray, it is *totally* polarized. i_p is the ***polarizing angle***.

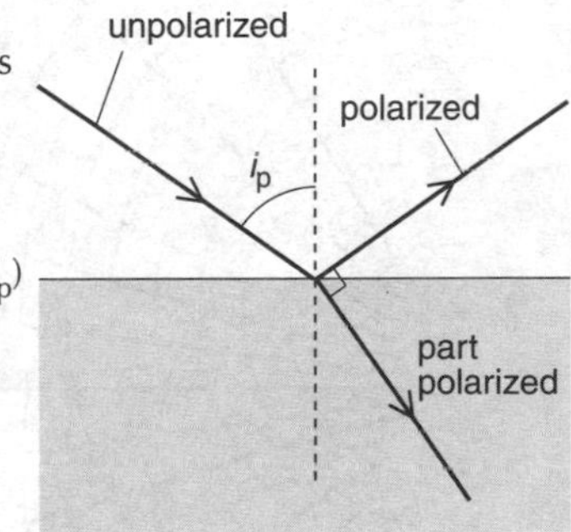

In this situation:

$$n = \sin i_p / \sin (90° - i_p)$$
$$= \sin i_p / \cos i_p$$

So $$n = \tan i_p$$

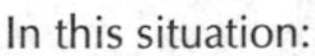

This result is called ***Brewster's law***.

Polaroid sunglasses reduce the glare from wet surfaces by blocking the reflected, polarized light.

Intensity

Waves transmit energy. If waves pass through a surface, their ***intensity*** (in W m^{-2}) is calculated like this:

$$\text{intensity} = \frac{\text{power crossing surface}}{\text{area of surface}}$$

Intensity is proportional to (amplitude)^2.

On the right, waves are radiating uniformly from a source of power output P. At a distance r from the source, the power is spread over an area $4\pi r^2$. So intensity $I = P/4\pi r^2$

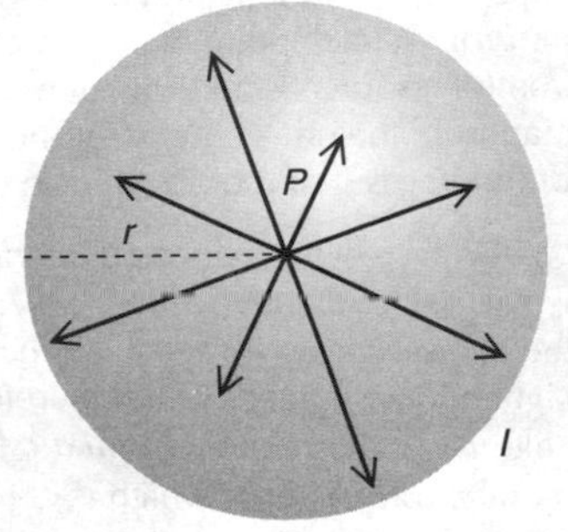

Note that $I \propto 1/r^2$. This is another example of an inverse square law.

C3 Combining waves

Superposition and interference

Two sets of waves can pass through the same point without affecting each other. However, they have a combined effect, found by adding their displacements (as vectors). This is known as the ***principle of superposition***.

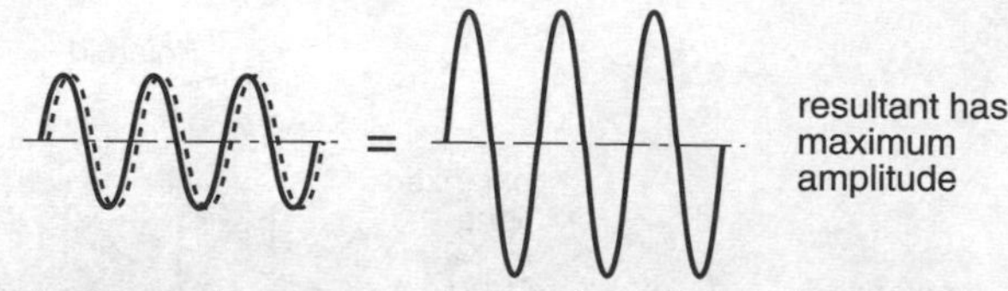

The waves above are *in phase* and reinforce each other. This is called ***constructive interference***.

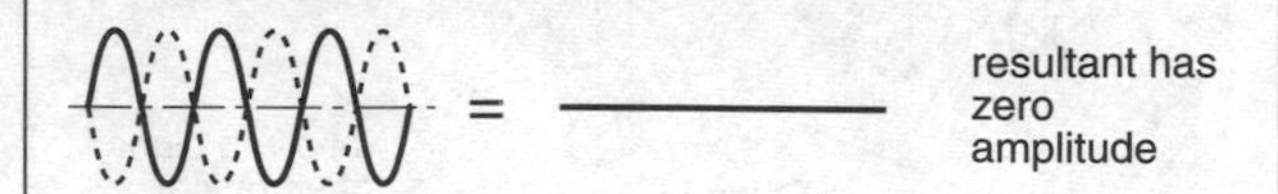

The waves above have a *phase difference* of $\frac{1}{2}$ cycle (180°) and cancel each other. This is called ***destructive interference***.

For interference to be observed:

- The sets of waves must be ***coherent***: there must be a constant phase difference between them. For this, they must have the same frequency.
- The sets of waves must have approximately the same amplitude and plane of vibration.

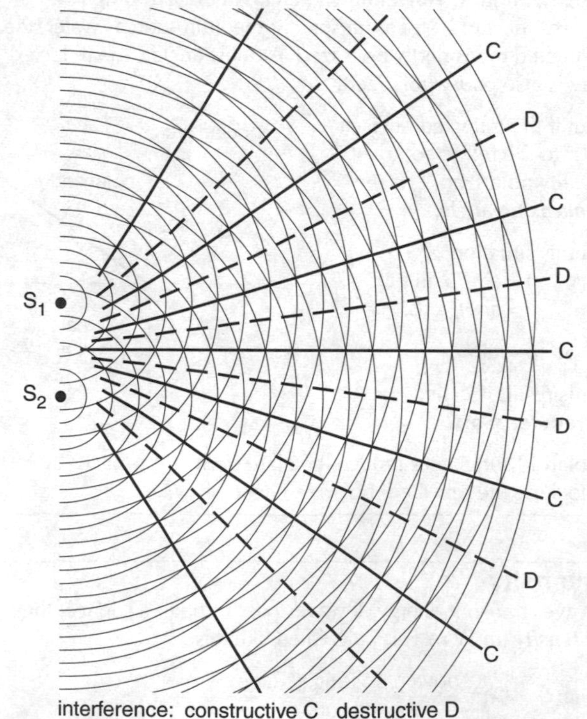

interference: constructive C destructive D

Above, waves from two coherent sources, S_1 and S_2, produce regions of reinforcing and cancelling called an ***interference pattern***. At each point of constructive interference, the path from one source is an exact number of wavelengths longer than from the other source (or the same length). The ***path difference*** is 0 or λ or 2λ, and so on.

Light waves will produce an interference pattern. However, waves from separate sources are not normally coherent, so the two sets of waves must originate from the same source. Light of one frequency (and therefore of one wavelength and colour) is called ***monochromatic light***. A laser emits monochromatic light which is coherent across its beam.

Double-slit experiment

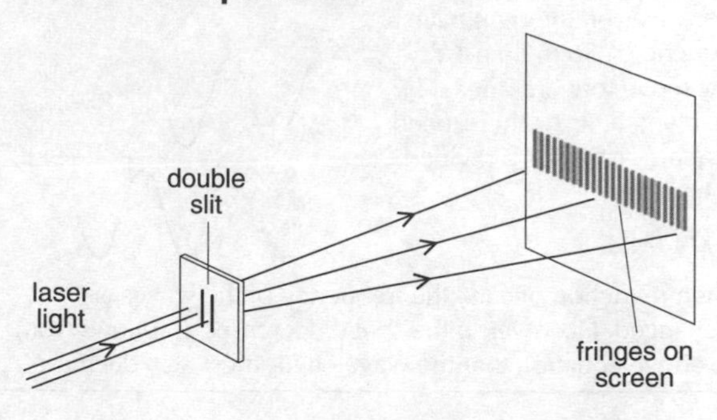

Above, light waves from a laser spread out from two slits (typically less than $\frac{1}{2}$ mm apart). The interference pattern produces a series of bright and dark ***fringes*** on the screen. The bright fringes are regions of constructive interference.

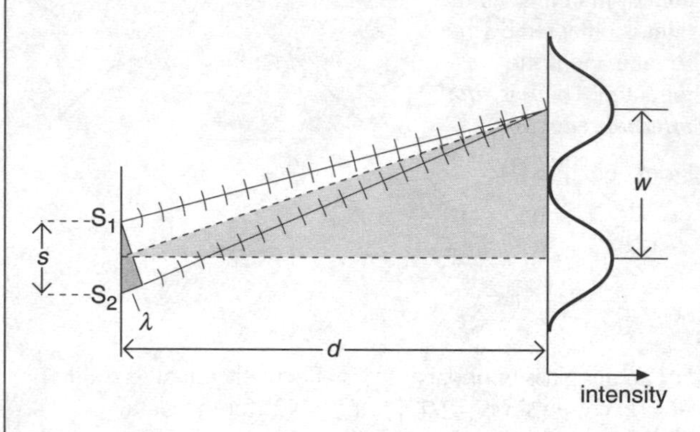

The first bright fringe occurs where the path difference is λ. For small angles, the shaded triangles above are similar, and the following equation applies:

$$\frac{\lambda}{s} = \frac{w}{D}$$

So $$w = \frac{\lambda D}{s}$$ w is the ***fringe spacing***

Note:

- The fringe spacing is increased if the slits are closer together or light of longer wavelength is used.
- By measuring w, D, and s, the wavelength of light can be found using the above equation. Light wavelengths range from 7×10^{-7} m (red) down to 4×10^{-7} m (violet).

Single-slit diffraction

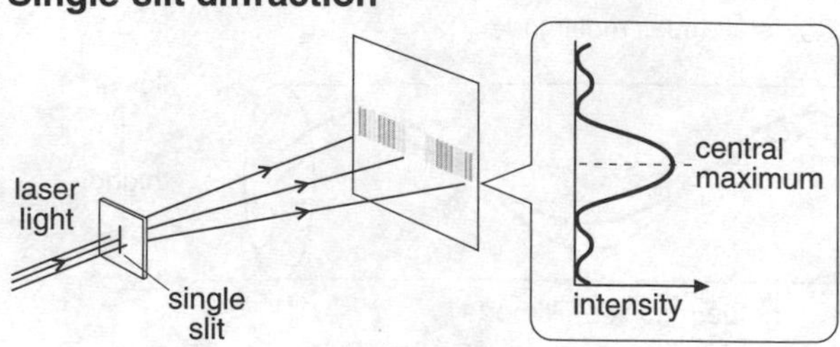

The spreading of light from a slit is an example of ***diffraction*** (see C1). Interference occurs between the different waves diffracted by the slit. The result is a pattern as above. The pattern becomes wider if the slit is made narrower or light of longer wavelength is used.

In optical instruments, diffraction limits the amount of detail in the image (see C4, C5).

Diffraction grating

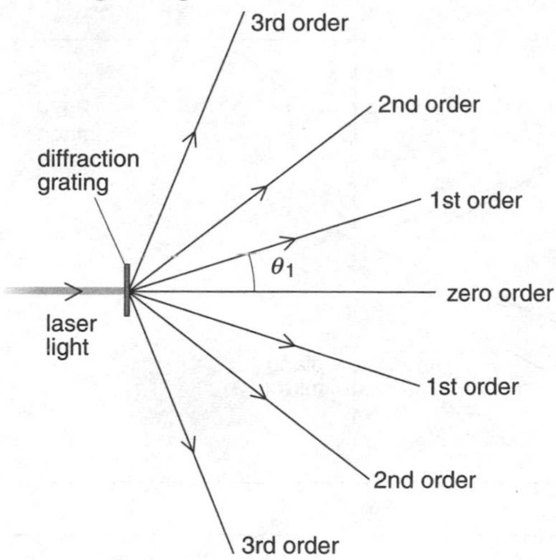

A ***diffraction grating,*** as above, has many slits (typically, 500 per mm). Constructive interference produces sharp lines of maximum intensity at set angles either side of a sharp, central maximum. In between, destructive interference gives zero or near-zero intensity. To identify the lines, they are each given an ***order number*** (0, 1, 2 etc.).

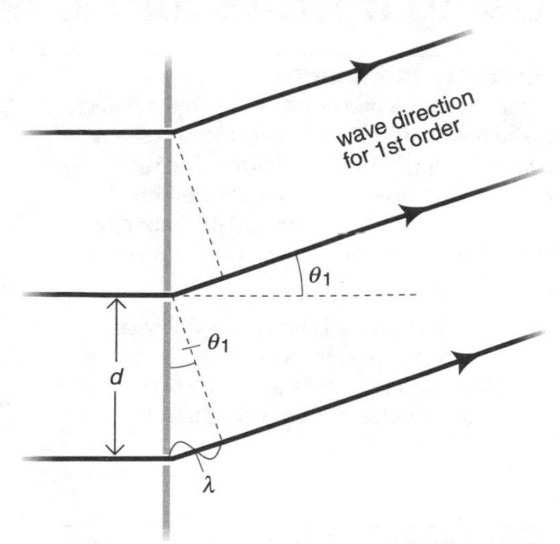

A close-up of part of the grating is shown above. d is the ***grating spacing***. θ_1 is the angle of the first order maximum. In this case, the path difference for any two adjacent slits is one wavelength, λ. From the triangles,

$$\sin\theta_1 = \frac{\lambda}{d} \qquad \text{So } d\sin\theta_1 = \lambda$$

For higher orders, the path differences are 2λ, 3λ etc. The following equation gives values of θ for all orders:

$$d\sin\theta = n\lambda$$

where n is the order number (0, 1, 2 etc.).

Note:
- If d, θ_1, and n are known, λ can be calculated.
- A longer wavelength gives a larger angle for each order.
- If the incoming light is a mixture of wavelengths (e.g. white), each order above zero becomes a spectrum.

Stationary waves in a stretched string

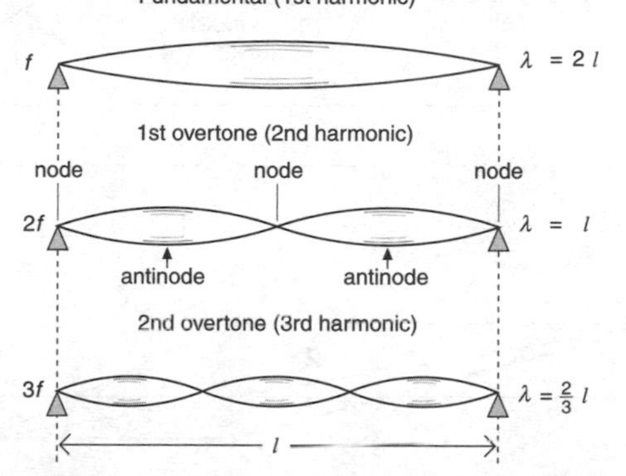

A stretched string can vibrate in various ***modes***, some of which are shown above. Each has a different frequency. The waves produced are known as ***stationary waves***. At ***nodes***, the amplitude of the oscillation is always zero. At ***antinodes***, it is always a maximum.

Stationary waves are produced by the superposition of two sets of progressive waves (of equal amplitude and frequency) travelling in opposite directions. For example, when a stretched string is vibrated, waves travel along the string, reflect from the ends, and are superimposed on waves travelling the other way.

On the right, you can see how a node is formed. As one wave moves to the right and the other to the left, the + and – displacements always cancel, so the resultant displacement is zero.

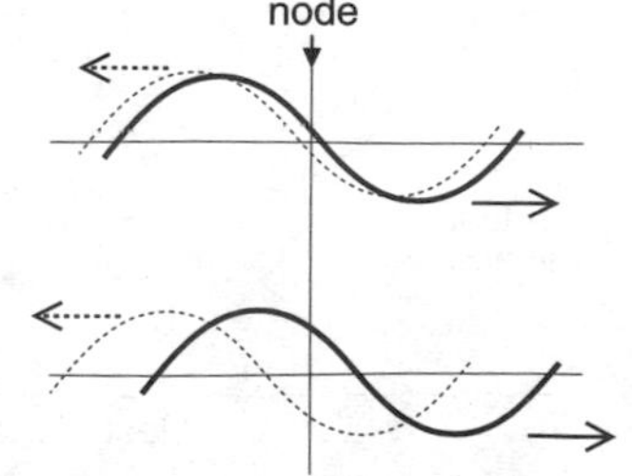

Frquency depends on tension T and the mass per unit length μ of the string.

$$f = \frac{1}{2l}\sqrt{\frac{T}{\mu}} \quad \text{for the fundamental frequency.}$$

Stationary waves in an air column

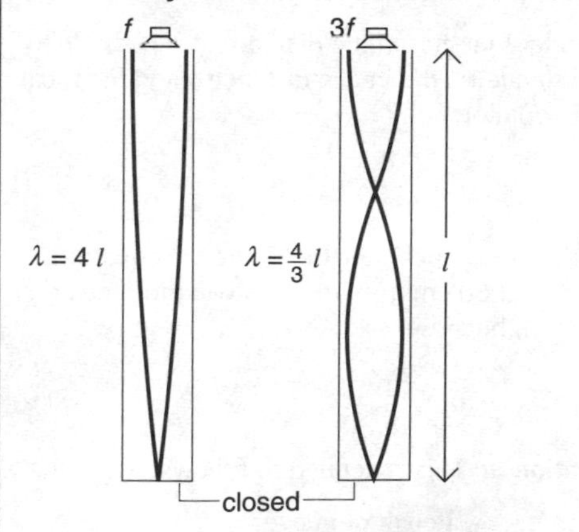

If a sound source is placed near the open end of a pipe, there are certain frequencies at which stationary waves are set up in the air column and the sound intensity reaches a maximum. This is another form of ***resonance*** (see also B13). Three examples are shown above.

Note:
- Sound waves are *longitudinal*. The 'waves' in each pipe are a *graphical representation* of the amplitude.
- Where the end of a pipe is open, there is an antinode. Where the end of a pipe is closed, there is a node.
- Knowing the frequency needed to produce resonance, and the wavelength from the pipe length, the speed of sound c can be calculated using $c = f\lambda$.

C4 Using mirrors and lenses – 1

Plane mirrors, and images

On the right, rays from one point on an object reflect from a plane (flat) mirror. To the eye, the rays seem to come from a point behind the mirror. The same process occurs with rays from other parts of the object. As a result, an image is seen. (For simplicity, ray diagrams like this usually only show the rays coming from one point.)

The image in a plane mirror is known as a ***virtual image*** because no rays pass through it. Some optical instruments bring rays to a focus to form a ***real image***. This type of image can be picked up on a screen.

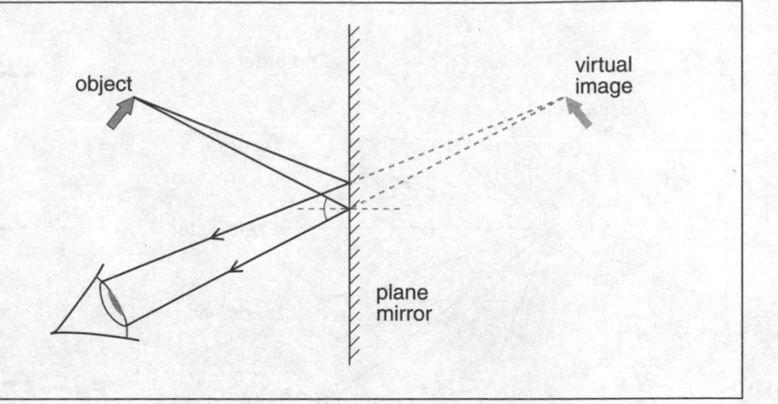

Convex and concave lenses

The diagrams on the right show how a convex and a concave lens each refract incoming rays parallel to the axis. Outgoing rays either converge towards or diverge from a ***principal focus***, F. Rays can come from either side, so there is another principal focus, F′, in an equivalent position on the opposite side of each lens. P is the ***optical centre***.

Note:

- In the rest of this unit, lenses are assumed to be thin, with no distorting effects. Diagrams will show refraction occuring at a line through the optical centre.

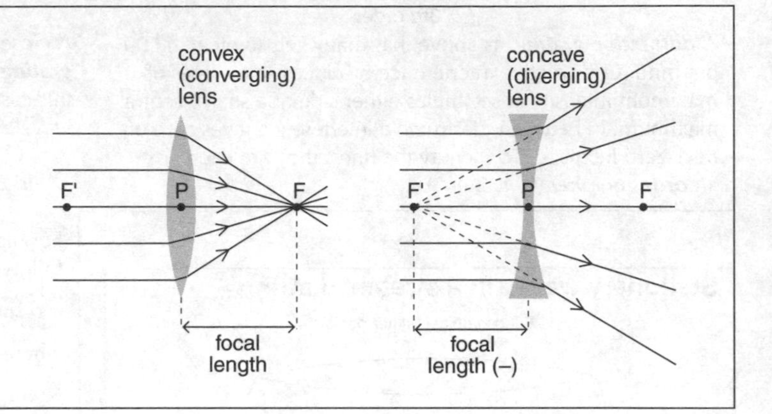

Convex lens equations

On the right, three rays have been used to show a convex lens forming a real, inverted image of an object:

1 A ray parallel to the axis is refracted through F.
2 A ray though P is undeviated (straight).
3 A ray through F′ is refracted parallel to the axis.

From pairs of similar triangles in the diagram, it is possible to link the object distance u, the image distance v and the focal length f with an equation:

$$\frac{1}{u}+\frac{1}{v}=\frac{1}{f} \qquad (1)$$

For example, for a lens of focal length 20 cm, the equation predicts that an object 60 cm from the lens will produce an image 30 cm from it, because

$$\frac{1}{60}+\frac{1}{30}=\frac{1}{20}$$

Linear magnification *m* This is defined as follows:

$$\text{linear magnification} = \frac{\text{height of image}}{\text{height of object}}$$

By comparing similar triangles in the diagram above right,

$$m=\frac{v}{u} \qquad (2)$$

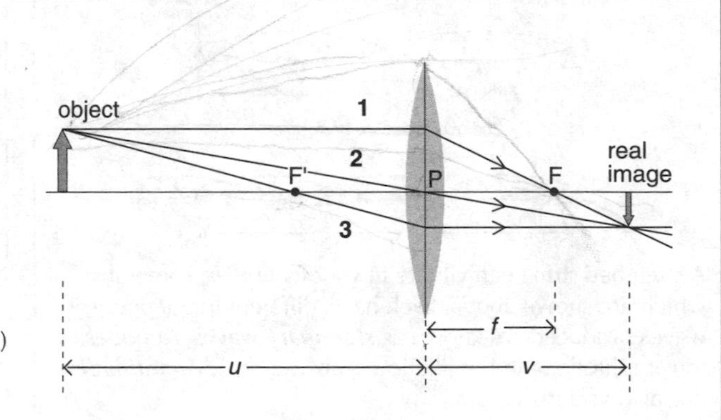

If the object above is moved towards F′, the image distance increases and the image gets larger. When u is $2f$, v is also $2f$, and the image and object are the same size (m = 1).

Convex lens as a magnifier

If u is less than f, a convex lens produces a virtual, upright, magnified image as on the right. Equation (1) applies, but the *virtual* image gives a *negative* value for v. For example, if f is 20 cm, and u is 15 cm, solving the equation gives v = –60 cm. This tells you that there is a virtual image 60 cm from the lens. Also, as v/u = –4, the image is four times the height of the object.

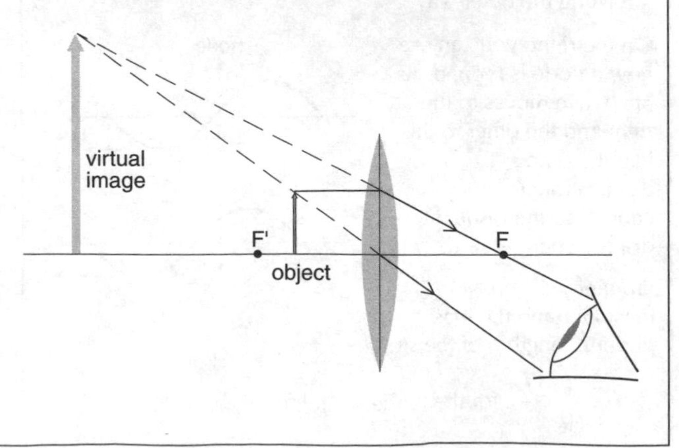

Lens and mirror equations

Concave mirrors are *converging* systems, with similar image-forming properties to convex lenses.

Convex mirrors are *diverging* systems, with similar image-forming properties to concave lenses.

Equations (1) and (2) on the opposite page apply to all thin lenses and mirrors. However, when using them, note:

- *Converging* systems have *positive* focal lengths.
- *Diverging* systems have *negative* focal lengths.

Focal point of a mirror

Light from a distant object converges toward the focus of a concave mirror.

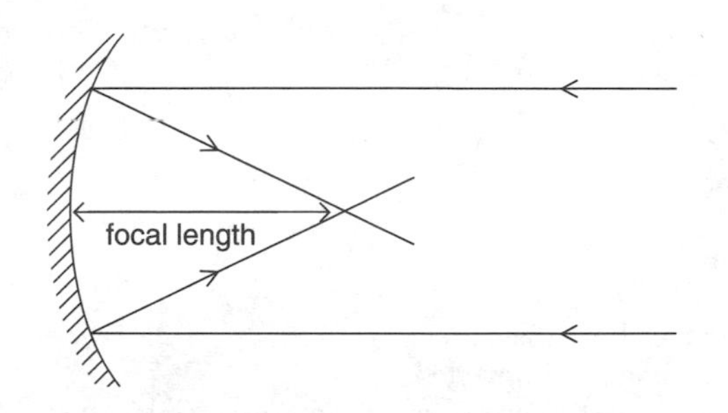

Light directed toward the focus of a spherical convex mirror is reflected as a parallel beam.

Light directed toward one focus of a hyperbolic convex mirror converges toward the other focus (see C5).

Spherical aberration

Light that is incident on a spherical lens or mirror a long way from its axis is brought to a focus at a different position from light incident closer to the axis.

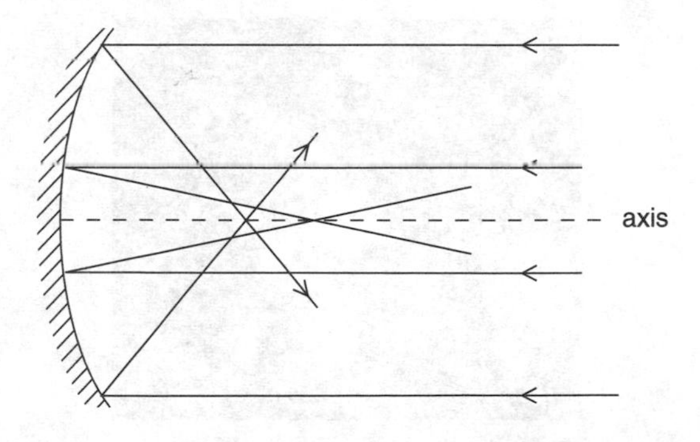

The result is a blurred image owing to the spherical nature of the lens or mirror.

Spherical aberration can be reduced by restricting the lens aperture. However, this has the effect of reducing the brightness of the image and making resolution worse.

Mirrors can be made parabolic, in which case all rays are brought to a focus at the same point and spherical aberration is eliminated.

Reflecting prisms

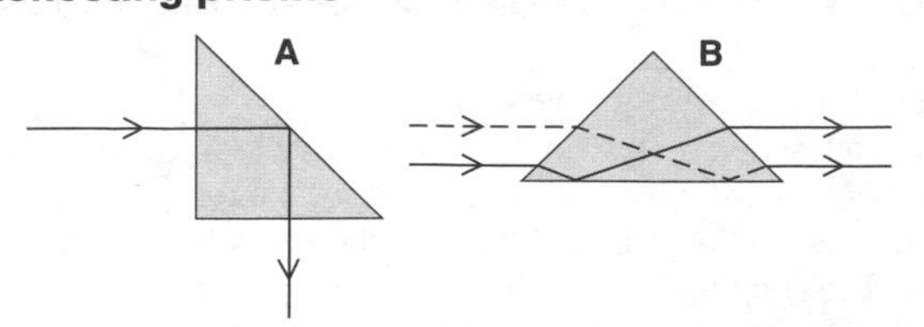

In prisms A and B above, total internal reflection occurs (see C2). Prism B inverts the beam. Some binoculars contain prisms like this to compensate for the inverting effect of the lens system. (Binoculars are two telescopes side by side.)

Problems with mirrors and lenses

A mirror that is silvered on the rear surface suffers from multiple images. One is produced by partial reflection at the front surface and another by reflection from the back surface. This can be overcome by using mirrors that are silvered on the front surface.

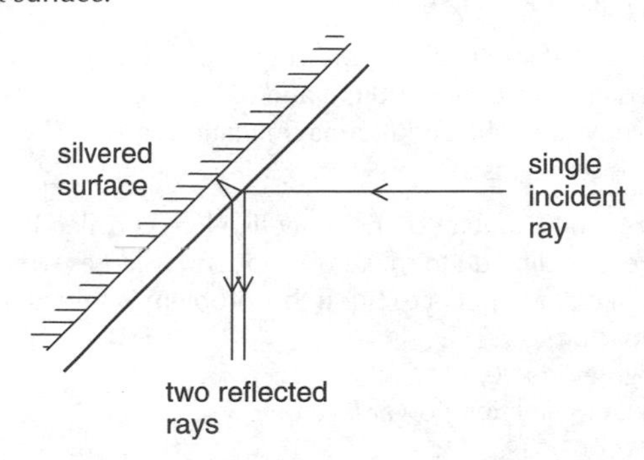

The ***reflecting prisms*** shown above use total internal reflection so that no silvering is needed. Because only one surface produces reflection there is only one image.

Chromatic aberration

Light of different colours refracts differently when passing through a lens. The image of an object emitting blue light is in a different position to that formed by one in the same position that is emitting red light.

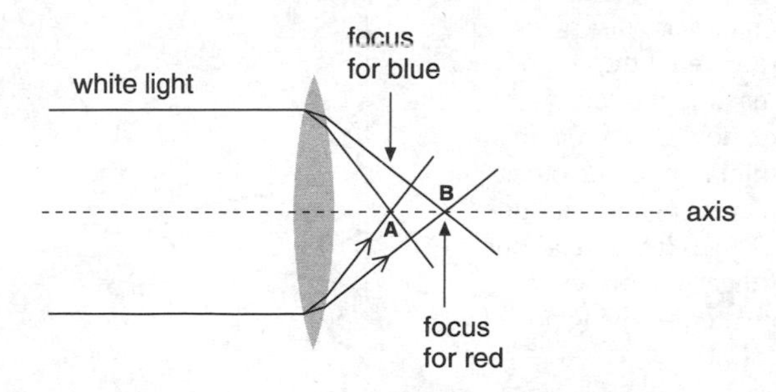

A screen placed at **A** would show an image with a blue edge and a screen at **B** would show an image with a red edge.

White objects emit all colours and there are therefore many images formed of an object, all in different positions. When trying to focus the image the result is a blurred image with coloured edges.

Problems due to chromatic aberration can be reduced by using combinations of converging and diverging lenses made from materials of different refractive indices.

Because mirrors do not depend on refraction there is no chromatic aberration with mirrors that are silvered on the front surface.

C5 Using mirrors and lenses – 2

Refracting telescopes

In the telescope on the right, the ***objective*** lens focuses light from a distant object, to form a real image just on the principal focus of the ***eyepiece*** lens. The eyepiece lens then forms a virtual, magnified image of this real image. Set like this, the telescope is in ***normal adjustment.***

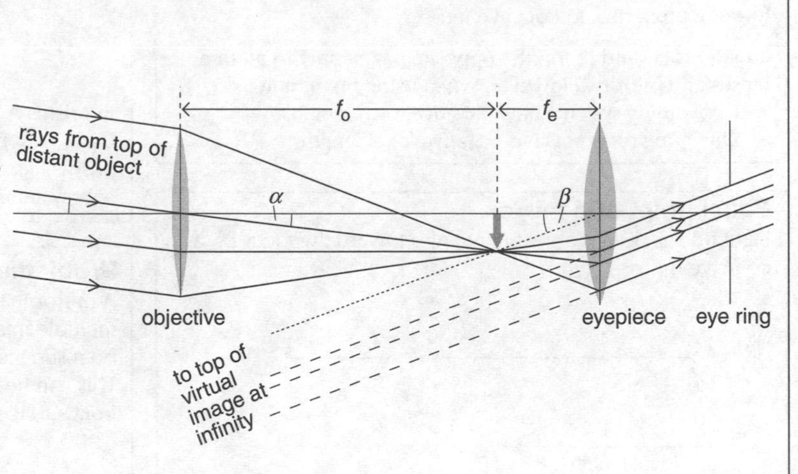

Angular magnification This is defined as the ratio of the two angles β and α in the diagram on the right:

$$\text{angular magnification} = \frac{\beta}{\alpha}$$

From triangles in the diagram, it can be shown that

$$\text{angular magnification} = \frac{f_o}{f_e}$$

For example, if the focal lengths of the objective and eyepiece lenses are 100 cm and 5 cm respectively, then the angular magnification (in the above setting) is 20.

Note, in the above diagram:

- The final image is inverted.
- For the widest angle of view, the eye is placed in the ***eye ring*** position (the position where the eyepiece forms a real image of the objective lens).
- By moving the eyepiece, the real image can be formed just inside its principal focus. The final, virtual image is then closer to the eye. However, it is more straining for the eye to look at this image for a long time.

Aperture The diameter of the objective lens is called its ***aperture***. It is difficult to make wide-aperture lenses which do not give a distorted image. But if this problem is solved, a wider aperture gives

1 a brighter image,
2 a greater resolving power (see below).

Resolving power This is $1/\theta_{min}$, where θ_{min} is the smallest angle between two points such that their images can just be *resolved* (seen separately and unmerged).

Diffraction limits the resolving power because a telescope's objective lens acts rather like a wide slit (see C3). As a result, the images of two 'point' stars will each appear as tiny, circular fringe systems with intensities as below.

According to the ***Rayleigh criterion***, two images are just resolved if the central maximum of one coincides with the first minimum of the other. Increasing the aperture D gives less spread-out fringe systems and a greater resolving power ($\approx D/\lambda$).

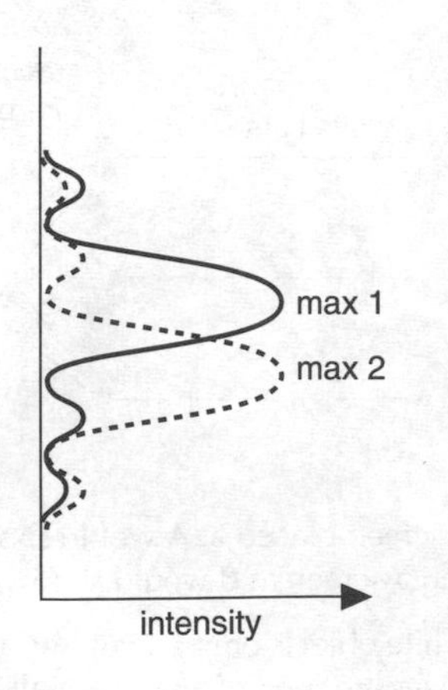

Airy's disc

When a distant point object is viewed with a telescope, the image observed is a disc of light caused by diffraction. This is called **Airy's disc**, named after the 19th century astronomer Sir George Airy.

The disc becomes smaller when larger-aperture telescopes are used. When two such objects are close together the diffraction discs merge so that the images are not resolved. Larger-aperture telescopes improve resolution.

Reflecting telescopes

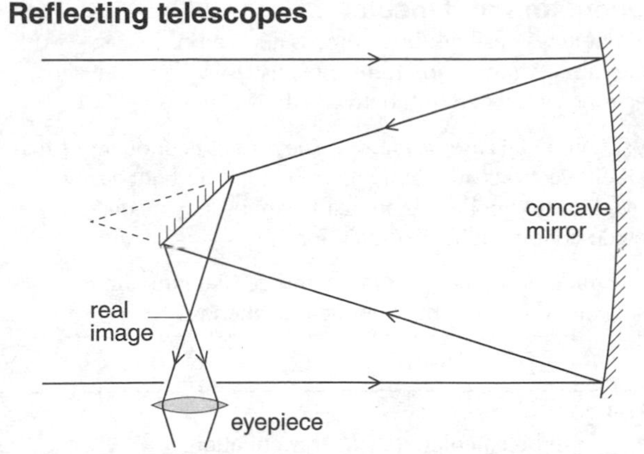

In large-aperture telescopes, the rays are brought to a focus by a concave mirror rather than an objective lens. With a mirror, support can be provided at the back to stop it flexing, and it is easier to reduce image distortion. Often, there is no eyepiece. The real image is formed either on a photographic plate, for later enlargement, or on a ***charge-coupled device (CCD)***, which stores it electronically.

Cassegrain arrangement

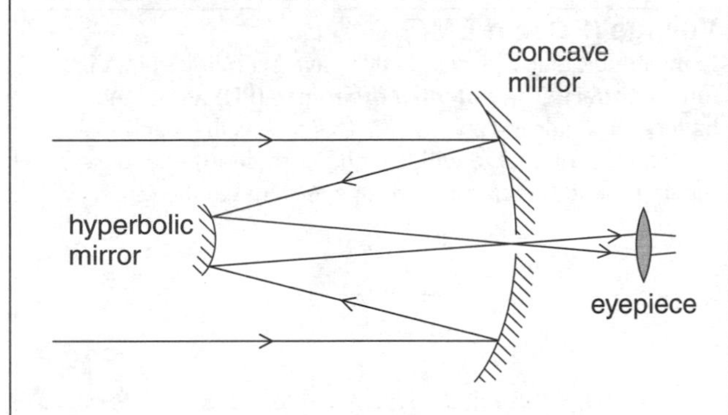

The convex mirror is hyperbolic and has one focus at the same point as that of the large concave mirror. The other focus is at the pole (centre). Light that passes through a small hole at the pole is then viewed through an eyepiece.

Merits of reflecting and refracting telescopes

As well as producing high magnification an astronomical telescope has to view faint objects so a large aperture is needed.

Small ***refracting telescopes*** are very convenient to use but to make large–aperture telescopes for astronomy a highly transparent material is needed that has few flaws to reflect the light. A large-diameter lens is heavy and can only be supported at its rim so that when mounted it suffers mechanical distortions which produce optical distortions in the images.

The large-aperture mirrors required in a ***reflecting telescope***

- are easier to construct since they can be supported anywhere behind it
- are silvered on the front surface so no refraction is involved
- be used with a wider range of wavelengths including those outside the optical region
- can be more easily constructed to eliminate aberrations than lenses.

These are amongst the reasons why large telescopes for astronomical and other purposes are usually reflecting rather than refracting.

Radio telescopes

Radio telescopes are reflecting telescopes. They are used to form images of distant objects due to the emission of microwaves or radio waves rather than visible light. The telescope scans the sky and builds up an image formed by the variation in intensity of radio waves coming from different directions.

Radio telescopes need large diameters

- to collect sufficient energy to form an image
- to produce adequate resolution.

Using the Rayleigh criterion on the opposite page, a radio telescope forming an image using 30 cm waves would need a diameter of about 3 km to produce the same resolution as that of a naked eye forming an optical image.

D1 Charges and circuits

Static electricity

If two materials are rubbed together, electrons may be transferred from one to another. As a result, one gains negative charge, while the other is left with an equal positive charge. If the materials are ***insulators*** (see right), the transferred charge does not readily flow away. It is sometimes called ***static electricity***.

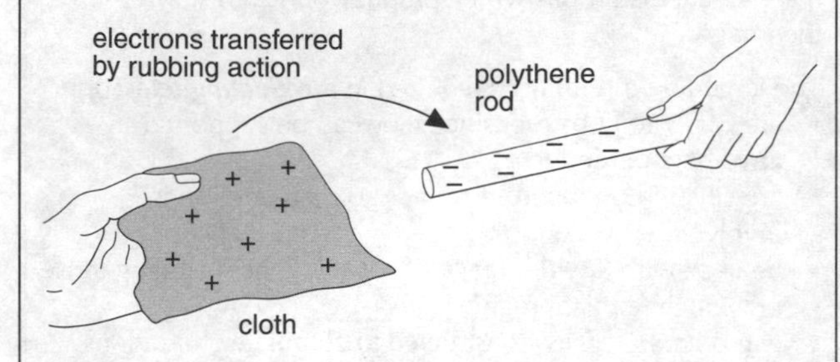

A charged object will attract an uncharged one. On the right, the charged rod has extra electrons. Being uncharged, the foil has equal amounts of – and + charge. The – charges are repelled by the rod and tend to move away, while the + charges are attracted. However, the force of attraction is greater because of the shorter distance.

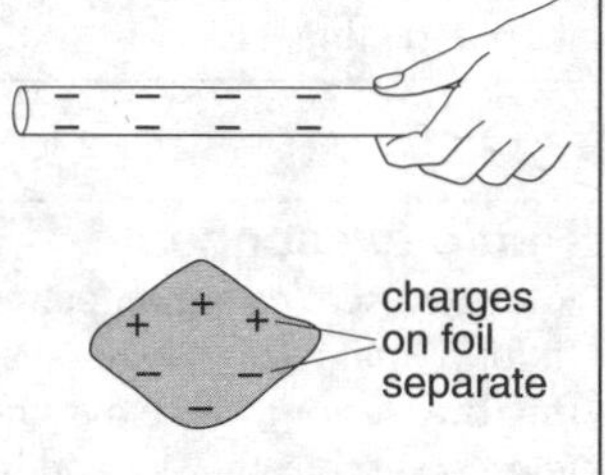

Charge which collects in one region because of the presence of charge on another object is called ***induced*** charge.

Current

+ –
cell
electron flow
conventional direction
A ammeter
bulb

In the circuit above, chemical reactions in the cell push electrons out of the negative (–) terminal, round the circuit, to the positive (+) terminal. This flow of electrons is called a ***current***.

An arrow in the circuit indicates the direction from the + terminal round to the –. Called the ***conventional direction***, it is the *opposite* direction to the actual electron flow.

The SI unit of current is the ***ampere*** (**A**).

A current of 1 A is equivalent to a flow of 6×10^{18} electrons per second. However, the ampere is not defined in this way, but in terms of its magnetic effect (see E6).

Current may be measured using an ***ammeter*** as above.

Conductors and insulators

Current flows easily through metals and carbon. These materials are good ***conductors*** because they have free electrons which can drift between their atoms (see F5).

Most non-metals are ***insulators***. They do not conduct because all their electrons are tightly held to atoms and not easily moved. Although liquids and gases are usually insulators, they do conduct if they contain ions.

Semiconductors, such as silicon and germanium, are insulators when cold but conductors when warm.

Charge

Charge can be calculated using this equation:

charge = current × time

The SI unit of charge is the ***coulomb*** (**C**).

For example, if a current of 1 A flows for 1 s, the charge passing is 1 C. (This is how the coulomb is defined.) Similarly, if a current of 2 A flows for 3 s, the charge passing is 6 C.

Voltage (PD and EMF)

In the circuit below, several cells have been linked in a line to form a ***battery***. The ***potential difference* (PD)** across the battery terminals is 12 volts (V). This means that each coulomb (C) of charge will 'spend' 12 joules (J) of energy in moving round the circuit from one terminal to the other.

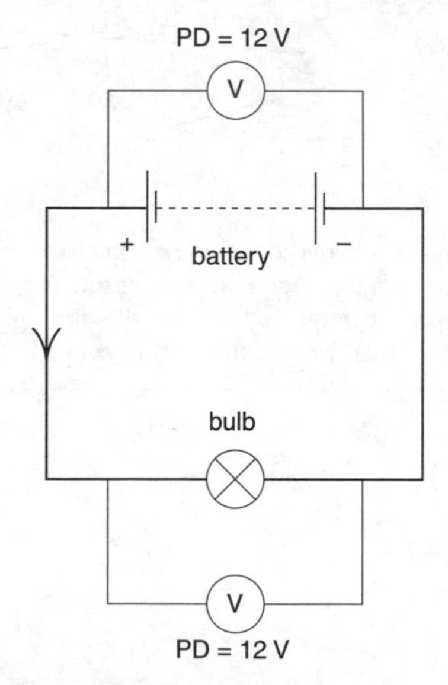

The PD across the bulb is also 12 V. This means that, for each coulomb pushed through it, 12 J of electrical energy is changed into other forms (heat and light energy).

PD may be measured using a ***voltmeter*** as shown above.

PD, energy, and charge are linked by this equation:

energy transformed = charge × PD

For example, if a charge of 2 C moves through a PD of 3 V, the energy transformed is 6 J.

The voltage produced by the chemical reactions inside a battery is called the ***electromotive force*** (**EMF**). When a battery is supplying current, some energy is wasted inside it, which reduces the PD across its terminals. For example, when a torch battery of EMF 3.0 V is supplying current, the PD across its terminals be might be only 2.5 V.

Ohm's law and resistance

If a conductor obeys ***Ohm's law***, then the current I through it is directly proportional to the PD V across it, provided the temperature is constant.

Metals obey Ohm's law. If a graph of I against V is plotted for a metal conductor at constant temperature, the result is as on the right. Expressed mathematically this is

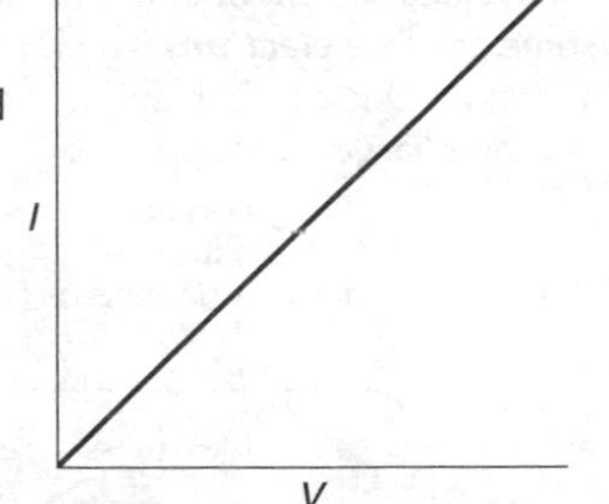

$$\frac{V}{I} = \text{constant}$$

The ***resistance*** R of a conductor is calculated like this:

$$\text{resistance} = \frac{\text{PD}}{\text{current}} \quad \text{In symbols} \quad R = \frac{V}{I}$$

The SI unit of resistance is the ***ohm*** (Ω).

For example, if a PD of 1 V causes a current of 1 A, then the resistance is 1 Ω. (This is how the ohm is defined.)

Similarly, if a PD of 12 V causes a current of 4 A, then the resistance is 3 Ω.

The resistance of a metal conductor (such as a wire) depends on various factors:

- **Length** A long wire has more resistance than a short one.
- **Cross-sectional area** A thin wire has more resistance than a thick one.
- **Temperature** A hot wire has more resistance than a cold one.
- **Type of material** A nichrome wire has more resistance than a copper wire of the same dimensions.

Note:
- While the resistance of a metal *increases* with temperature, that of a semiconductor *decreases*.

Resistance components

Heating elements If a conductor (such as a wire) has resistance, then electrical energy is changed into heat when a current passes. This effect is used in heating elements.

Resistors These are components specially designed to provide resistance. In electronic circuits, they are needed so that other components are supplied with the correct current.

Variable resistors These have a control for varying the length of resistance material through which the current passes.

Thermistors These components have a resistance which changes considerably with temperature (e.g. high when cold, low when hot). They contain semiconducting materials.

Light-dependent resistors (LDRs) These have a high resistance in the dark but a low resistance in the light.

Diodes These have an extremely high resistance in one direction but a low resistance in the other. In effect, they allow current to flow in one direction only.

Symbols

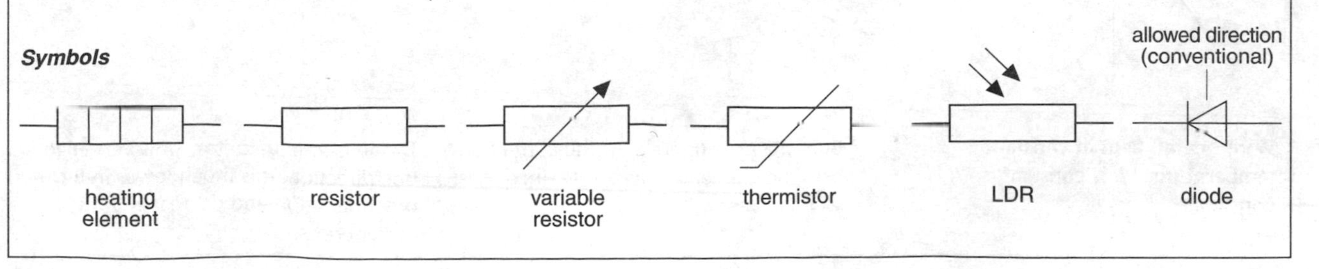

Circuit rules

Resistors in series The current through the battery and each resistor is the same. However, the voltage aross the battery is shared by the resistors.

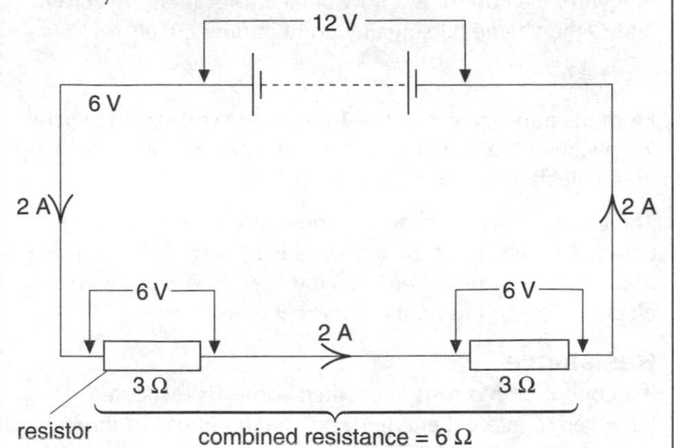

Resistors in parallel The voltage across the battery and each resistor is the same. However, the current from the battery is shared by the resistors.

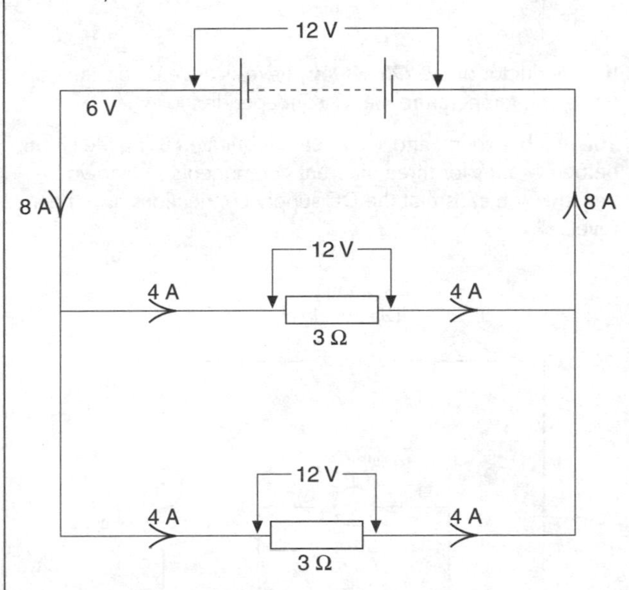

Power

Power P is calculated with this equation:

$$\text{power} = \text{PD} \times \text{current} \qquad \text{In symbols} \quad P = VI$$

For example, if there is a PD of 12 V across a resistor and a current of 3 A through it, then the power = 12 × 3 = 36 W (watts). In other words, the resistor is changing 36 joules of electrical energy into heat every second.

Alternative forms of the power equation are

$$P = I^2R \qquad P = \frac{V^2}{R}$$

D2 Current and resistance

Current, charge, and electrons

In a wire, the current is a flow of electrons. If I is the current and Q the charge passing any point in time t, then

$$Q = It$$

From the above, a current of 1 ampere (A) means that charge is flowing at the rate of 1 coulomb (C) per second. The charge on an electron is $e = -1.60 \times 10^{-19}$ C.

Therefore, 1 C is the charge carried by $1/e$ electrons, i.e. 6.24×10^{18} electrons. So a flow of 6.24×10^{18} electrons per second gives a current of 1 A. However, as e is negative, an electron flow to the *right* is a current to the *left*.

Resistance

If a conductor has resistance, then energy is *dissipated* (changed to internal energy) when a current passes through.

The PD V (in V) across a conductor, current I (in A) through it, and its resistance R (in Ω) are linked by this equation:

$$R = \frac{V}{I}$$

If a conductor obeys ***Ohm's law***, its resistance is constant for any given temperature (i.e. R is independent of V).

The link between I and V can be investigated using the circuit below. Graphs for three different components are shown. (A negative V means that the DC supply connections have been reversed.)

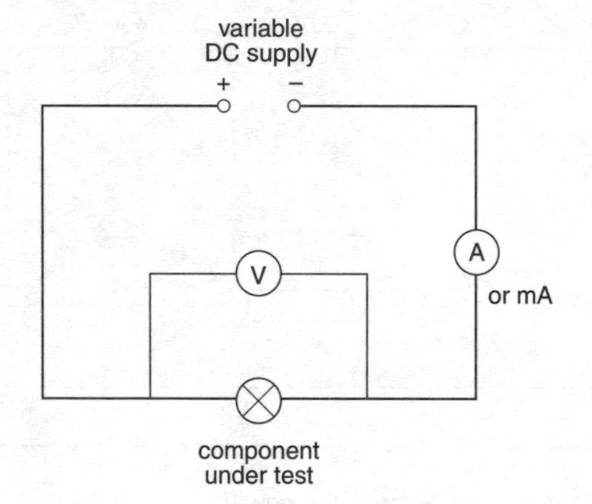

Current and drift speed

Most electrons are bound to their atoms. However, in a metal, some are ***free electrons*** which can move between atoms. When a PD is applied, and a current flows, the free electrons are the ***charge carriers***.

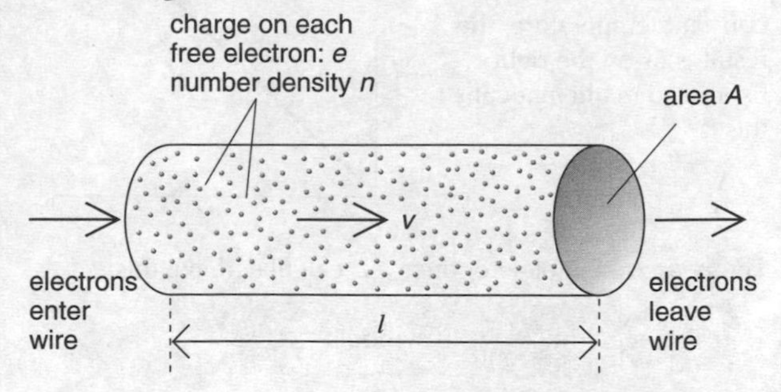

In the wire above, free electrons (each of charge e) are moving with an average speed v. n is the ***number density*** of free electrons: the number per unit volume (per m^3).

In the wire the number of free electrons = nAl
So total charge carried by free electrons = $nAle$

As time = distance/speed:
time taken for all the free electrons to pass through A
$= l/v$

As current I = charge/time $\quad I = \dfrac{nAle \times v}{l}$

$$\therefore \quad I = nAev$$

v is called the ***drift speed***. Typically, it can be less than a millimetre per second for the current in a wire.

The ***current density*** J is the current per unit cross-sectional area (per m^2).

$$J = \frac{I}{A} \quad \text{so} \quad J = nev$$

Note:

- The number density of free electrons is different for different metals. For copper, it is $8 \times 10^{28}\ m^{-3}$.
- When liquids conduct, ions are the charge carriers. The above equations apply, except that e and n must be replaced by the charge and number density of the ions.

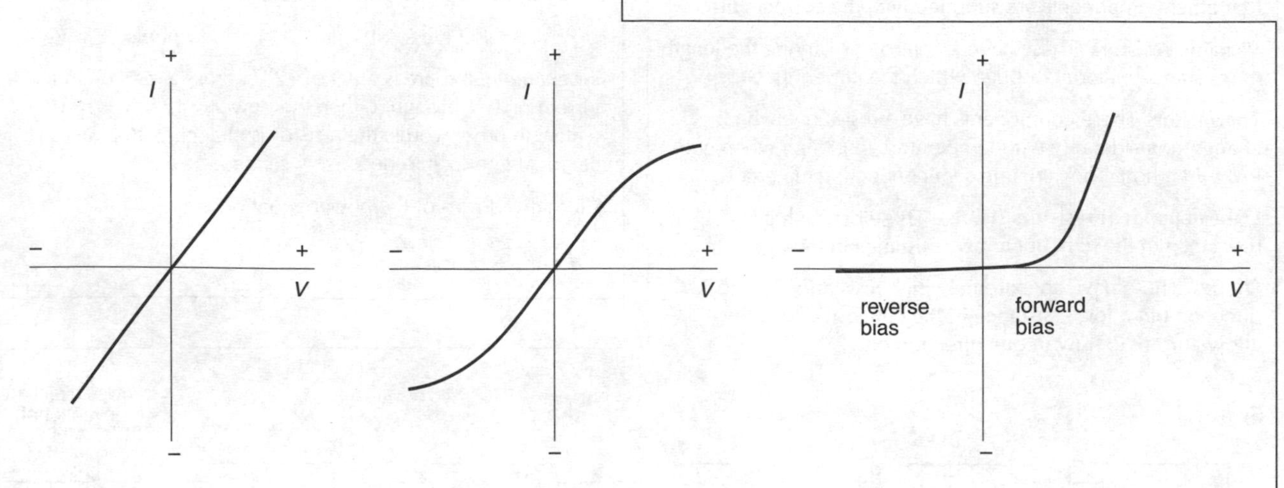

Wire (metal) kept at constant temperature V/I is constant, so R is constant.

Bulb filament (metal) As the current rises the filament heats up. V/I increases, so R increases.

Diode (semiconductor) R is very high in one direction. It is much lower in the other direction and decreases as the current rises.

Resistance and temperature

A conducting solid is made up of a ***lattice*** of atoms. When a current flows, electrons move through this lattice.

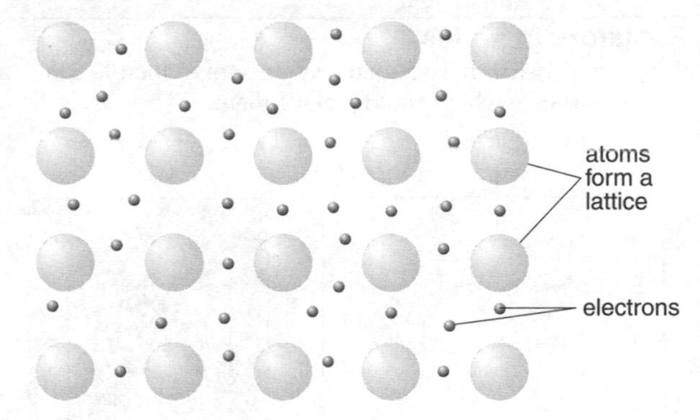

Metals When free electrons drift through a metal, they make occasional collisions with the lattice. These collisions are inelastic and transfer energy to the lattice as internal energy. That is why a metal has resistance. If the temperature of a metal rises, the atoms of the lattice vibrate more vigorously. Free electrons collide with the lattice more frequently, which increases the resistance.

Semiconductors (e.g. silicon) At low temperature, the electrons are tightly bound to their atoms. But as the temperature rises, more and more electrons break free and can take part in conduction. This easily outweighs the effects of more vigorous lattice vibrations, so the resistance decreases. At around 100–150 °C, ***breakdown*** occurs. There is a sudden fall in resistance – and a huge increase in current. That is why semiconductor devices are easily damaged if they start to overheat.

The conduction properties of a semiconductor can be changed by ***doping*** it with tiny amounts of impurities. For example, a diode can be made by doping a piece of silicon so that a current in one direction increases its resistance while a current in the opposite direction decreases it.

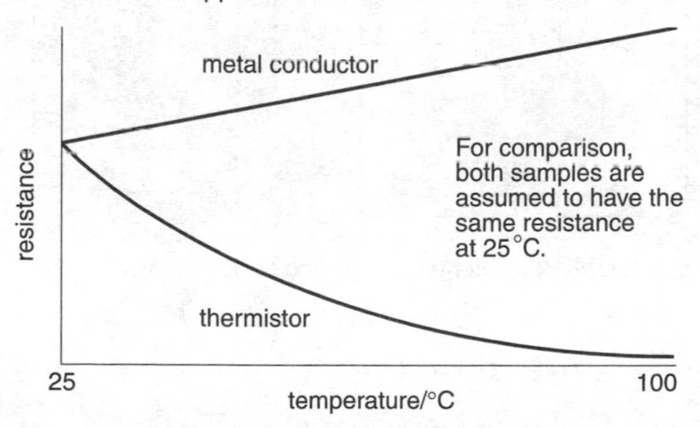

The graphs above are for a typical metal conductor and one type of thermistor. The thermistor contains semiconducting materials.

Superconductivity

When some metals are cooled towards absolute zero, a ***transition temperature*** is reached at which the resistance suddenly falls to zero. This effect is called ***superconductivity***. It occurs when there is no interaction between the free electrons and the lattice, and is explained by the quantum theory. Some specially developed metal compounds have transition temperatures above 100 K.

If an electromagnet has a superconducting coil, a huge current can be maintained in it with no loss of energy. This enables a very strong magnetic field to be produced.

Energy transfer

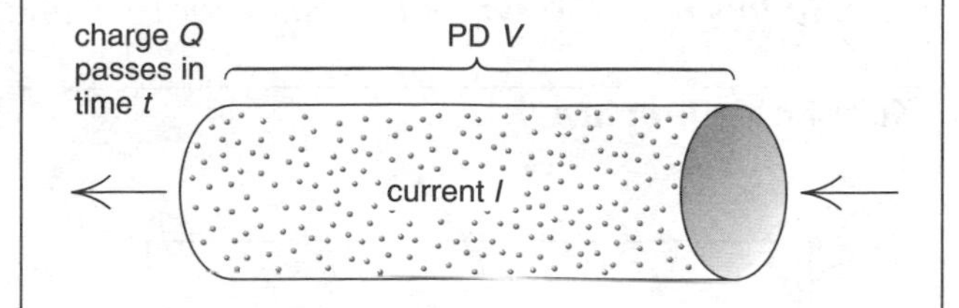

Above, charge Q passes through a resistor in time t. Work W is done by the charge, so energy W is transformed – the electrons lose electrical potential energy and the lattice gains internal energy (it heats up).

W, Q, and V are linked by this equation (see also E1):

$$W = QV$$

But $Q = It$, so $\quad W = VIt \qquad (1)$

Applying $V = IR$ to the above equation gives

$W = I^2Rt \quad$ and $\quad W = V^2t/R \qquad (2)$

For example, if a current of 2 A flows through a 3 Ω resistor for 5 s, $W = 2^2 \times 3 \times 5 = 60$ J. So the energy dissipated is 60 J. *Double* the current gives *four* times the energy dissipation.

Note:

- Equation (1) can be used to calculate the total energy transformation whenever electrical potential energy is changed into other forms (e.g. KE and internal energy in an electric motor). Equations (2) are only valid where *all* the energy is changed into internal energy. Similar comments apply to the power equations which follow.

As power $P = W/t$, it follows from (1) and (2) that

$P = VI \qquad P = I^2R \qquad P = V^2/R$

Resistivity

The resistance R of a conductor depends on its length l and cross-sectional area A:

$$R \propto \frac{l}{A}$$

This can be changed into an equation by means of a constant, ρ, known as the ***resistivity*** of the material:

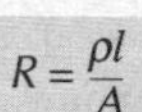

$$R = \frac{\rho l}{A}$$

With this equation, the resistance of a wire can be calculated if its dimensions and resistivity are known.

Resistivities, in Ω m	
copper: 1.55×10^{-8}	aluminium: 2.50×10^{-8}

Conductance and conductivity

If a PD V is applied across a conductor, and a current I flows, then $V = IR$. However, as V is the cause of the current and I is the effect, it is more logical to write this as:

$$I = \frac{1}{R} \times V$$

$1/R$ is called the ***conductance***. $1/\rho$ is the ***conductivity***.

D3 Analysing circuits

Note: in this unit, the symbol E stands for EMF and not electric field strength.

Kirchhoff's first law

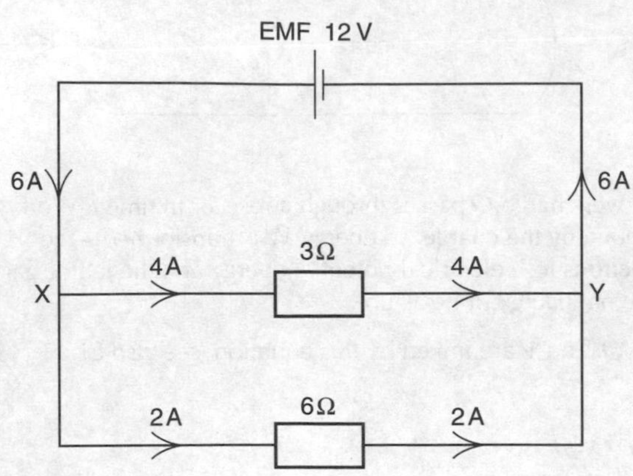

The currents at junctions X and Y above illustrate a law which applies to all circuits:

total current out of junction = total current into junction

This is known as ***Kirchhoff's first law***. It arises because, in a complete circuit, charge is never gained or lost. It is conserved. So the total rate of flow of charge is constant.

Kirchhoff's second law

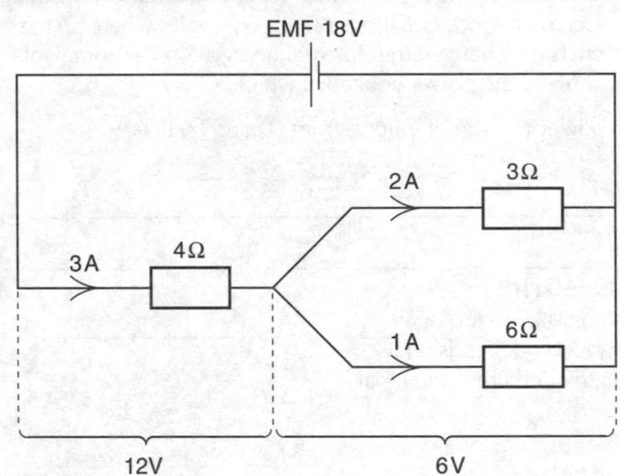

The arrangement above is called 'a circuit'. But, really, there are *two* complete circuits through the battery. To avoid confusion, these will be called *loops*.

In the circuit above, charge leaves the battery with electrical potential energy. As the charge flows round a loop, its energy is 'spent' – in stages – as heat. The principle that the total energy supplied is equal to the total energy spent is expressed by ***Kirchhoff's second law***.

Round any closed loop of a circuit, the algebraic sum of the EMFs is equal to the algebraic sum of the PDs (i.e. the algebraic sum of all the *IR*s).

Note:

- From the law, it follows that if sections of a circuit are in parallel, they have the same PD across them.
- 'Algebraic' implies that the direction of the voltage must be considered. For example, in the circuit on the right, the EMF of the right-hand battery is taken as *negative* (–4 V) because it is opposing the current. Therefore:

  algebraic sum of EMFs = 18 + (–4) = +14 V
  algebraic sum of *IR*s = (2 × 3) + (2 × 4) = +14 V

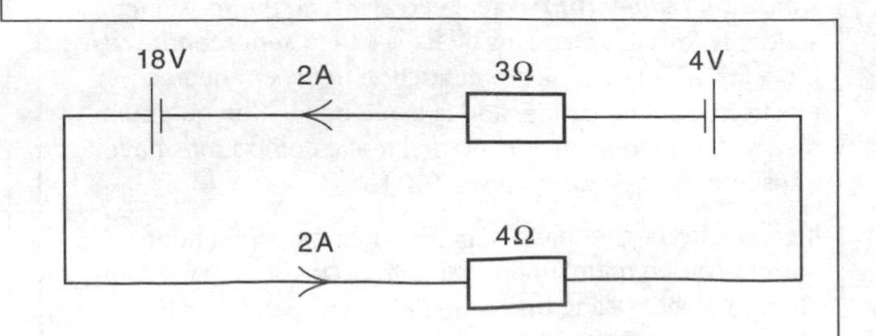

Resistors in series

If R_1 and R_2 below have a total resistance of R, then R is the single resistance which could replace them.

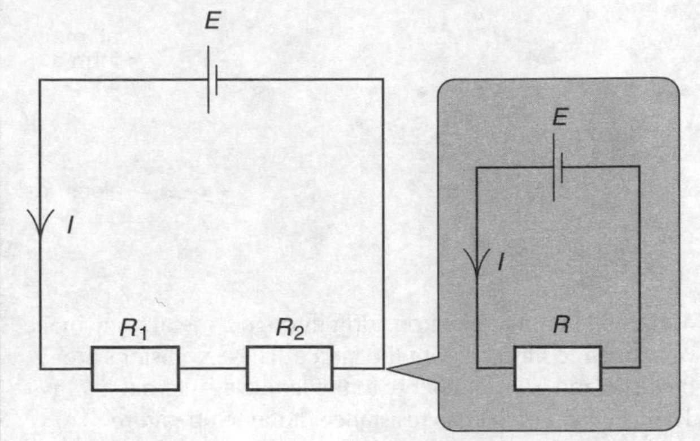

From Kirchhoff's first law, all parts of the circuit have the same current I through them.

From Kirchhoff's second law $E = IR$ and $E = IR_1 + IR_2$.

So $IR = IR_1 + IR_2$

$\therefore \quad R = R_1 + R_2$

For example, if $R_1 = 3\ \Omega$ and $R_2 = 6\ \Omega$, then $R = 9\ \Omega$.

Resistors in parallel

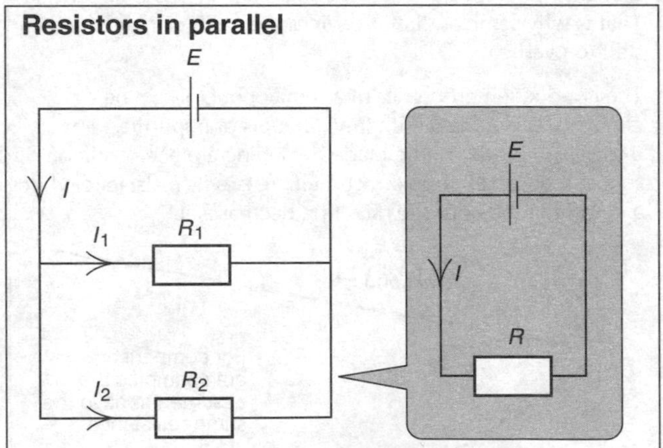

From Kirchhoff's second law (applied to the various loops):

$E = IR$ and $E = I_1R_1$ and $E = I_2R_2$

From Kirchhoff's first law $I = I_1 + I_2$.

So $\dfrac{E}{R} = \dfrac{E}{R_1} + \dfrac{E}{R_2}$

$\therefore \quad \dfrac{1}{R} = \dfrac{1}{R_1} + \dfrac{1}{R_2}$

For example, if $R_1 = 3\ \Omega$ and $R_2 = 6\ \Omega$,
$1/R = 1/3 + 1/6 = 1/2$. So $R = 2\ \Omega$.

Internal resistance

On the opposite page, it was assumed that each battery's output PD (the PD across its terminals) was equal to its EMF. In reality, when a battery is supplying current, its output PD is *less* than its EMF. The greater the current, the lower the output PD. This reduced voltage is due to energy dissipation in the battery. In effect, the battery has ***internal resistance***. Mathematically, this can be treated as an additional resistor in the circuit.

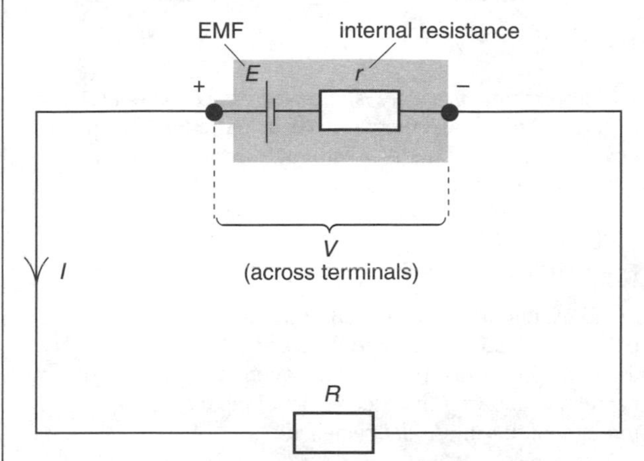

The battery above is supplying a current I to an external circuit. The battery has a constant internal resistance r.

From Kirchhoff's second law $\quad E = IR + Ir$

But $V = IR$, so $\quad E = V + Ir$

So $\quad V = E - Ir \quad (1)$

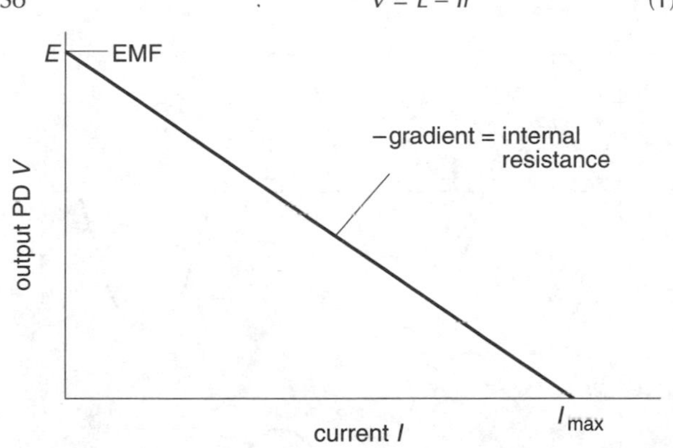

The graph above shows how V varies with I. Unlike earlier graphs, V is on the vertical axis.

Note:

- When I is zero, $V = E$. In other words, when a battery is in ***open circuit*** (no external circuit), the PD across its terminals is equal to its EMF
- When R is zero, V is zero. In other words, when the battery is in ***short circuit*** (its terminals directly connected), its output PD is zero. In this situation, the battery is delivering the maximum possible current, I_{max}, which is equal to E/r. Also, the battery's entire energy output is being wasted internally as heat.
- As $I_{max} = E/r$, it follows that $r = E/I_{max}$. So the gradient of the graph is numerically equal to the internal resistance of the battery.

If both sides of equation (1) are multiplied by I, the result is $VI = EI - I^2r$. Rearranged, this gives the following:

EI	=	VI	+	I^2r
power released by chemical action		power delivered to external circuit		power dissipated inside battery

Potential divider

A ***potential divider*** or ***potentiometer*** like the one below passes on a fraction of the PD supplied to it.

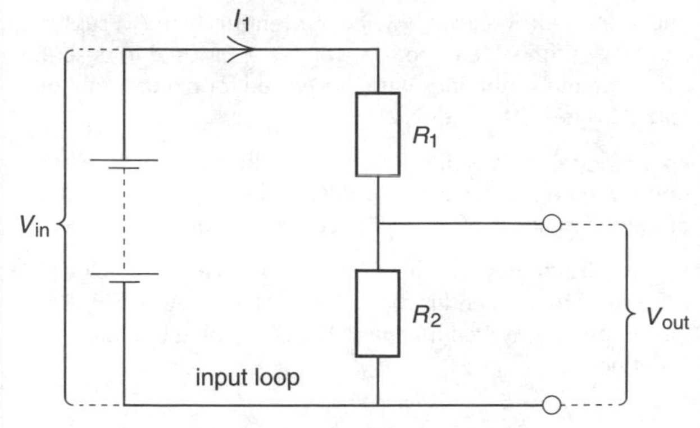

In the input loop above, the total resistance = $R_1 + R_2$.

So $\quad I = V_{in}/(R_1 + R_2)$

But $V_{out} = IR_2$, $\quad$ so $V_{out} = \left(\dfrac{R_2}{R_1 + R_2}\right)V_{in}$

For example, if R_1 and R_2 are both 2 kΩ, then $R_2/(R_1 + R_2)$ works out at 1/2, so V_{out} is a half of V_{in}.

Note:

- The above analysis assumes that no external circuit is connected across R_2. If such a circuit is connected, then the output PD is reduced.

In electronics, a potential divider can change the signals from a sensor (such as a heat or light detector) into voltage changes which can be processed electrically. For example, if R_2 is a thermistor, then a rise in temperature will cause a fall in R_2, and therefore a fall in V_{out}. Similarly, if R_2 is a ***light-dependent resistor*** (**LDR**), then a rise in light level will cause a fall in R_2, and therefore a fall in V_{out}.

Potential dividers are not really suitable for high-power applications because of energy dissipation in the resistors.

Balanced PDs

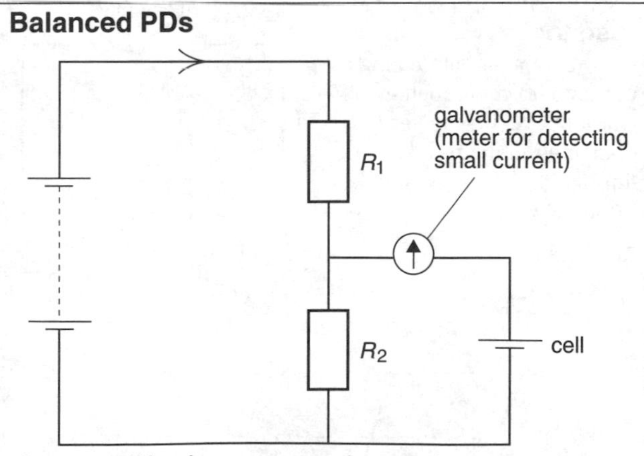

Above, a cell has been connected across the output of a potential divider, and the values of R_1 and R_2 adjusted so that the galvanometer reads zero. This happens when the PD across R_2 exactly balances the cell's output PD.

The above method can be used to compare the EMFs of different cells. It has several advantages.

- The cell is effectively in open circuit, so the PD across the cell's terminals is equal to its EMF
- As the meter is only being used to test for zero current, it does not need an accurately calibrated scale.
- A very sensitive meter can be used.

D4 Alternating current

AC terms

The graph below shows how the current from an AC supply varies with time. Here are some of the terms used to describe AC. Note the similarities with those used for circular motion and SHM (see B11 and B12).

Frequency f This is the number of cycles per second. The unit is the hertz (Hz). For example, in the UK, the frequency of the AC mains is 50 Hz (50 cycles per second).

Angular frequency ω This is equivalent to angular velocity in circular motion and is measured in rad s^{-1}. As in SHM, it has no direct physical meaning, but is useful in equations. It is defined as follows:

$$\omega = 2\pi f$$

Peak current I_0 This is the maximum current during the cycle. It is the amplitude of the waveform in the graph below.

The current I at any instant is related to the peak current by the following equation (given in two alternative versions):

$I = I_0 \sin \omega t$ and $I = I_0 \sin 2\pi ft$

The graph is an example of a ***sinusoidal*** waveform.

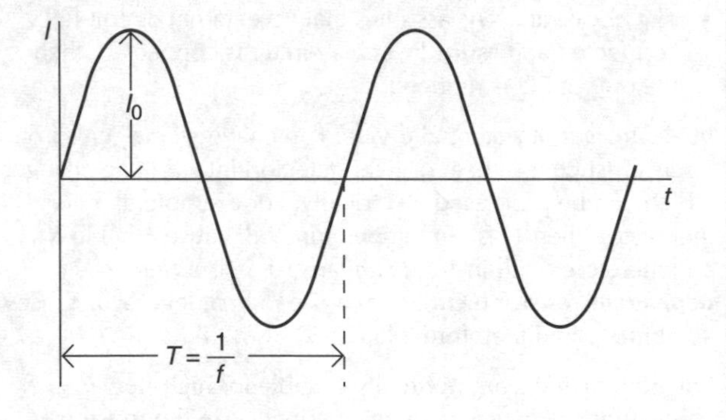

Power dissipated in a resistor

On the right, an alternating current I flows through a resistance R. The power P dissipated in the resistor varies through the cycle. At any instant $P = I^2R$.

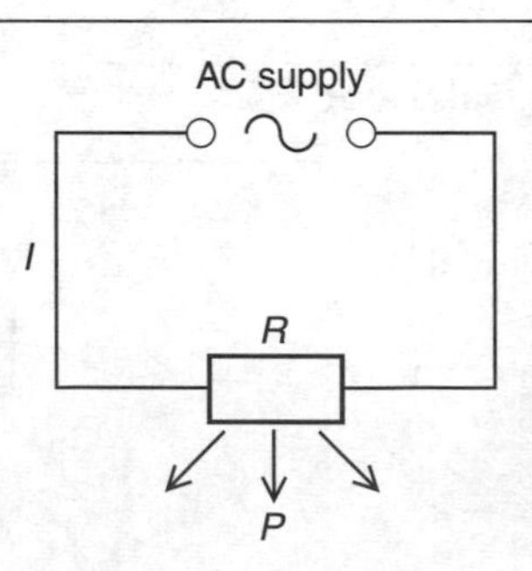

But $I = I_0 \sin \omega t$

So $P = I_0^2 R \sin^2 \omega t$

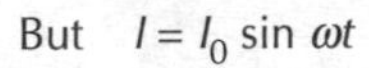

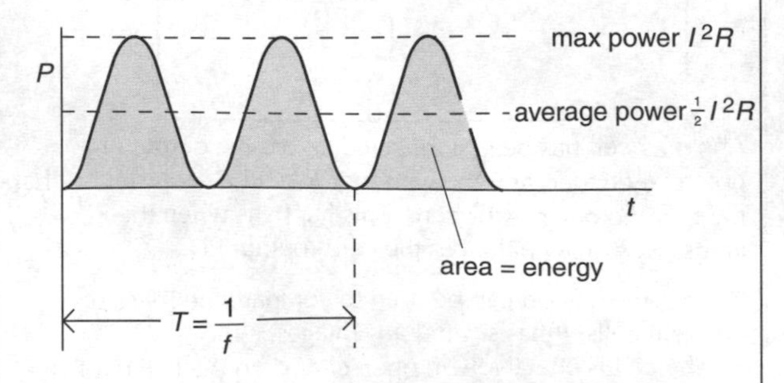

The graph shows how P varies with t. P is always positive because all values of $\sin^2 \omega t$ lie between 0 and 1. Note:

average power $= \frac{1}{2}$ maximum power $= \frac{1}{2} I_0^2 R$

RMS voltage and current

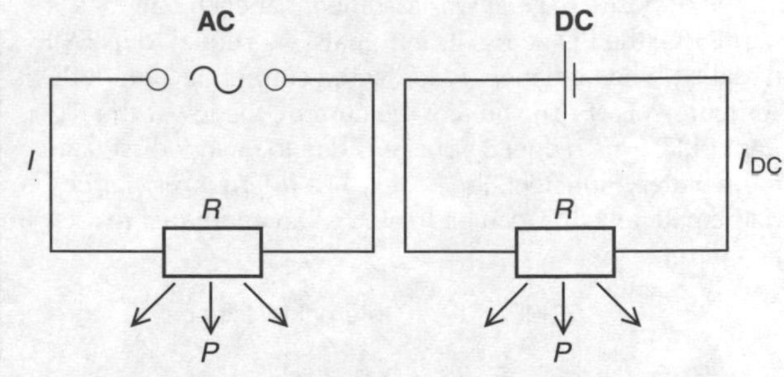

The current from a battery is one-way ***direct current*** **(DC)**. In the DC circuit above, there is a steady current I_{DC} such that power is dissipated in the resistor at exactly the same rate as in the AC circuit. So

$$I_{DC}^2 R = \frac{1}{2} I_0^2 R$$

From which it follows that $I_{DC} = I_0/\sqrt{2}$.

In the AC circuit, the ***root mean square (RMS) current*** is equal to I_{DC}. So it is equal to the steady current which gives the same power dissipation as the alternating current. (It is so named because it is the square root of the mean of I^2 throughout the sinusoidal cycle.) So:

$$I_{RMS} = \frac{I_0}{\sqrt{2}}$$

Similarly, for the alternating PD of peak value V_0,

$$V_{RMS} = \frac{V_0}{\sqrt{2}}$$

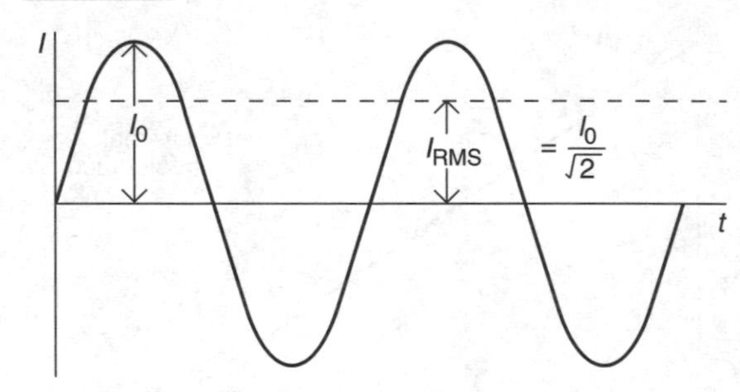

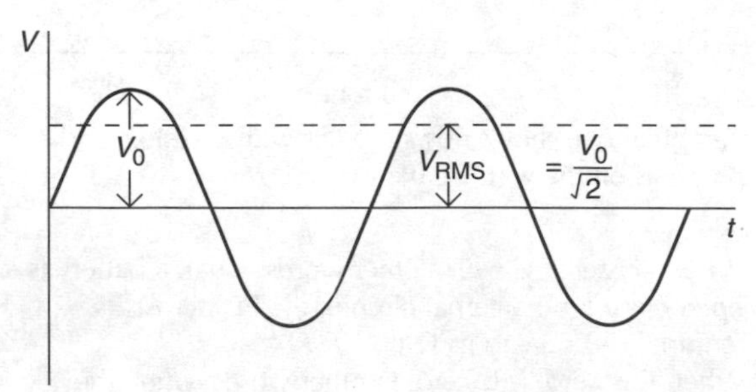

The graphs above show how the alternating V across a resistor, and I through it, vary with time. The resistor is assumed to obey Ohm's law.

Note:

- When the PD is zero, the current is zero. And when the PD is at its peak, the current is at its peak. In other words, the PD and current are ***in phase***.
- $R = \frac{V}{I} = \frac{V_0}{I_0} = \frac{V_{RMS}}{I_{RMS}}$
- The average power dissipated $= \frac{1}{2} V_0 I_0 = V_{RMS} I_{RMS}$

The UK AC mains voltage of 230 V is a RMS value. The peak voltage $= 230 \times \sqrt{2} = 325$ V. In doing calculations on power dissipation, RMS values are normally used because the factor $\frac{1}{2}$ does not need to be included in the working.

Transformers

Transformers increase or decrease alternating voltage (see E5 for diagram and principle). Provided there is no flux leakage from the core:

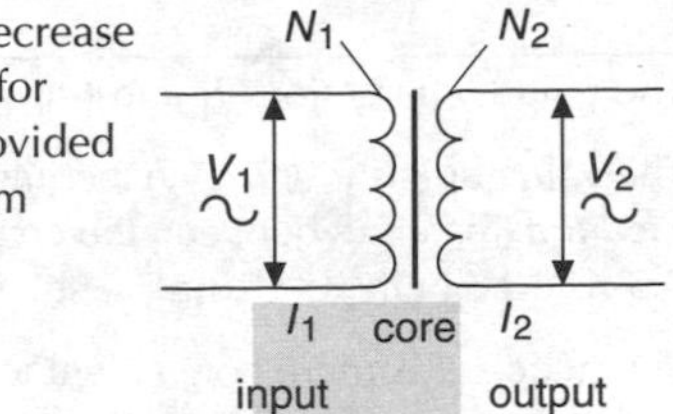

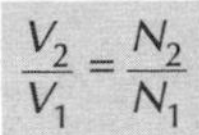

$$\frac{V_2}{V_1} = \frac{N_2}{N_1}$$

In an ideal transformer, the power output is equal to the power input. In practice, a transformer wastes some power (as heat). Two reasons for this are:

- coil resistance
- the changing flux induces eddy currents (see E7) in the core. The core is laminated (layered) to reduce these.

For a practical transformer $V_2 I_2 = eV_1 I_1$

where e is the efficiency (typically over 0.95).

Supplying AC for the mains

waste gases

steam

step-up transformer

burning fuel heats boiler

turbine

alternator

cooled water

Mains AC is generated in power stations. The layout of a typical fuel-burning station is shown above. The power is fed into a distribution network called the ***Grid***.

Power is sent across country through overhead lines at very high voltage (typically 400 000 V). The voltage is increased to this level by transformers and then reduced again at the far end. As power = voltage × current, transmitting at a higher voltage means a lower current and, therefore, less power wasted as heat because of line resistance.

Impedance

In electrical circuits, all components have ***impedance***. In other words, they oppose the flow of current when a voltage is applied across them. Impedance is measured in ohms (Ω).

There are two types of impedance. In DC circuits, ***resistance*** is the only type that has an effect. However, in AC circuits, components may also have ***reactance***. Reactance causes a phase difference between the alternating current and voltage. If a component only has resistance, the current and voltage are in phase.

Rectification

Low voltage DC equipment can use mains power, provided the voltage is reduced, and the AC is converted to DC. Turning AC into DC, is called ***rectification***.

Half-wave rectification The circuit below reduces the alternating voltage, and then uses a diode to block the backwards part of the alternating current. The DC output is in half-wave pulses with zero PD between.

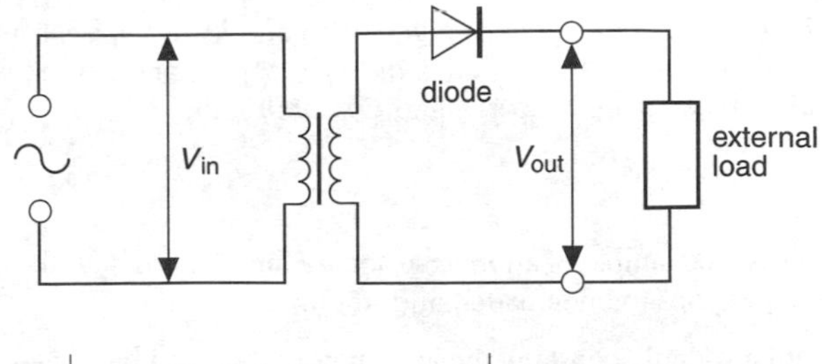

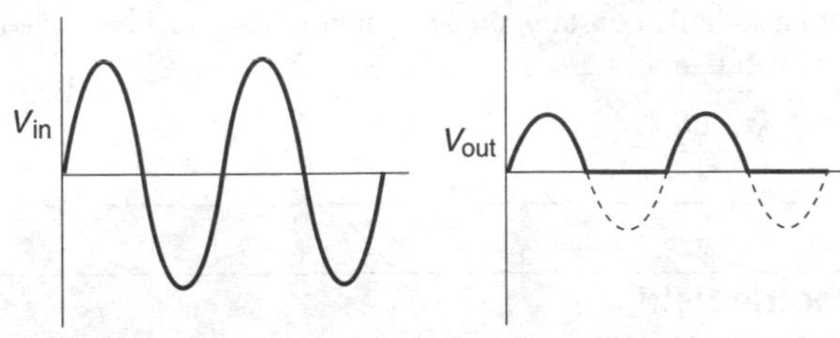

Full-wave rectification The four diodes in the circuit below form a ***bridge rectifier***. In one half of the AC cycle, diodes A conduct and diodes B block. In the other half, diodes B conduct and diodes A block. The effect is to reverse the backwards parts of the AC.

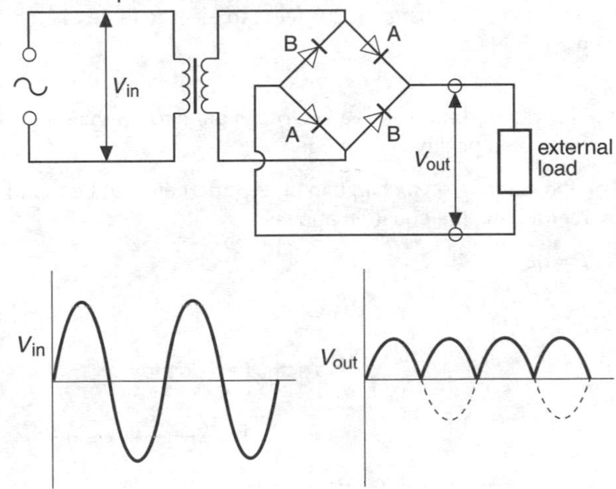

Smoothing The output from a rectifier circuit can be smoothed by connecting a large capacitor across it, as below. The capacitor charges up during the forward peaks, then releases its charge in between.

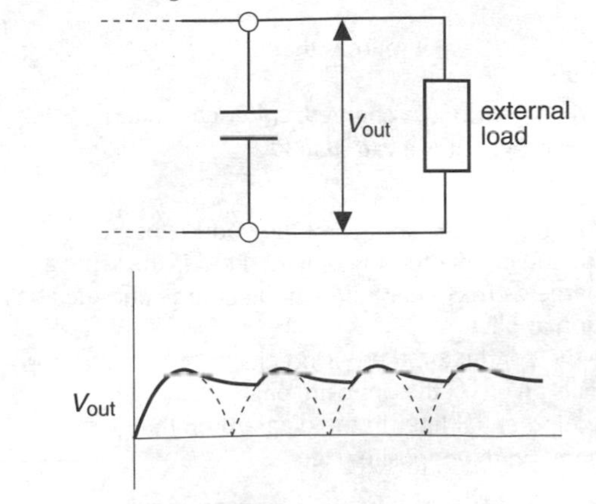

E1 Electric charges and fields

Electric force

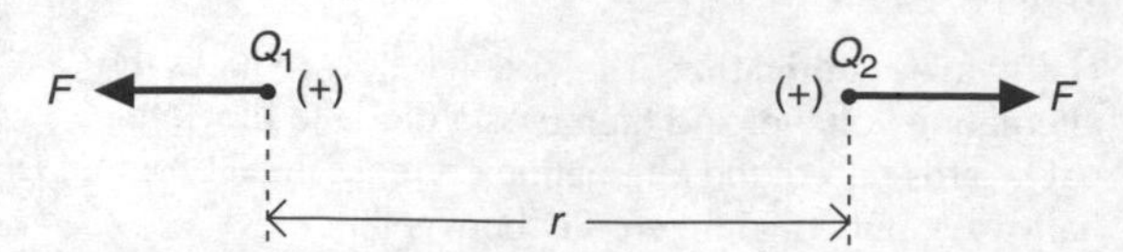

Charges attract or repel each other with an ***electric force***. If point charges Q_1 and Q_2, are a distance r apart, and F is the force on each, then according to ***Coulomb's law***:

$$F \propto \frac{Q_1 Q_2}{r^2}$$

This is an example of an ***inverse square law***. If r doubles, the force F drops to one quarter, and so on.

With a suitable constant, the above proportion can be turned into an equation:

$$F = \frac{kQ_1 Q_2}{r^2}$$

The unit of charge for Q_1 and Q_2 is the coulomb (C).

The value of k is found by experiment. It depends on the ***medium*** (material) between the charges. For a vacuum, k is 8.99×10^{-9} N m^2 C^{-2}, and is effectively the same for air.

In practice, it is more convenient to use another constant, ε_0, and rewrite the equation on the left in the following form:

In a vacuum $$F = \frac{Q_1 Q_2}{4\pi\varepsilon_0 r^2}$$

ε_0 is called the ***permittivity of free space***. Its value is 8.85×10^{-12} C^2 N^{-1} m^{-2}.

Note:
- Although '4π' complicates the above equation, it simplifies others derived from it.
- In the above equation, if, say, Q_1 and Q_2 are *like* charges (e.g. – and –), then F is *positive*. So a positive F is a force of *repulsion*. Similarly, it follows that a negative F is a force of *attraction*.

Electric field

If a charge feels an electric force, then it is in an ***electric field***. If a charge q feels a force F, then the ***electric field strength*** E is defined like this:

$$E = \frac{\text{electric force}}{\text{charge}} \qquad \text{In symbols} \quad E = \frac{F}{q} \qquad (1)$$

For example, if a charge of 2 C feels an electric force of 10 N, then E is 5 N C^{-1}.

Note:
- Electric field strength is a vector. Its direction is that of the force on a positive (+) charge.

The force acting on a charge in an electric field can be found by rearranging the equation above:

$$F = qE$$

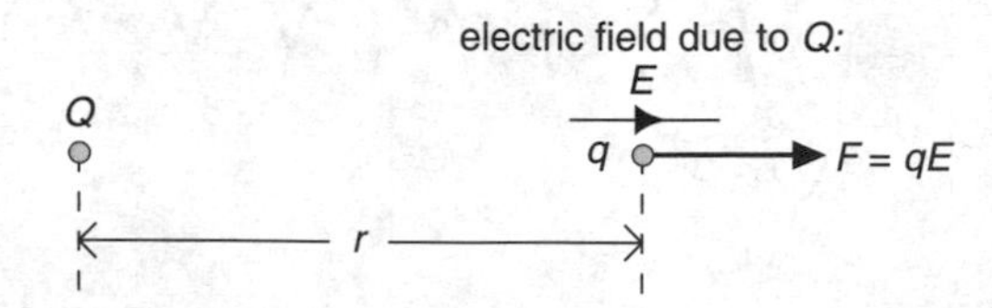

Above, charge Q produces an electric field which acts on a small charge q. As $F = qE$ and

$$F = \frac{Qq}{4\pi\varepsilon_0 r^2} \qquad \text{it follows that} \quad E = \frac{Q}{4\pi\varepsilon_0 r^2} \qquad (2)$$

The electric field round a charged, spherical conductor is shown on the right. It is a ***radial*** field.

Note:
- The charge is on the surface of the conductor.
- Outside the conductor, the electric field is the same as if all the charge were concentrated at the centre, and the above equation applies.
- Inside the conductor, there is no electric field. The reason for this is given on the opposite page.
- The equipotential lines in the diagram on the right are explained on the opposite page.

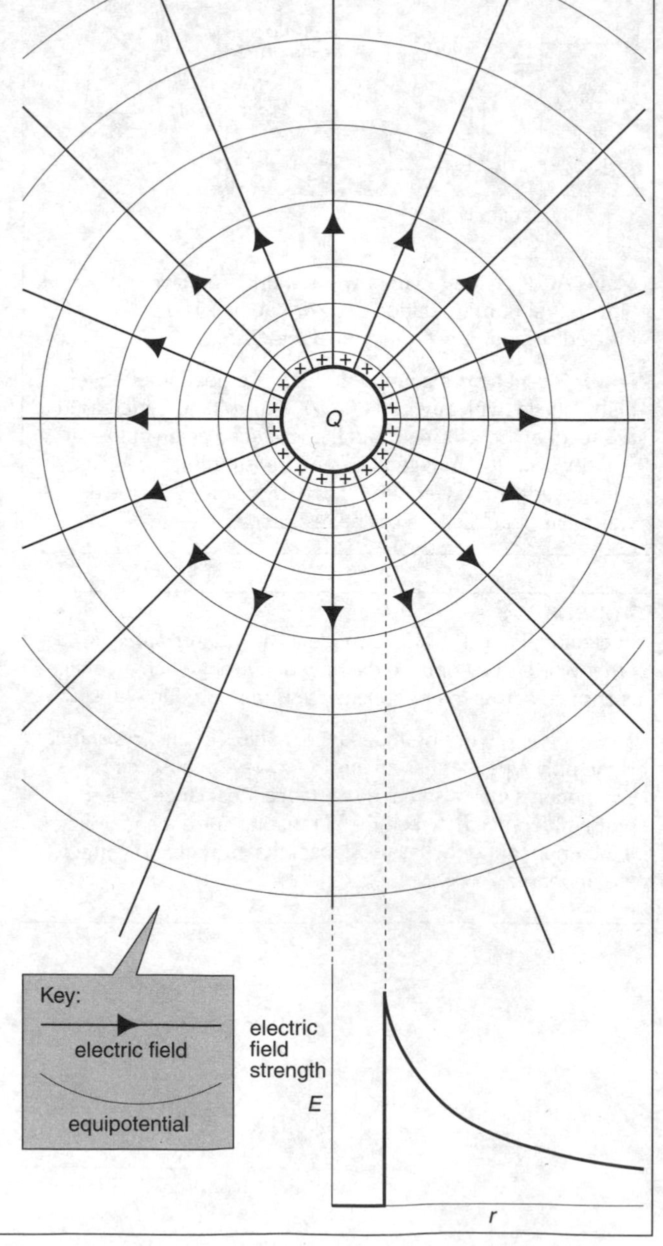

Electric potential

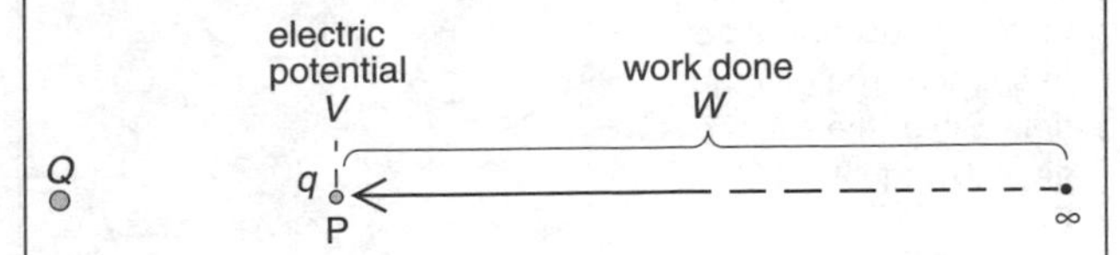

Above, charge Q causes an electric field. A small charge q has been moved through this field, from an infinite distance (where the electric force is zero), to point P.

The ***electric potential*** V (at point P) is defined as follows:

$$V = \frac{W}{q} \quad (3)$$

where W is the work done in moving a charge q from infinity (∞) to point P.

The SI unit of electric potential is the ***volt*** (V). For example, if 1000 J of work is done in moving a charge of +2 C from ∞ to P, then the potential at P = 1000/2 = 500 V.

Note:
- Electric potential is a scalar.
- At infinity, the electric potential is *zero*.
- Elsewhere, the electric potential due to *positive* charge is *positive*. Similarly, the electric potential due to a *negative* charge is *negative*.
- Equation (3) can also be used to find the work done in moving a charge q between two points. In this case, V is the ***potential difference*** **(PD)** between the points.

Linking potential and field strength

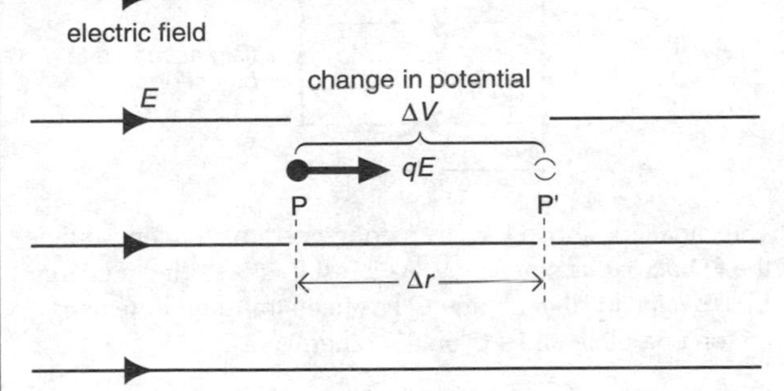

Above, work ΔW is done on a small charge q in moving it from P to P′ in a uniform electric field E. So, from (3):

$$\Delta W = q\Delta V \quad (3)$$

This equation gives the work done *on* the charge.

So work done *by* charge = $-\,q\Delta V$

But work done by charge = force × distance moved = $qE\Delta r$

So $\quad qE\Delta r = -\,q\Delta V$

Therefore $\quad E = -\dfrac{\Delta V}{\Delta r} \quad (4)$

In calculus notation, there is a more general version of this equation which also applies to non-uniform fields:

$$E = -\frac{\mathrm{d}V}{\mathrm{d}r}$$

Note:
- In the above equations, the minus sign indicates that E is in the direction of *decreasing* potential.

Electric potential in a radial field

A radial field is shown on the opposite page. Provided r is not less than the radius of the sphere,

$$E = \frac{Q}{4\pi\varepsilon_0 r^2} \qquad \text{Also, } E = -\frac{\mathrm{d}V}{\mathrm{d}r}$$ (see bottom panel)

Combining these and using calculus gives

$$V = \frac{Q}{4\pi\varepsilon_0 r}$$ (This also applies to a point charge.)

Note:
- In the diagram on the opposite page, each ***equipotential line*** is a line joining points of equal potential.
- Inside the charged conductor, all points are at the same potential, so the potential gradient (see bottom panel) is zero. From this it follows that E is also zero, so there is no electric field inside the conductor.

Comparing electric and gravitational fields

For particles of similar size, electric forces are very much stronger than gravitational ones. For example, electric forces hold atoms together to form solids.

Electric and gravitational fields have similar features. That is why the equations is this unit have a similar form to those in E3. However, comparing equivalent equations, a minus sign may be present in one but absent from the other. This arises because of the differing force directions.

Gravity is always a force of attraction. Mass is always positive and it produces a gravitational field which is directed *towards* it.

Electric charges may attract or repel. However, if a charge is positive, then it produces an electric field which is directed *away* from it.

The metal plates on the right have a small test charge q between them. The charge feels a force F. So, from equation (1),

$$E = \frac{F}{q}$$

There is a potential difference V between the plates. From the graph on the right, and equation (4), the potential gradient is $-\,V/d$. So

$$E = \frac{V}{d} \quad (5)$$

The constant potential gradient means that the electric field is uniform.

The above equations show different aspects of electric field strength. If, say, E is 10 N C^{-1}, you can think of this either as a force of 10 N per coulomb or a potential drop of 10 V per metre. (E can be expressed in N C^{-1} or in V m^{-1}.)

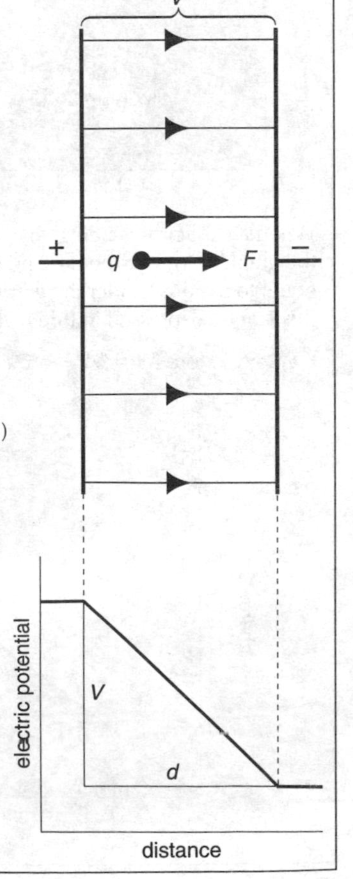

E2 Capacitors and fields

Capacitance

Capacitors store small amounts of electric charge. For information on their practical uses, see D4, H16, and H17.

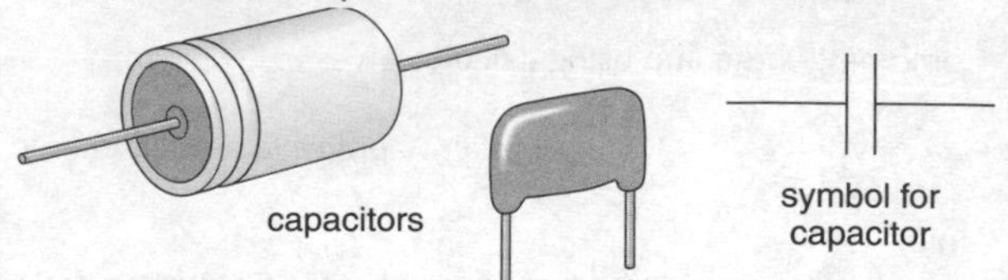

A capacitor can be charged by connecting a battery across it. The higher the PD V, the greater the charge Q stored. Experiments show that $Q \propto V$. Therefore, Q/V is a constant.

The ***capacitance*** C of a capacitor is defined as follows:

$$\text{capacitance} = \frac{\text{charge}}{\text{PD}} \qquad \text{In symbols} \quad C = \frac{Q}{V}$$

The higher the capacitance, the more charge is stored for any given PD.

Capacitance is measured in C V^{-1}, known as a ***farad*** (F). However, a farad is a very large unit, and the μF (10^{-6} F) is more commonly used for practical capacitors.

Energy stored by a capacitor

Work must be done to charge up a capacitor. Electrical potential energy is stored as a result.

If a charge of 2 C is moved through a *steady* PD of 10 V, then, using equation (3) in E1,

work done $W = QV = 2 \times 10 = 20$ J.

So the stored energy is 20 J. Numerically, this is the area under the graph below.

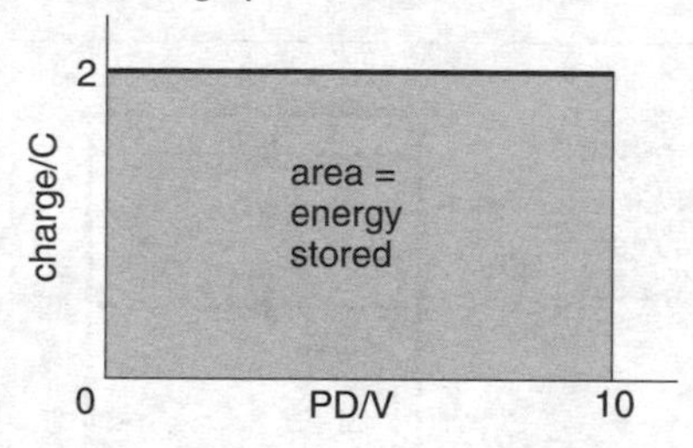

When a capacitor is being charged, Q and V are related as in the graph below. As before, the energy stored is numerically equal to the area under the graph, which is $\frac{1}{2}QV$. As $C = Q/V$, this can be expressed in three ways:

$$\text{energy stored} = \tfrac{1}{2}QV = \tfrac{1}{2}CV^2 = \tfrac{1}{2}Q^2/C$$

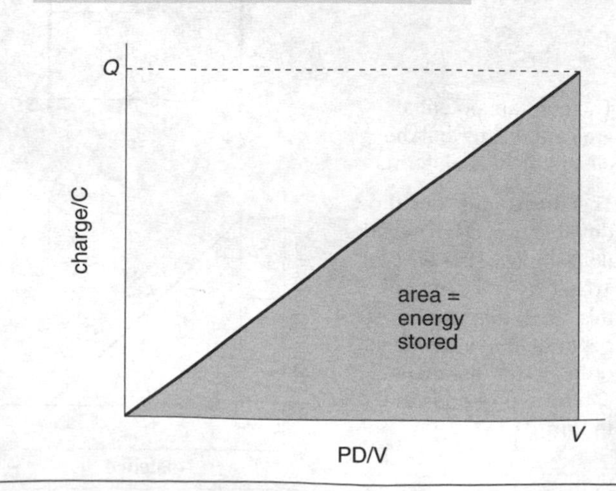

Electric field near a charged plate

On the right, a metal sphere has a charge Q uniformly distributed over its surface. The electric field E near the surface is given by equation (2) in E1:

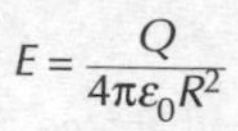

$$E = \frac{Q}{4\pi\varepsilon_0 R^2}$$

But the surface area of the sphere $A = 4\pi R^2$. So

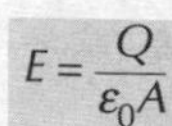

$$E = \frac{Q}{\varepsilon_0 A} \qquad (1)$$

charge Q
surface area A

This equation also applies to a flat, charged metal plate of surface area A.

Parallel plate capacitor

The simplest form of capacitor is made up of two parallel metal plates, separated by an air gap.

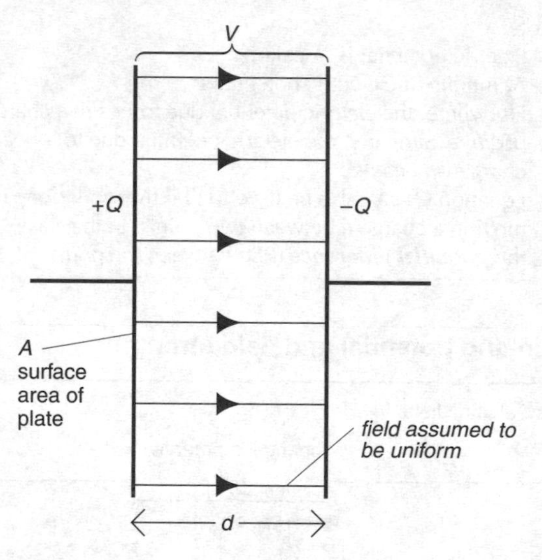

The capacitor above has been connected to a battery so that the PD across its plates is V. As a result, it is storing a charge Q. (This means that charge Q has been transferred, leaving $-Q$ on one plate and $+Q$ on the other.)

From equation (1) above $\quad E = Q/\varepsilon_0 A$
From equation (5) in B13 $\quad E = V/d$
From these, it follows that $\quad Q/V = \varepsilon_0 A/d$

But $C = \dfrac{Q}{V}$ $\quad$ So $C = \dfrac{\varepsilon_0 A}{d}$ $\quad$ (2)

From the above equation, note:

- $C \propto A$, so a larger plate area gives a higher C.
- $C \propto 1/d$, so a smaller plate separation gives a higher C.

Dielectric

If the gap between the capacitor plates is filled with a material such as polythene, the capacitance is increased. Any insulating material which has this effect is called a ***dielectric***.

The ***relative permittivity*** ε_r of the dielectric is the factor by which the capacitance is increased. For polythene, $\varepsilon_r = 2.3$. With a dielectric present, equation (2) becomes

$$C = \frac{\varepsilon_r \varepsilon_0 A}{d} \qquad (3)$$

Discharge of a capacitor

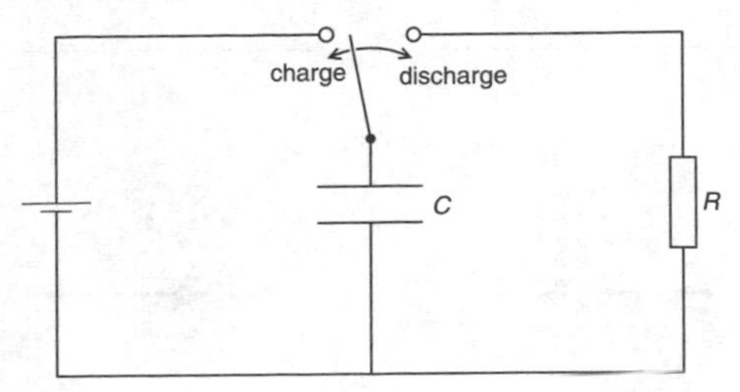

The capacitor above is charged from a battery and then discharged through a resistance R.

Graph A shows how, during discharge, the charge Q decreases with time t, according to the following equation:

$$Q = Q_0 e^{-t/RC} \qquad \text{where } e = 2.718$$

RC is called the ***time constant***. (It equals the time which the charge would take to fall to zero if the initial rate of loss of charge were maintained.) Increasing R or C gives a higher time constant, and therefore a slower discharge.

The gradient of the graph at any time t is equal to the current at that time.

Graph A **Graph B**

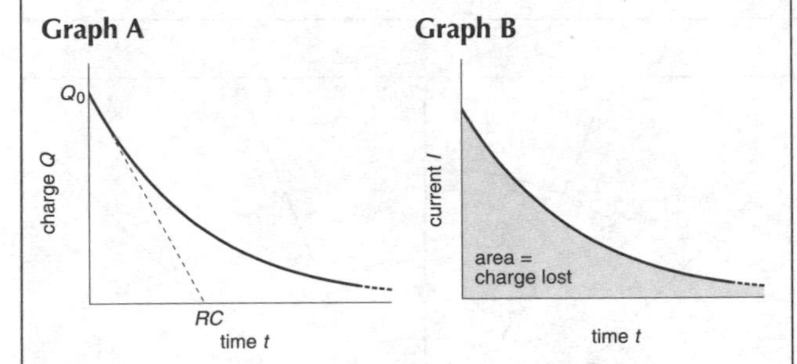

Graph B shows how the current decreases with time. The area under the graph is numerically equal to the charge lost.

Note:

- Each graph is an ***exponential decay curve***, with the same characteristics as a radioactive decay curve. A ***half-life*** can be calculated in the same way (see G2).

Practical capacitors

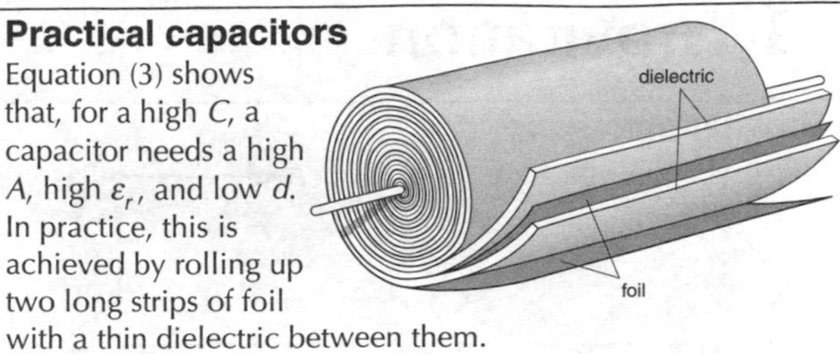

Equation (3) shows that, for a high C, a capacitor needs a high A, high ε_r, and low d. In practice, this is achieved by rolling up two long strips of foil with a thin dielectric between them.

In ***electrolytic capacitors***, the dielectric is formed by the chemical action of a current. This gives a very thin dielectric, and a very high capacitance. But the capacitor must always be used with the same plate positive, or the chemical action is reversed.

Capacitors have a ***maximum working voltage*** above which the dielectric breaks down and starts to conduct.

Charging a capacitor

The capacitor below is charged through a resistance R. The graph shows how the charge builds up.

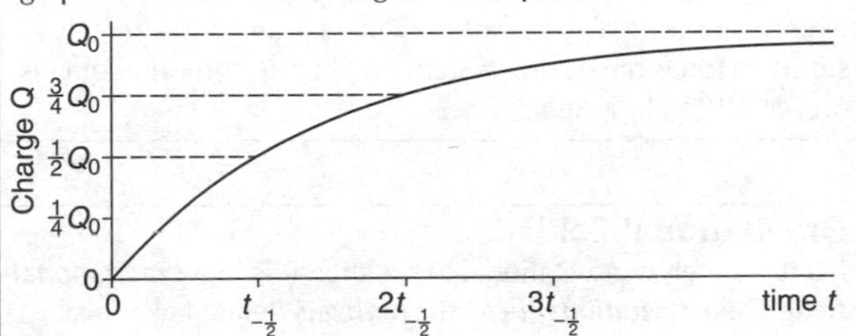

The equation for the charging process is

$$Q = Q_0(1 - e^{-t/CR})$$

The charge reaches a maximum value of Q_0 which is equal to VC.

The charging current starts at a maximum value of V/R and falls to a lower value in the same way as it does when the capacitor discharges.

Voltage–time graphs

Since the voltage across a capacitor is proportional to the charge on it the variations of voltage with time are the same shape as the charge–time graphs. The equations for calculating the voltage at any time are similar to those for charge, substituting V for Q.

Capacitors in series

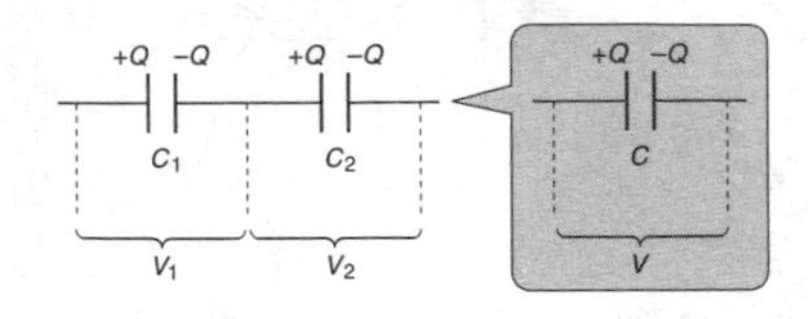

If C_1 and C_2 have a total capacitance of C, then C is the single capacitance which could replace them.

Two capacitors in series store only the same charge Q as a single capacitor. So $V = Q/C$, $V_1 = Q/C_1$, and $V_2 = Q/C_2$.

But $V = V_1 + V_2$ So $\dfrac{Q}{C} = \dfrac{Q}{C_1} + \dfrac{Q}{C_2}$

or $$\frac{1}{C} = \frac{1}{C_1} + \frac{1}{C_2}$$

For example, if $C_1 = 3\ \mu F$ and $C_2 = 6\ \mu F$, then $1/C = 1/3 + 1/6 = 1/2$. So $C = 2\ \mu F$.

Capacitors in parallel

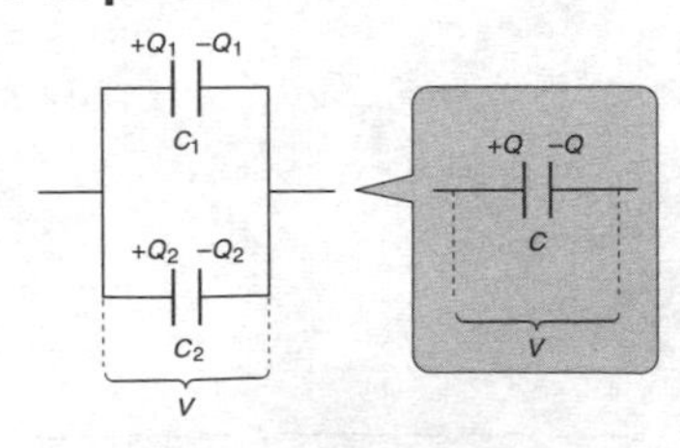

If C_1 and C_2 have a total capacitance of C, then C is the single capacitance which could replace them.

Capacitors in parallel each have the same PD across them. So $Q = CV$, $Q_1 = C_1V$, and $Q_2 = C_2V$.

Together, the capacitors act like a single capacitor with a larger plate area. So $Q = Q_1 + Q_2$

$$\therefore \quad CV = C_1V + C_2V$$

and $$C = C_1 + C_2$$

For example, if $C_1 = 3\ \mu F$ and $C_2 = 6\ \mu F$, then $C = 9\ \mu F$.

E3 Gravitation

Gravitational force

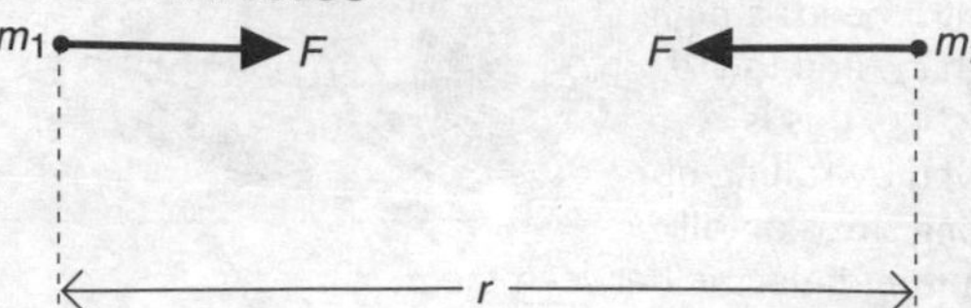

All masses attract each other with a ***gravitational force***. If point masses m_1 and m_2, are a distance r apart, and F is the force on each, then according to ***Newton's law of gravitation***

$$F \propto \frac{m_1 m_2}{r^2}$$

With a suitable constant, the above proportion can be turned into an equation:

$$F = \frac{G m_1 m_2}{r^2}$$

G is called the ***gravitational constant***. It is found by experiment using large laboratory masses and an extremely sensitive force-measuring system. In SI units, the value of G is 6.67×10^{-11} N m^2 kg^{-2}.

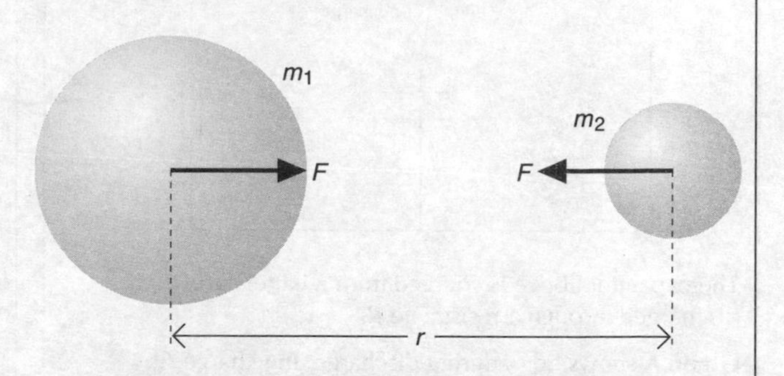

The equation on the left is also valid for spherical masses of uniform density, with centres r apart, as above.

Note:

- Newton's law of gravitation is an example of an ***inverse square law***. If the distance r doubles, the force F drops to one quarter, and so on.
- Gravitational forces are always forces of *attraction*.
- Gravitational forces are extremely weak, unless at least one of the objects is of planetary mass or more.

Gravitational field

If a mass feels a gravitational force, then it is in a ***gravitational field***. The ***gravitational field strength*** g is defined like this:

$$g = \frac{\text{gravitational force}}{\text{mass}}$$

In symbols $g = \dfrac{F}{m}$

For example, if a mass of 2 kg feels a gravitational force of 10 N, then g is 5 N kg^{-1}.

Note:

- Gravitational field strength is a vector.
- g is a variable and can have different values. The symbol g above does not imply the particular value of 9.81 N kg^{-1} near the Earth's surface.
- The force acting on a mass in a gravitational field can be found by rearranging the equation above: $F = mg$.

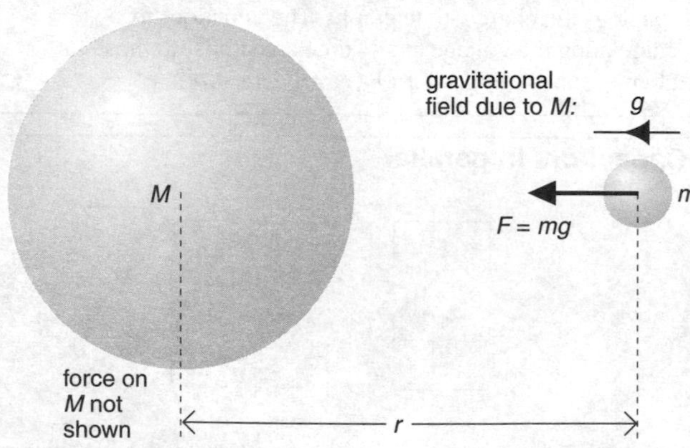

Above, mass M produces a gravitational field which acts on mass m.

As $F = \dfrac{GMm}{r^2}$ and $F = mg$

it follows that $$g = \frac{GM}{r^2}$$

It is equally true to say that m produces a gravitational field which acts on M. Either way, mass × gravitational field strength gives a force of the same magnitude, GMm/r^2.

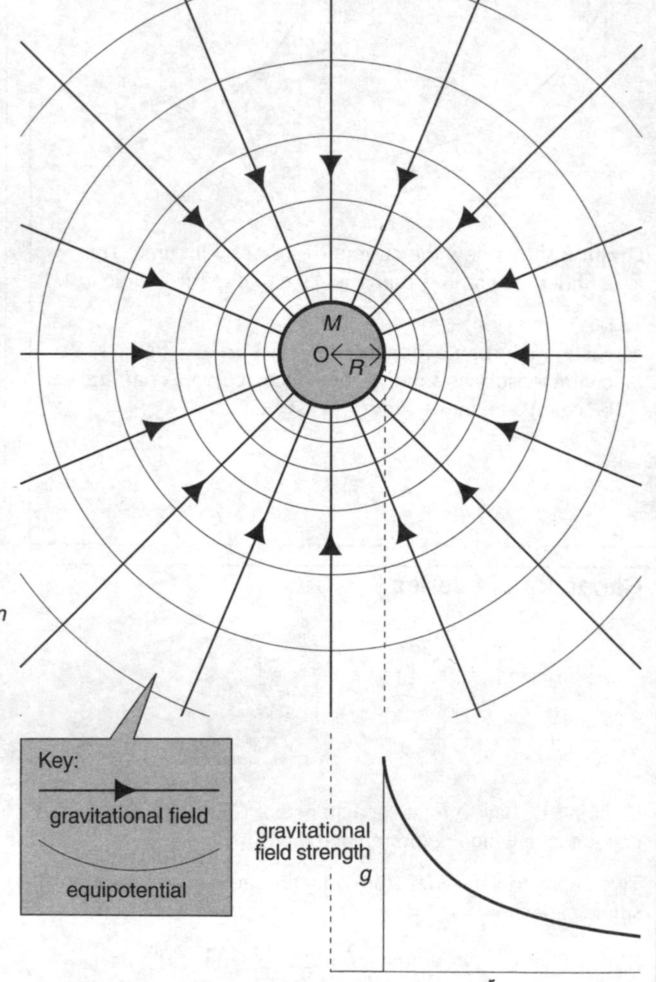

Note:
The gravitational field around a spherical mass is shown above. It is called a ***radial*** field because of its shape.

- Inside the mass, the equation on the left does not apply. g falls to zero at the centre.
- Equipotential lines are explained on the next page.

The Earth's gravitational field strength

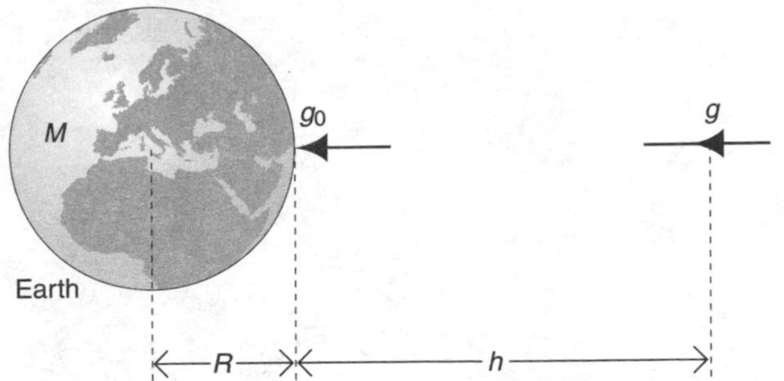

At the surface If M is the Earth's mass, R is its radius, and g_0 is the gravitational field strength at its surface, then

$$g_0 = \frac{GM}{R^2} \qquad (1)$$

Note:

- g_0 is 9.81 N kg^{-1}. It is more commonly known as g (without the $_0$). Here however, the $_0$ has been added to distinguish it from other possible values of g.
- Using measured values of g_0, R, and G in the above equation, the Earth's mass M can be calculated. With R known, the Earth's average density can also be found.

Above the surface In this case, $g = GM/r^2$. From this and equation (1), the following result is obtained:

$$g = \frac{g_0 R^2}{r^2}$$

So as the distance from the Earth increases, g decreases.

Gravitational potential

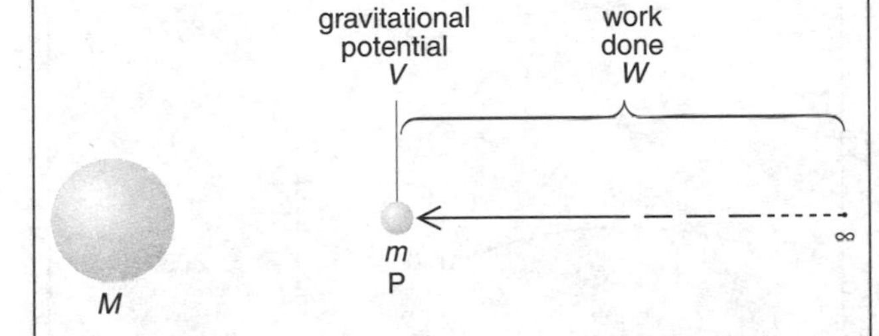

Work must be done to move a mass against a gravitational field. Above, mass M causes a gravitational field. Mass m has been moved through this field, from an infinite distance (where the gravitational force is zero) to point P.

The ***gravitational potential*** V (at point P) is defined as follows:

$$V = \frac{W}{m} \qquad (2)$$

where W is the work done in moving a mass m from infinity (∞) to point P.

Note:

- Like energy, gravitational potential is a scalar.
- At infinity, the gravitational potential is *zero*.
- Elsewhere, the gravitational potential is *negative*. This is because gravity is a force of attraction. Work is done *by* the mass as it is pulled from ∞ to P, so *negative* work is done *on* it. For example, if 1000 J of work is done by a 2 kg mass when it moves from ∞ to P, then –1000 J of work is done on it. So $V = -1000/2 = -500$ J kg^{-1}.

Linking potential and field strength

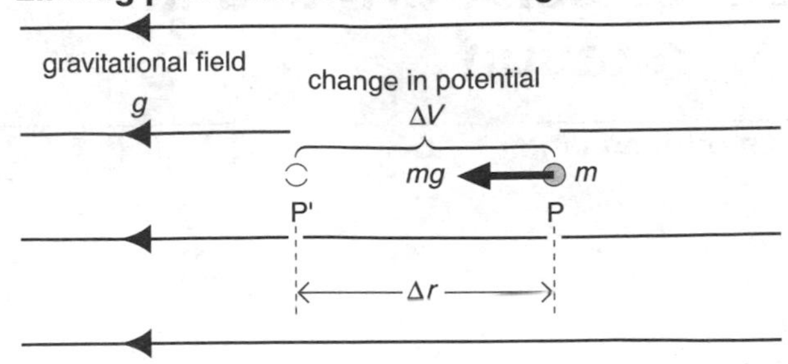

Above, work ΔW is done on a small mass m in moving it from P to P' in a uniform gravitational field g. So, from (2),

$$\Delta W = m\Delta V \qquad (3)$$

This equation gives the work done *on* the mass.

So work done *by* mass $= -m\Delta V$

But work done *by* mass = force × distance moved $= mg\Delta r$

So $mg\Delta r = -m\Delta V$

Therefore $g = -\dfrac{\Delta V}{\Delta r}$

In calculus notation, there is a more general version of this equation which also applies to non-uniform fields:

$$g = -\frac{dV}{dr}$$

Note:

- In the above equations, the minus sign indicates that g is in the direction of decreasing potential.

Gravitational potential in a radial field

A radial field is shown on the opposite page. Provided r is not less than the radius of the sphere:

$$g = \frac{GM}{r^2}$$

Also, $g = -\dfrac{dV}{dr}$

Combining these, and using calculus, gives $V = -\dfrac{GM}{r}$

Note:

- $V \propto 1/r$. So if the distance r doubles, the gravitational potential V halves, and so on. (Inside the mass, this does not apply.)
- In the diagram on the opposite page, each ***equipotential line*** is a line joining points of equal potential.
- In the case of the Earth, the gravitational potential V_0 at the surface is $-GM/R$, where R is the radius.

  As $g_0 = \dfrac{GM}{R^2}$

  it follows that $V_0 = -g_0 R$.

Escape speed

This is the speed, v_{esc}, at which an object must leave a planet's surface to completely escape its gravitational field (i.e. be 'thrown' to infinity). For this, the object must be given enough KE to do the work necessary to move it from the surface (where $V_0 = -g_0 R$) to infinity (where $V = 0$).
The required gain in potential is $g_0 R$. So, from (3), the amount of work to be done is $mg_0 R$. Therefore

$$\tfrac{1}{2} m v_{esc}^2 = m g_0 R \qquad \therefore \quad v_{esc} = \sqrt{2 g_0 R}$$

The escape speed from the Earth is 11.2×10^3 m s^{-1}.

E4 Circular orbits and rotation

An orbit equation

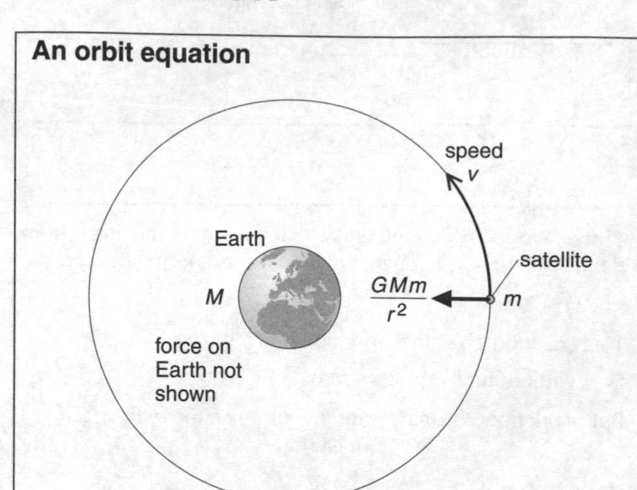

Above, a satellite is in a circular orbit around the Earth. The gravitational force on the satellite provides the centripetal force needed for the circular motion. So

$$\frac{GMm}{r^2} = \frac{mv^2}{r} \qquad (1)$$

Note:

- The equation can be used to find the speed v needed for an orbit of any given radius r.
- As m cancels from each side of the equation, the speed needed does not depend on the mass of the satellite.

A value for *GM* To use the above equation, you do not need to know the Earth's mass M. Instead, a value for GM can be found using equation (1) in E3. Rearranged, this gives

$$GM = g_0R^2 \qquad (2)$$

where R is the Earth's radius (12.8×10^9 m), and g_0 is the gravitational field strength at its surface (9.81 N kg^{-1}).

Period of orbit

The period T is the time taken for one orbit.
Equation (1) can be rewritten using a different version of the equation for centripetal force:

$$\frac{GMm}{r^2} = mr\omega^2 \qquad (3)$$

Cancelling m, then substituting $\omega = 2\pi/T$ (see B11), and rearranging gives the following link between T and r:

$$\frac{T^2}{r^3} = \frac{4\pi^2}{GM} \qquad (4)$$

As $4\pi^2/GM$ is a constant, T^2/r^3 has the same value for all satellites. So as r increases, the period gets longer.

Weightlessness

An astronaut in a satellite is in a state of free fall. Her acceleration towards the Earth is exactly the same as that of the satellite, so the floor of the satellite exerts no forces on her. As a result, she experiences exactly the same sensation of weightlessness as she would in zero gravity. However, she is not really weightless. A few hundred kilometres above the Earth, the gravitational force on her is almost as strong as it is down on the surface.

Geostationary orbit

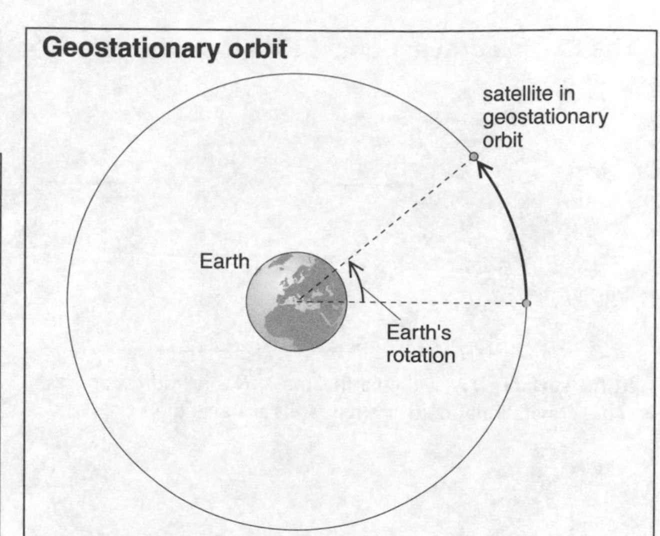

If a satellite is in a ***geostationary orbit***, then viewed from Earth, it appears to be in a fixed position in the sky. This is because the period of its orbit exactly matches the period of the Earth's rotation (24 hours). Communications satellites are normally in geostationary orbits.

Equation (3) can be used to calculate the value of r needed for a geostationary orbit. ω is found using $\omega = 2\pi/T$, with $T = 24 \times 3600$ s. GM is found using equation (2). r works out at 4.23×10^7 m.

Sun, planets, and moons

	Distance from Sun/ $\times 10^7$ km	Period of orbit/ days
Mars	22.8	687.0
Earth	15.0	365.3
Venus	10.8	224.7
Mercury	5.8	88.0
Sun		

Planets are natural satellites of the Sun. Most are in approximately circular orbits, so equation (4) also applies to them (except that M in the equation now becomes the mass of the Sun).

So (period of planet)$^2 \propto$ (distance from Sun)^3

This is called ***Kepler's third law***.

T^2/r^3 has the same value for all the planets around the Sun. If this value is known from astronomical data, equation (4) can be used to calculate the mass of the Sun (M).

Moons are natural satellites of planets. So the above method can also be used to find the mass of a planet.

Equivalent quantities

Motion can be *translational* (e.g. linear) or *rotational*. Quantities used in measuring linear motion all have their rotational equivalents, as shown below.

Linear quantity	Symbol	Rotational quantity	Symbol
displacement	s	angular displacement	θ
velocity	v	angular velocity	ω
acceleration	a	angular acceleration	α
force	F	torque	T
mass	m	moment of inertia	I

Note:
- In physics, the symbol T may stand for torque, period or tension. Above, it represents torque.
- Moment of inertia is the property of a body which resists angular acceleration. It is explained below.
- The equivalent quantities above do not have the same units.

Equivalent equations

The equations used when dealing with linear motion all have their equivalents in rotational motion, as shown below.

Linear equation		Rotational quantity	
velocity	$v = \frac{\theta}{t}$	angular velocity	$\omega = \frac{\theta}{t}$
acceleration	$a = \frac{v}{t}$	angular acceleration	$\alpha = \frac{\omega}{t}$
force	$F = ma$	torque	$T = I\alpha$
work done	$W = Fs$	work done	$W = T\theta$
KE	$E = \frac{1}{2}mv^2$	KE	$E = \frac{1}{2}I\omega^2$
power	$P = Fv$	power	$P = T\omega$
momentum	$p = mv$	angular momentum	$L = I\omega$

Note:
- The equations above are in simplified, non-calculus form. They assume uniform changes, and uniform forces and torques.
- Any rotational equation can be found by taking the linear equation and replacing the symbols with their rotational equivalents.
- Whether the motion is linear or rotational, work and energy are measured in joules. (Similarly power is measured in watts.)

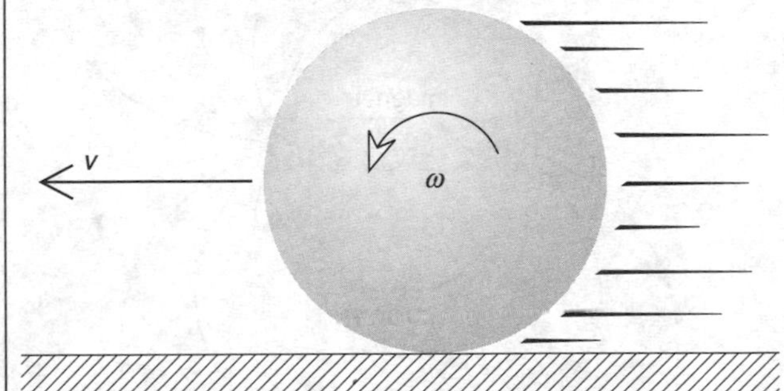

The rolling ball above has both linear *and* rotational motion. In this case,

total KE $= \frac{1}{2}mv^2 + \frac{1}{2}I\omega^2$

Moment of inertia

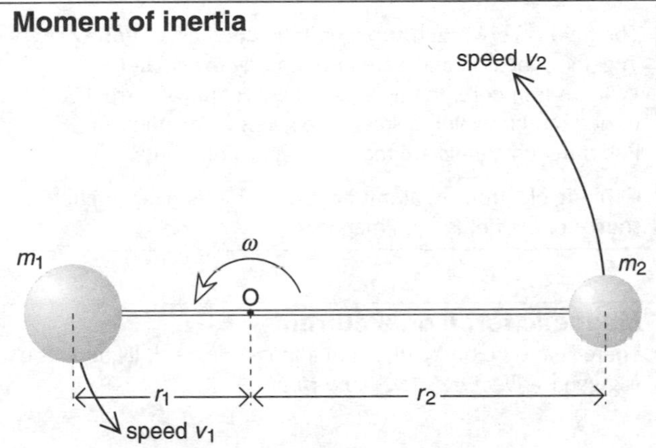

The dumb-bell above is rotating about an axis through its centre of mass O. (The bar of the dumb-bell has no mass.) All parts of the dumb-bell have the same angular velocity ω. The rotational KE of the system is the sum of the linear KEs of the two masses. So

rotational KE $= \frac{1}{2}m_1v_1^2 + \frac{1}{2}m_2v_2^2$

But $v_1 = \omega r_1$ and $v_2 = \omega r_2$. So the above equation can be rewritten:

rotational KE $= \frac{1}{2}(m_1r_1^2 + m_2r_2^2)\omega^2$

Comparing this with the equation for rotational KE in the right-hand table at the top of the page shows that

moment of inertia $I = m_1r_1^2 + m_2r_2^2$

This principle applies to all objects – for example, the one on the right. The moment of inertia is the sum of the mr^2 terms for all the particles in an object. Expressed mathematically:

moment of inertia $I = \Sigma(mr^2)$

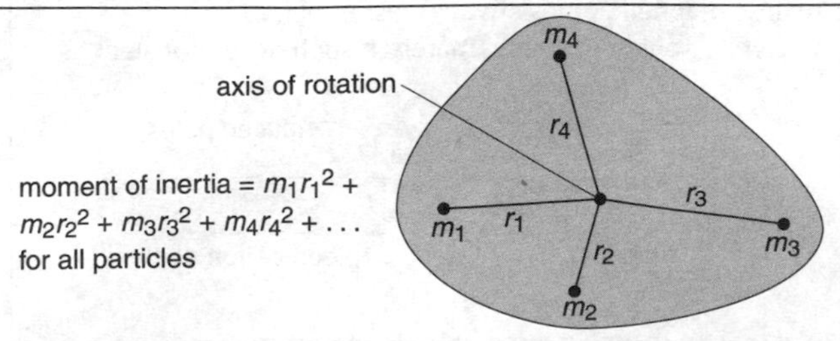

moment of inertia $= m_1r_1^2 + m_2r_2^2 + m_3r_3^2 + m_4r_4^2 + \ldots$ for all particles

Note:
- An object's moment of inertia depends on which axis it is being rotated about. The more spread out the mass, the higher the moment of inertia.
- From this, it follows that objects with the same mass can have different moments of inertia.

E5 Magnets and currents

Magnets and fields

Strongly magnetic materials, such as iron and steel, are called ***ferromagnetic*** materials. They feel forces from other magnets and can be made into magnets.

The forces from a magnet seem to come from ***magnetic poles*** near its ends. In reality, every particle in a magnet acts like a tiny magnet. The two poles are the combined effect of all these tiny magnets lined up.

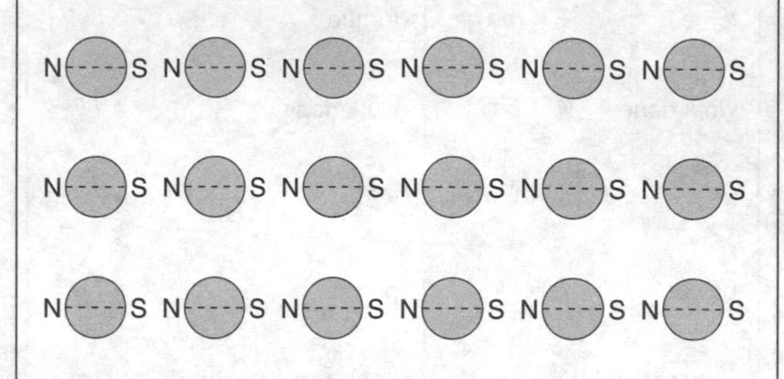

There are two types of pole: ***north*** (**N**) and ***south*** (**S**):
Unlike poles attract. Like poles repel.

The Earth is a weak magnet. As a result, a freely suspended bar magnet turns until it ends are pointing roughly north–south – which is how the two types of pole got their names. However, the north end of the Earth is, magnetically, a south pole because it attracts the north pole of a magnet.

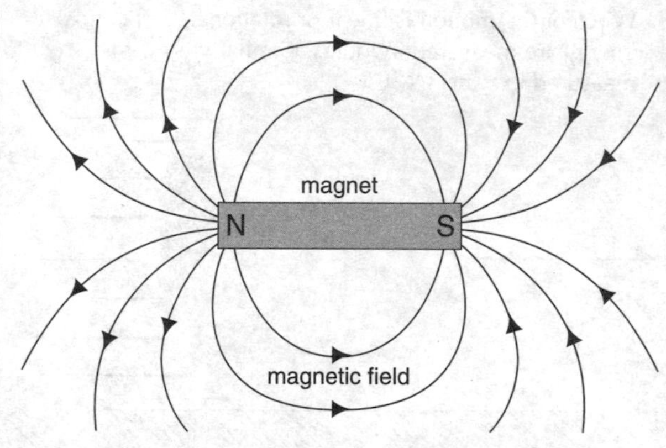

The region around a magnet where magnetic forces act is sometimes called a ***magnetic field***. However, see E6 for a more precise definition of a magnetic field. In diagrams, a magnetic field is represented by ***field lines***. The stronger the field, the closer the lines. The direction of the field is the direction in which a 'free' N pole would move (though in reality, magnetic poles always exist in pairs).
When a strongly magnetic material, such as iron or steel, is

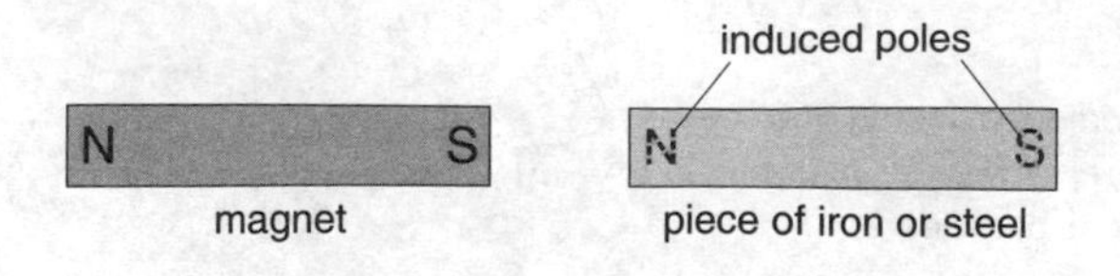

placed in a magnetic field, it becomes magnetized, as shown above. This is called ***induced magnetism***. A magnet attracts iron or steel because the direction of the induced magnetism means that two unlike poles are close together.

When a magnetic field is removed, iron quickly loses its magnetism, but steel becomes permanently magnetized.

Magnetic fields from currents

A current has a magnetic field around it. The greater the current, the stronger the field. The field round a current-carrying wire is shown on the right. The field direction is given by ***Maxwell's screw rule***. Imagine a screw moving in the conventional direction. The field turns the same way as the screw.

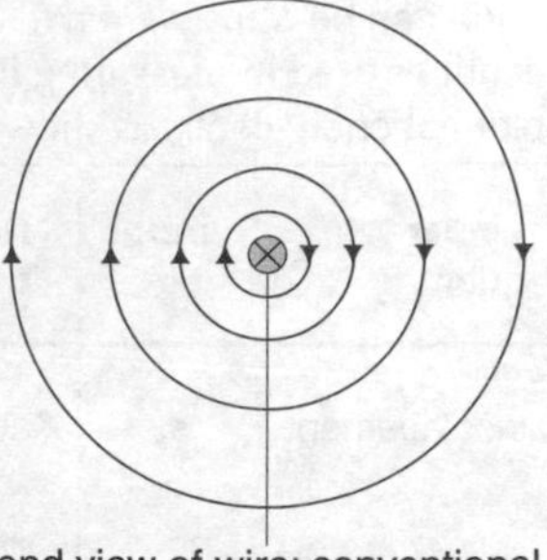
end view of wire: conventional current direction into paper

The field round a current-carrying coil (below) is similar to that round a bar magnet. The field direction can be worked out using either the screw rule or the ***right-hand grip rule***. Imagine gripping the coil with your right hand so that your fingers curl the same way as the conventional direction. Your thumb is then pointing to the N pole.

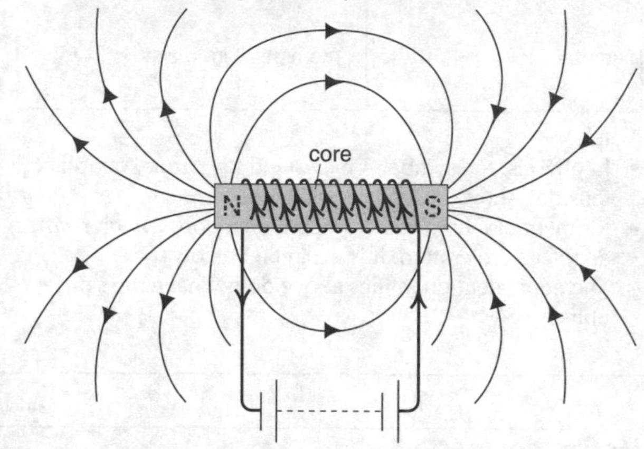

The field is very much stronger if the coil has an iron ***core***. Together, the coil and core form an ***electromagnet***. With an iron core, the field vanishes when the current is turned off. However, a steel core keeps its magnetism. Permanent magnets are made using this principle.

Orbiting electrons in atoms are tiny currents. They are the source of the fields from magnets.

Magnetic force on a current

There is a force on a current in a magnetic field. Its direction is given by ***Fleming's left-hand rule***:

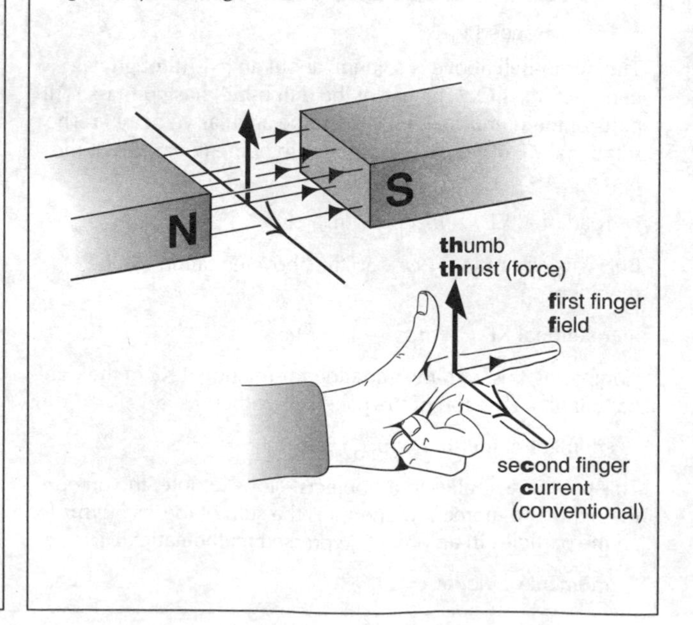

A simple electric motor

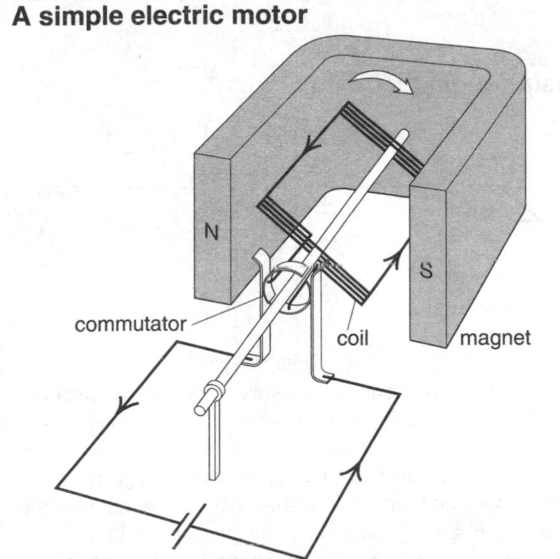

The electric motor above has a coil which can rotate in a magnetic field. When a current passes through the coil, the left side is pushed up, and the right side is pushed down, so there is a turning effect. A switching mechanism called a ***commutator*** keeps the left- and right-side current directions the same, whatever the position of the coil. So the turning effect is always the same way.

Generating AC

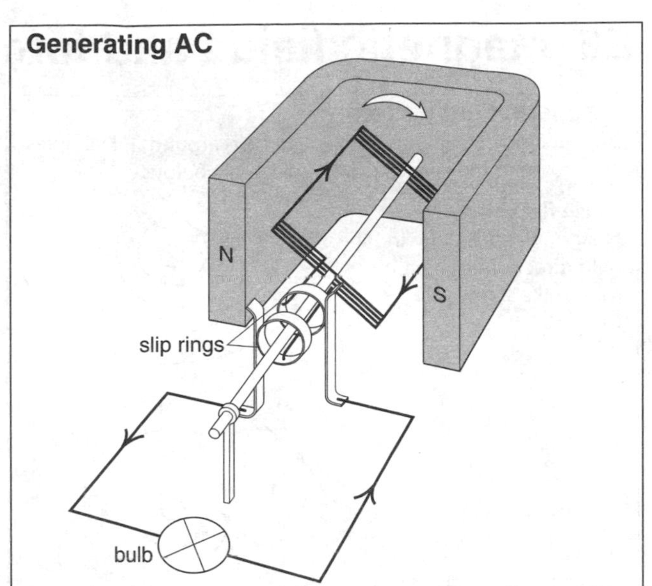

In the generator above, a coil is rotated in a magnetic field. This induces an EMF in the coil, so a current flows. The current keeps changing direction as the coil faces first one way and then the other: it is ***alternating current*** (**AC**). A generator which produces AC is called an ***alternator***.

'One way' current from a battery is ***direct current*** (**DC**).

Currents from magnetic fields

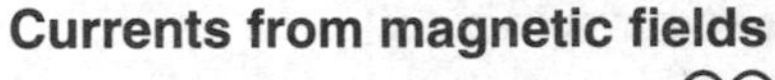

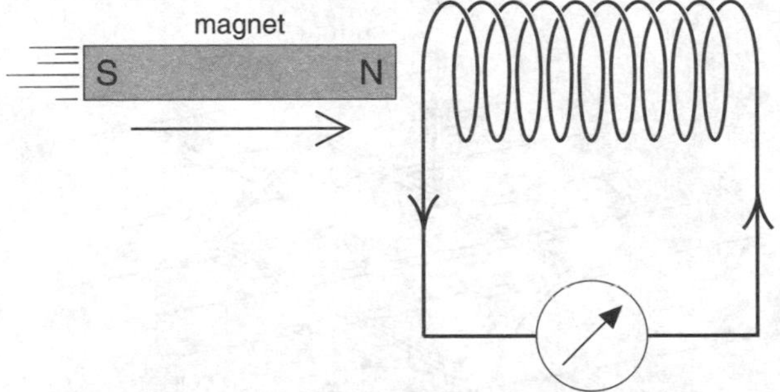

If one end of a bar magnet is moved into a coil, as above, an EMF (voltage) is generated in the coil. This effect is called ***electromagnetic induction***. The EMF makes a current flow in the circuit.

The induced EMF (and the current) is increased if:

- the magnet is moved faster,
- there are more turns on the coil,
- a magnet giving a stronger field is used.

If the magnet is moved out of the coil, the EMF is reversed. If the magnet is stationary, the EMF is zero.

Varying a magnetic field can have the same effect as moving a magnet. Below, an EMF is induced in the right-hand coil whenever the electromagnet is switched on or off. The current flows one way at switch-on and the opposite way at switch-off.

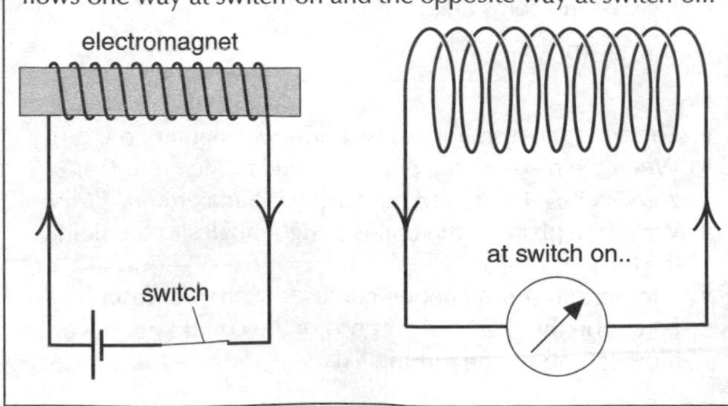

Transformers

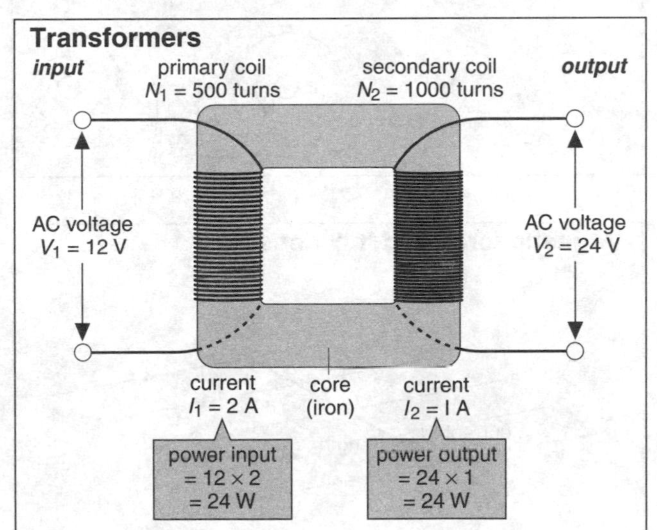

In a ***transformer***, an alternating current in the ***primary*** (input) coil creates a changing magnetic field in the core, which induces an alternating voltage in the ***secondary*** (output) coil. A ***step-up*** transformer, as above, increases the voltage. A ***step-down*** transfomer reduces it.

There is a link between the output and input voltages (V_2 and V_1) and the numbers of turns (N_2 and N_1) on the coils:

$$\frac{V_2}{V_1} = \frac{N_2}{N_1}$$

For the transformer in the diagram,

$$\frac{N_2}{N_1} = \frac{1000}{500} = 2 = \frac{24}{12} = \frac{V_2}{V_1}$$

If no power is wasted in the coils or core,

power output = power input

$$V_2 I_2 = V_1 I_1$$

This means that a transformer which *increases* voltage will *decrease* current, and vice versa.

E6 Magnetic fields and forces

Magnetic field patterns

Any electric current has a magnetic field around it. Examples are shown in the cross-sectional diagrams below.

Single flat coil
Near any section of wire, the field direction is given by Maxwell's screw rule (see E5).

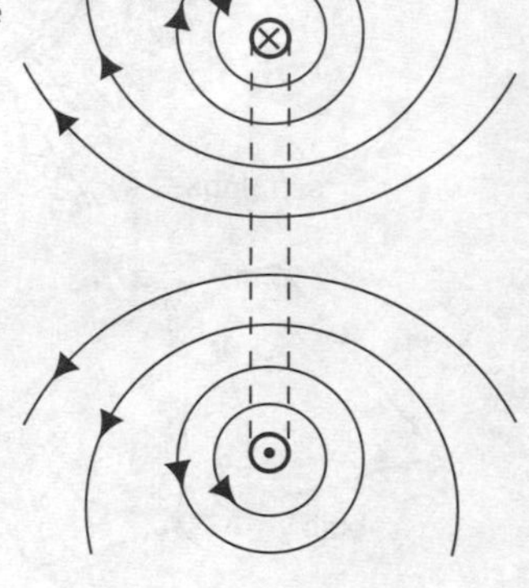

Opposing flat coils
A ***neutral point*** (N on the right) is the point where one magnetic field cancels another.

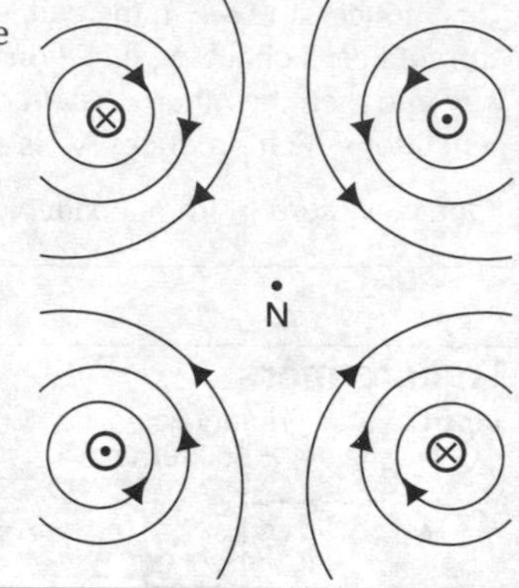

Magnetic force and flux density

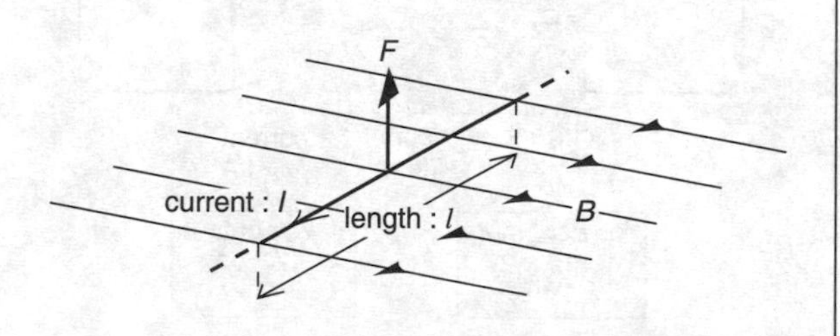

Above, a current-carrying wire is at right-angles to a uniform magnetic field. The field exerts a force on the wire. The direction of the force is given by Fleming's left-hand rule (see E5). The size of the force depends on the current I, the length l in the field, and the strength of the field. This effect can be used to define the magnetic field strength, known as the ***magnetic flux density***, B:

$$F = BIl \qquad (1)$$

B is a vector. The SI unit of B is the ***tesla*** (T). For example, if the magnetic flux density is 2 T, then the force on 2 m of wire carrying a current of 3 A is $2 \times 2 \times 3 = 12$ N.

If a wire is not at right angles to the field, then the above equation becomes

$$F = BIl \sin \theta$$

where θ is the angle between the field and the wire. As θ becomes less, the force becomes less. When the wire is *parallel* to the field, $\sin \theta = 0$, so the force is zero.

Measuring magnetic flux density

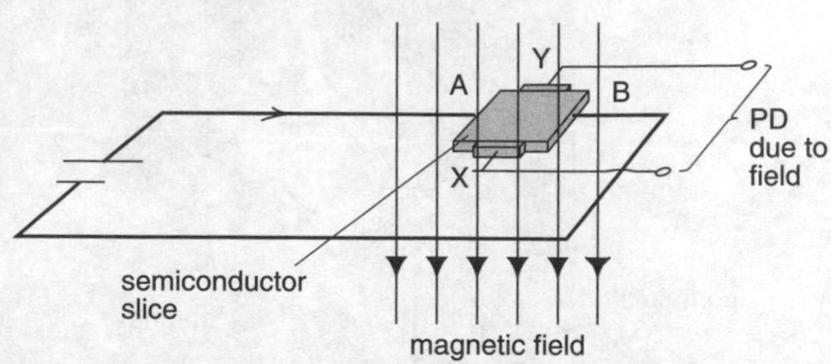

Magnetic flux density can be measured using a ***Hall probe.*** This contains a small slice of semiconductor material, as shown above. When a current is passed between A and B, the drifting charge carriers (see D2) experience a sideways force from the magnetic field. This pushes them sideways, causing a PD between X and Y which can be measured with a sensitive meter. The higher the magnetic flux density, the greater the force on the charge carriers and the higher the PD. A semiconductor is used because the charge carriers move faster than in a metal, for any given current, and this gives a higher PD.

Before it can be used, a Hall probe must be calibrated by finding the meter reading produced by a field of known B.

Torque on a current-carrying coil

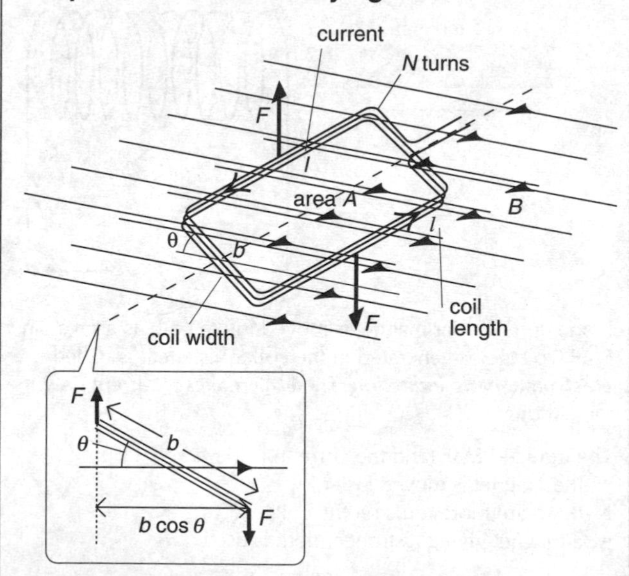

Above, applying equation (1), there is a total upward force F of $BIl \times N$ on one edge of the coil and the same downward force on the other. These form a couple of torque T:

$T = Fb \cos \theta$ (see B5) So $T = BIlNb \cos \theta$

But $lb = A$, the coil's area.

So $$T = BIAN \cos \theta$$

Note:

- The equation also applies to a non-rectangular coil.
- When the plane of the coil is parallel to the field, θ is zero, so $\cos \theta = 1$, and the torque is a maximum, $BIAN$.
- When the plane of the coil is at right-angles to the field, the torque is zero.
- T for torque should not be confused with T for tesla.
- Forces on the near and far ends of the coil do not have moments about the turning axis.

The Biot–Savart law, and μ_0

On the right, a short length δl of thin wire, carrying a current I, causes a magnetic flux density δB at P. According to the ***Biot–Savart law***

$$\delta B \propto \frac{I\delta l \sin\theta}{r^2}$$

With a suitable constant, this can be turned into an equation:

$$\delta B = \frac{kI\delta l \sin\theta}{r^2}$$

For a vacuum (and effectively for air), the value of k is 10^{-7} T m A^{-1}. However, in practice, another constant, μ_0, is used, and the above equation is rewritten as follows:

$$\delta B = \frac{\mu_0 I\delta l \sin\theta}{4\pi r^2} \qquad (2)$$

μ_0 is called the ***permeability of free space***. Its value is $4\pi \times 10^{-7}$ T m A^{-1}. This is not found by experiment. It is a *defined* value, linked with the definition of the ampere (see the panel right).

Calculating magnetic flux density

Using equation (2) and calculus, it is possible to derive equations for B near wires and inside coils carrying a current I.

***B* near an infinitely long, thin, straight wire** At a distance a from such a wire (as on the right)

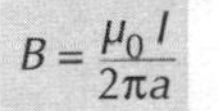

$$B = \frac{\mu_0 I}{2\pi a}$$

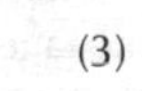

(3)

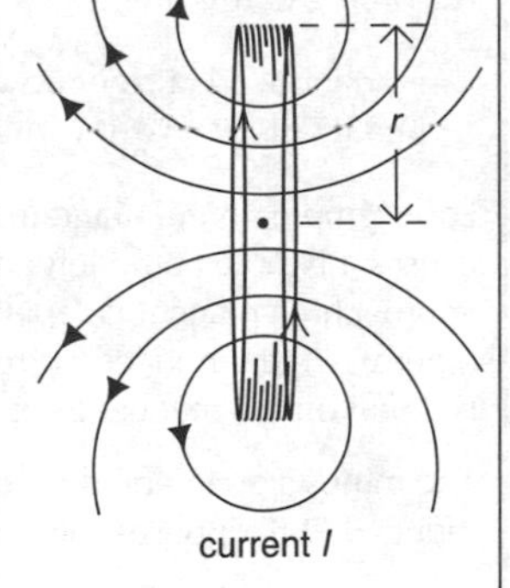

***B* at the centre of a thin coil** On the axis of such a coil, of N turns and radius r (as on the right),

$$B = \frac{\mu_0 NI}{2r}$$

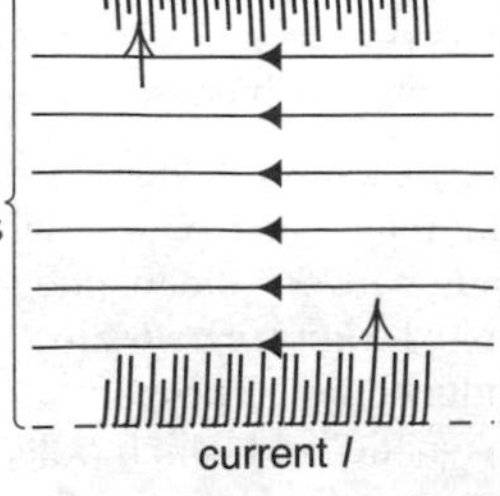

***B* inside an infinitely long solenoid** (coil) The field inside such a solenoid (as on the right) is uniform. If n is the number of turns *per unit length* (per m), then

$$B = \mu_0 nI$$

This equation is a reasonable approximation for any solenoid which is at least ten times longer than it is wide.

Force between two current-carrying wires

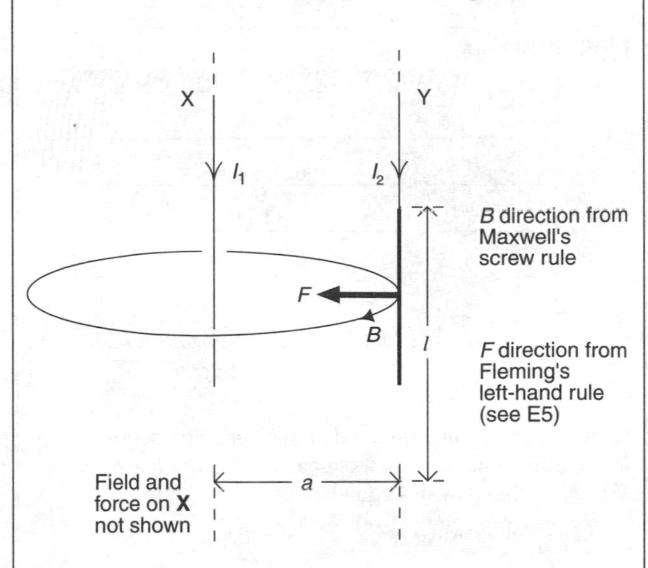

X and Y above are two infinitely long, straight wires in a vacuum. The current in X produces a magnetic field, whose flux density is B at Y. As a result, there is a force on Y. F is the force acting on length l.

From equation (3) $\qquad B = \dfrac{\mu_0 I_1}{2\pi a}$

From equation (1) $\qquad F = BI_2 l$

$$\therefore \quad F = \frac{\mu_0 I_1 I_2 l}{2\pi a} \qquad (4)$$

Note:

- The above equation gives the force of X on Y. Working out the force of Y on X gives exactly the same result.
- If the two currents are in the *same* direction (as above), then the wires *attract* each other. If the two currents are in *opposite* directions, then the wires *repel*.

Defining the ampere

The SI unit of current is defined as follows:

One ampere is the current which, flowing through two infinitely long, thin, straight wires placed one metre apart in a vacuum, produces a force of 2×10^{-7} newtons on each metre length of wire.

Using the various factors in equation (4), the above definition can be expressed in the following form (for simplicity, units have been omitted):

If $I_1 = I_2 = 1$, $a = 1$, and $l = 1$, then $F = 2 \times 10^{-7}$.

Substituting these in equation (4) gives $\mu_0 = 4\pi \times 10^{-7}$.

The above definition is not a practical way of fixing a standard ampere. This is done by measuring the force between two current-carrying coils.

***B* inside a solenoid with a core** The value of B is changed by a core. For example, with a pure iron core, B is increased by a factor of about 1000 (depending on the temperature). The previous equation then becomes:

$$B = \mu_r \mu_0 nI$$

μ_r is called the ***relative permeability*** of the material. So, for pure iron, μ_r is about 1000.

An electromagnet is a solenoid with a core of high μ_r.

E7 Electromagnetic induction

Magnetic flux

flux Φ (6 Wb)
flux density B (3 T)
area A (2 m^2)
solenoid

Above, there is a uniform field of magnetic flux density B inside a solenoid of cross-sectional area A. The ***magnetic flux*** Φ is defined by this equation:

flux = flux density × area $\qquad \Phi = BA$

The SI unit of magnetic flux is the ***weber*** (**Wb**). For example, if B is 3 T and A is 2 m^2, then Φ is 6 Wb.

Note:

- Field lines do not really exist, but they can help you visualize what magnetic flux means. In the diagram above, each field line represents a flux of 1 Wb. There are 6 lines altogether, so the flux is 6 Wb. But there are 3 lines per m^2, so the flux density is 3 T.
- With flux, 'density' implies 'per m^2' and not 'per m^3'.
- 1 tesla = 1 weber per metre^2 i.e. 1 T = 1 Wb m^{-2}.

Faraday's law

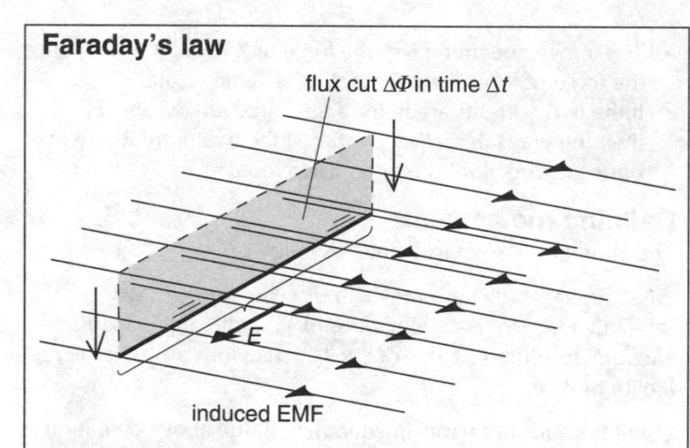

Above, a conductor is moving at a steady speed through a magnetic field. It cuts through flux $\Delta\Phi$ in time Δt. As a result, an EMF E is induced in the conductor. According to ***Faraday's law of electromagnetic induction***, the induced EMF is proportional to the rate of cutting flux:

$$\text{induced EMF} \propto \frac{\text{flux cut}}{\text{time taken}} \qquad E \propto \frac{\Delta\Phi}{\Delta t}$$

With a constant, this can be turned into an equation. The constant is 1 because of the way the units are defined. So

$$\text{induced EMF} = \frac{\text{flux cut}}{\text{time taken}} \qquad E = \frac{\Delta\Phi}{\Delta t}$$

For example, if 6 Wb of flux are cut in 2 s, E is 3 V.

There is a calculus version of the above equation which also applies if flux is not cut at a steady rate:

$$E = -\frac{d\Phi}{dt} \qquad (1)$$

The significance of the minus sign is explained on the right.

Flux change and flux linkage

Changing flux has exactly the same effect as cutting flux, and the induced EMF is calculated in the same way.

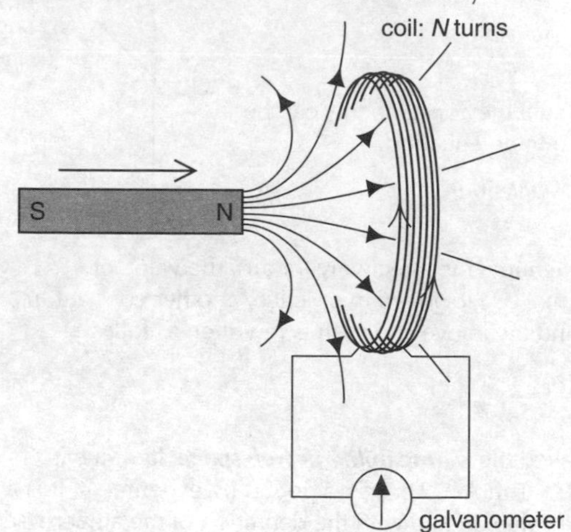

Above, a magnet is moved into a coil. If the flux through the coil changes (at a steady rate) by $\Delta\Phi$ in time Δt, then an EMF of $\Delta\Phi/\Delta t$ is induced *in each turn*. But there are N turns in series. So, the *total* induced EMF E is as follows:

$$E = N\frac{\Delta\Phi}{\Delta t}$$

For example, if the flux changes by 6 Wb in 2 s, and the coil has 100 turns, then the total induced EMF is 300 V.

If there is a flux Φ through a coil of N turns, then $N\Phi$ is called the ***flux linkage***. $N\Delta\Phi$ is the *change* in flux linkage (600 Wb turns, in the previous example). So, the previous equation can be written as follows:

$$\text{induced EMF} = \frac{\text{change in flux linkage}}{\text{time taken}}$$

Lenz's law

According to this law,

> if an induced current flows, its direction is always such that it will oppose the change in flux which produced it.

For example, in the diagram above, the induced current causes a N pole at the left end of the coil so that the approaching magnet is repelled. If the magnet is moved the other way, the induced current direction reverses so that there is a pull on the magnet to oppose its motion.

The minus sign in equation (1) comes about because the induced EMF opposes the flux change.

Lenz's law follows from the law of conservation of energy. Energy must be transferred to produce an induced current. So work must be done to make the change which causes it.

Eddy currents When the aluminium disc on the right is spun between the magnetic poles, ***eddy currents*** are induced in it. These set up a magnetic field which pulls on the poles and opposes the motion. So the disc quickly comes to a halt. Electromagnetic braking systems use this effect.

N

Calculating an induced EMF

The conductor on the right is cutting magnetic flux. It moves a distance of 4 m in 6 s, at a steady speed. From the data supplied, the PD across the ends of the conductor can be calculated. (The PD is equal to the induced EMF E.)

area of flux cut = $3 \times 4 = 12\ \text{m}^2$

$\therefore$ flux cut = $BA = 2 \times 12 = 24$ Wb

$\therefore$ PD = $E = \Delta\Phi/\Delta t = 24/6 = 4$ V

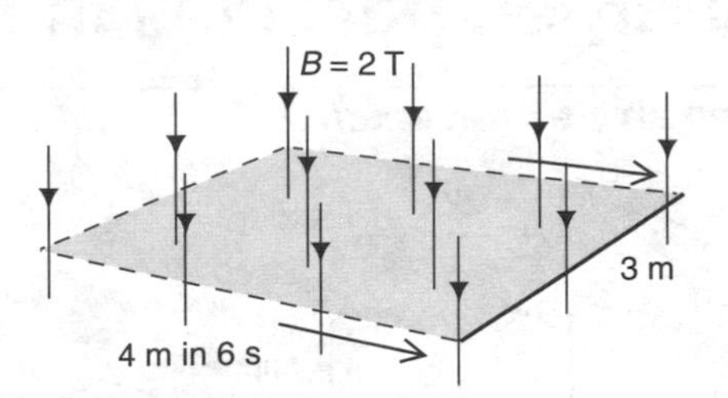

Induced EMF = $B \times$ area cut per second
= BLv

where L is the length of the wire and v is its velocity.

Induced EMF in a rotating coil

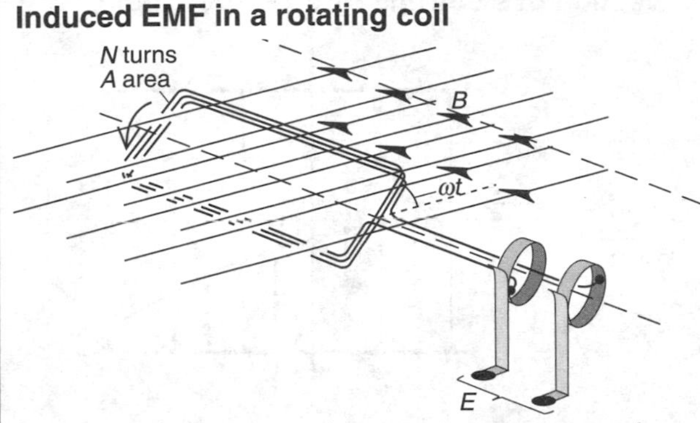

Above, a coil of N turns and area A is rotating at a steady angular velocity ω in a magnetic field of flux density B. In time t, the coil has turned through an angle ωt. For a vertical coil, the flux linkage would be BAN. However, as the coil is an angle to the ωt to the field: flux linkage = $BAN \sin \omega t$.

The induced EMF E can be found by working out the rate of change of flux linkage, using calculus. The result is

$$E = BAN\omega \cos \omega t$$

The graph shows how E varies with t. The ***peak*** (maximum) EMF, E_0, occurs when the coil is horizontal and $\cos \omega t = 1$.

So $E_0 = BAN\omega$

Increases in B, A, N, and ω all give an increased EMF.

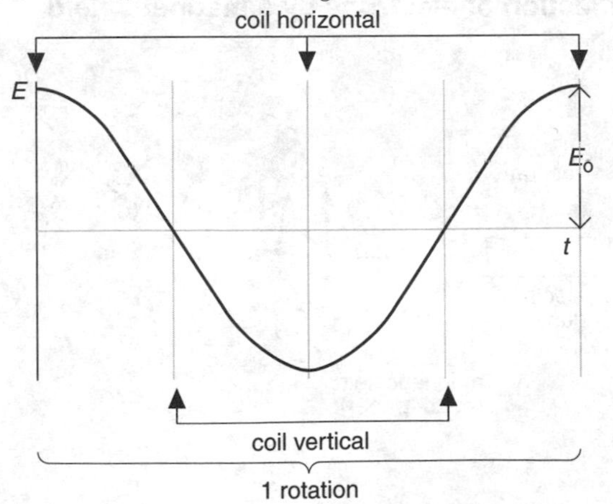

The output from an alternator (see E5) is as shown in the graph above. The current is ***alternating current*** (**AC**).

Induced EMF in an electric motor (See E5 for motor diagram.) When the coil of a motor is turning, it is cutting magnetic flux. As a result, an EMF is induced which, by Lenz's law, opposes the current. This is known as a ***back EMF***. At switch-on, when the coil is stationary, there is no back EMF, so the current is high. As the motor speeds up, the back EMF rises and the current drops.

For higher B, the coil of a motor is normally wound on an iron-based *armature*. This has a laminated (layered) structure to reduce eddy currents, which waste energy.

Self-induction

When the circuit below is switched on, the current through the coil rises as shown in the graph. The rising current causes a changing flux which induces a back EMF in the coil. By Lenz's law, this opposes the rising current. The effect is called ***self-induction***. The coil is an ***inductor***.

If a back EMF E_b is induced in a coil when the current rises by ΔI in time Δt (at a steady rate) then the ***self-inductance*** L of the coil is given by this equation:

$$\text{back EMF} = L \times \frac{\text{current change}}{\text{time taken}} \qquad E_b = L\frac{\Delta I}{\Delta t} \qquad (2)$$

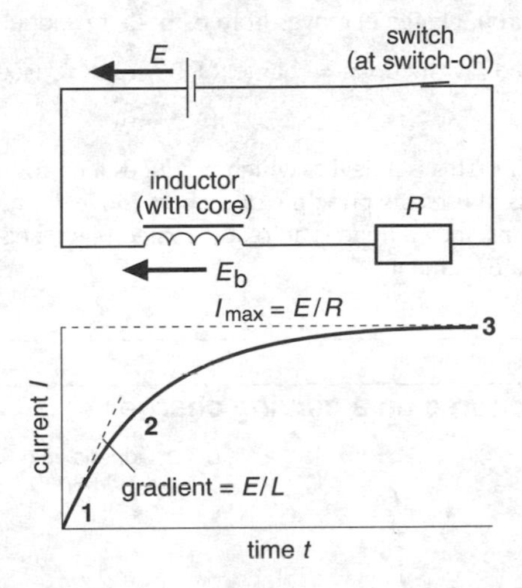

The unit of self-inductance is the ***henry*** (H). For example, if current is rising at the rate of 2 A s^{-1} in a coil of self-inductance 3 H, then the back EMF is 6 V.

1 At switch-on As the current is zero, the back EMF must balance the supply EMF. So $E_b = E$, and from equation (2), the initial rate of rise of current,

$$\frac{\Delta I}{\Delta t} = \frac{E}{L}$$

2 After switch-on Kirchhoff's second law (see D3) gives the equation linking E_b and I:

$$E - E_b = IR$$

3 At maximum current E_b is zero. So

$$I = \frac{E}{R}$$

Stored energy When current I flows through an inductor, energy is stored by its magnetic field:

$$\text{stored energy} = \tfrac{1}{2}LI^2$$

At switch-off, the energy is released – often as heat and light from a spark across the switch contacts, caused by the high EMF induced when the current falls rapidly to zero.

E8 Charged particles in motion

Producing an electron beam

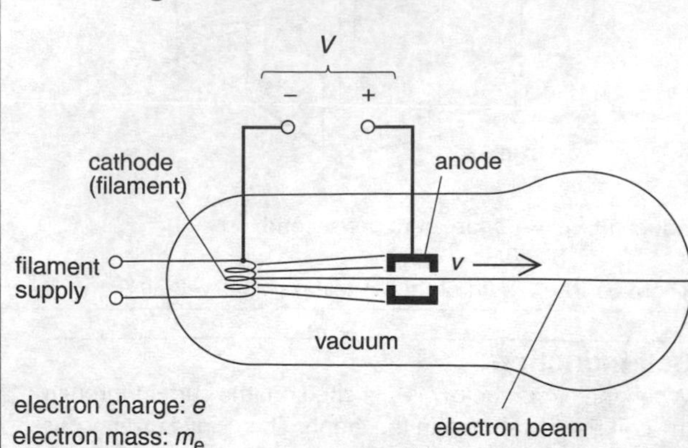

electron charge: e
electron mass: m_e

In the vacuum tube above, electrons are given off by a hot tungsten filament. The effect is called ***thermionic emission.*** The electrons gain kinetic energy (KE) as they are pulled from the ***cathode*** (–) to the ***anode*** (+). Some pass through the hole in the anode and emerge as a narrow beam, at speed v. Electrons in a beam are sometimes called ***cathode rays*** because they come from the cathode.

As an electron (charge e) moves from cathode to anode,

KE gained = work done = charge × PD = eV (see E1)

So $$\tfrac{1}{2} m_e v^2 = eV$$

Electron gun This is a device which produces a narrow beam of electrons. It uses the principle described above. In many electron guns, the cathode is an oxide-coated plate, heated by a separate filament.

Electronvolt (eV) This is a unit often used for measuring particle energies. 1 eV is the energy gained by an electron when moving through a PD of 1 V.

If an electron (1.60×10^{-19} C) moves through 1 V,
KE gained = charge × PD = $1.60 \times 10^{-19} \times 1$ J

So $$1 \text{ eV} = 1.60 \times 10^{-19} \text{ J}$$

Deflection of electrons by an electric field

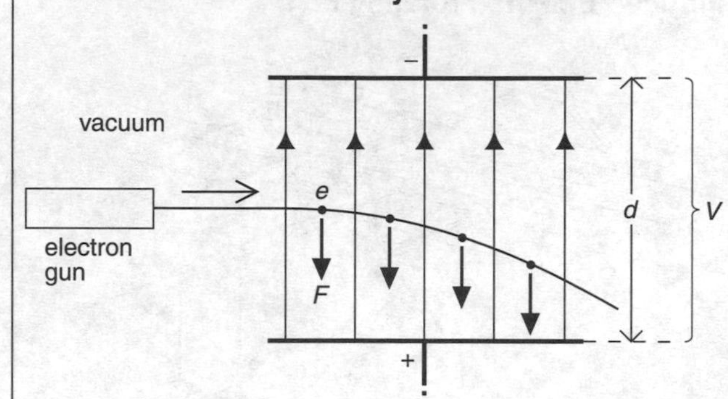

Above, electrons pass between two horizontal plates. The electric field strength between the plates is V/d (see E1).

force F on electron = electric field strength × charge

So $$F = \frac{Ve}{d}$$

Note:
- The force on the electron does not depend on its speed.
- The force is always in line with the electric field.
- The path of the electrons is a *parabola* (just as it is for the thrown ball in B3).

Magnetic force on a moving charge

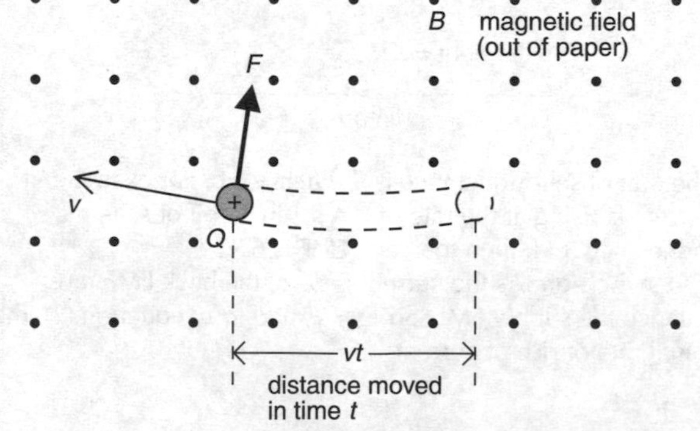

Above, a particle of charge Q, is moving at a steady speed through a uniform magnetic field. In time t, it travels a distance vt. So it is equivalent to a current Q/t in a wire of length vt. According to equation (1) in E6, the force on a current-carrying wire is BIl. Applying this to the above:

force $F = B \times \frac{Q}{t} \times vt$

So $$F = BQv \qquad (1)$$

Note:
- As the speed increases, the force increases.
- The direction of the force is given by Fleming's left-hand rule (see E5).
- If the particle is travelling at an angle θ to the field, then the above equation becomes $F = BQv \sin \theta$.

Deflection of electrons by a magnetic field

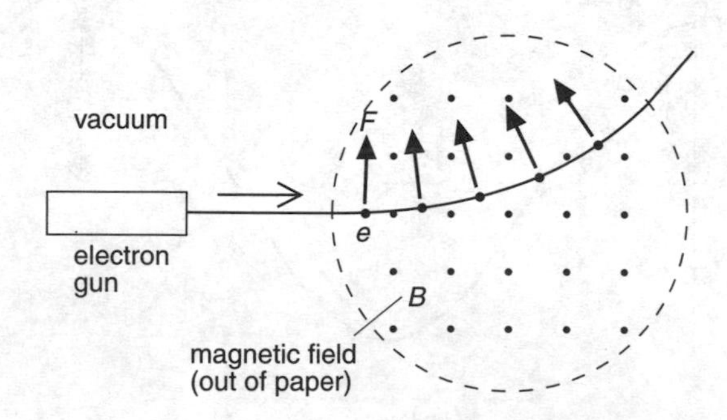

Above, electrons travel at right angles to a uniform magnetic field. The force F on an electron is found by putting Q equal to e in equation (1) on the left. So

$$F = Bev$$

Note:
- The force on the electron increases with speed.
- The force is always at right-angles to the direction of motion, as predicted by Fleming's left-hand rule. But in applying this rule, remember that the electron has negative charge, so electron motion to the *right* represents a conventional current to the *left.*
- The path of the beam is *circular* (see next page).

Circular motion in a magnetic field

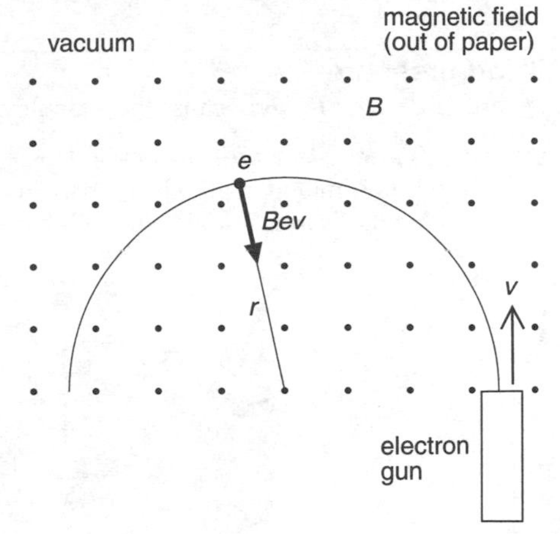

Above, electrons leave the electron gun at speed v. They move in a circle because the magnetic force Bev supplies the necessary centripetal force (see B11). So

$$Bev = \frac{m_e v^2}{r}$$

So $$r = \frac{m_e v}{eB} \quad (2)$$

Note:
- The radius of the circle is proportional to the speed.
- Increasing B decreases the radius.

Specific charge of an electron This an electron's charge per unit mass, e/m_e. Its value is -1.8×10^{11} C kg^{-1}. Methods of measuring it make use of the above equation.

Ion beams If atoms lose electrons, they become positive ions. These particles also have circular paths in a magnetic field. If Q is the charge on a particle, and m the mass, then equation (2) can be written in this more general form:

$$r = \frac{mv}{QB}$$

Note:
- The radius depends on the specific charge (Q/m). This idea is used in a ***mass spectrometer*** to separate nuclei with different specific charges.

Speed selection

An ion beam may contain ions with a range of speeds. Ions of only one speed can be selected by the method shown below. This uses the principle that the magnetic force on an ion depends on its speed, while the electric force does not:

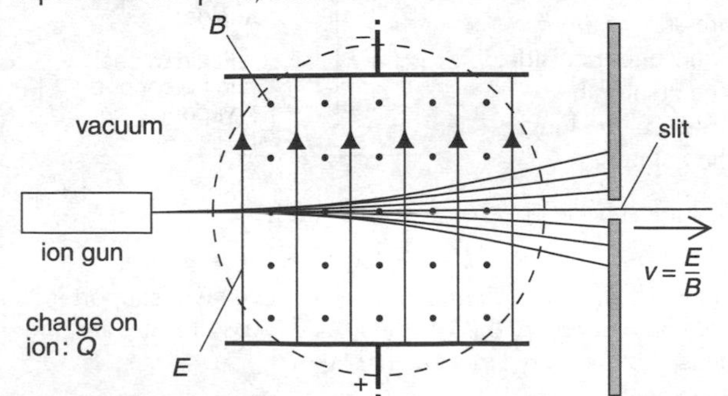

The magnetic field produces an upward force BQv. The electric field produces a downward force EQ. The only ions to pass through the slit are those for which these forces are equal. So, if $EQ = BQv$, the selected speed $v = E/B$. Faster ions are deflected upwards (because BQv is more); slower ions are deflected downwards.

Cathode ray oscilloscope (CRO)

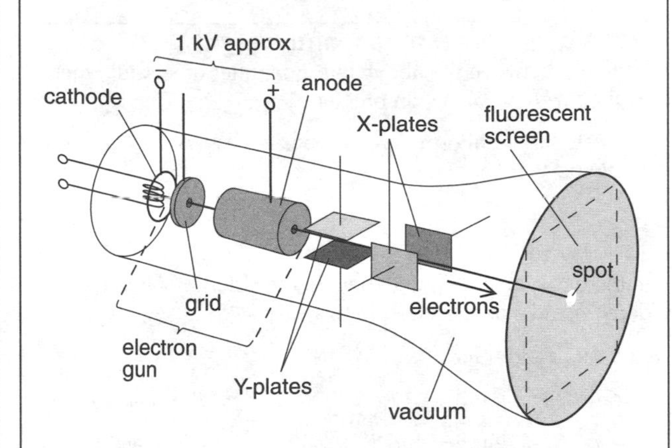

In a CRO, as above, a beam of electrons is used to draw a graph on a screen. The screen is coated with a ***fluorescent*** material, so that it glows where the electrons strike it.

The electrons come from an ***electron gun***. The anode is designed so that it focuses the beam. The flow of electrons and, therefore, the brightness of the spot, is controlled by making the ***grid*** negative relative to the cathode.

The ***Y-plates*** and ***X-plates*** are used to move the beam up and down and side to side. The beam is deflected whenever there is a PD across a set of plates.

Displaying an alternating voltage The line below has two components. Across the Y-plates, an alternating PD, coming from an external source via an amplifier in the CRO, moves the spot up and down. Across the X-plates, a ***time base*** circuit in the CRO changes the PD so that the spot moves from left to right at a steady rate – until it reaches the edge of the screen and flicks back. The result is a graph of PD against time, drawn over and over again.

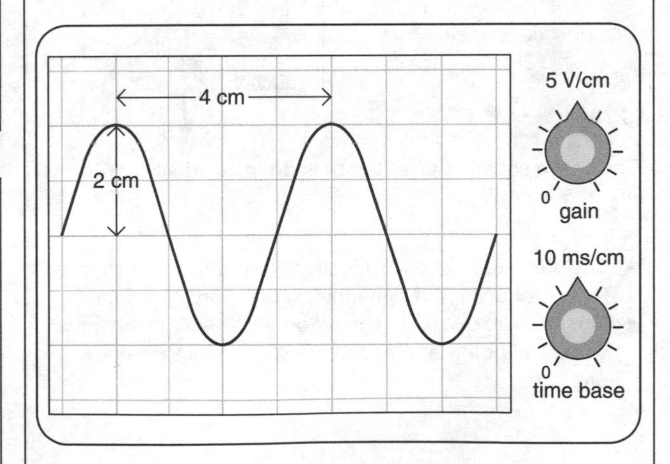

Measuring peak alternating PD The setting on the gain control (above) means that 5 V across the Y-input terminals cause the spot to move 1 cm vertically. On the screen, the amplitude of the wave trace is 2.0 cm. So in this case, the peak voltage is $2.0 \times 5 = 10$ V.

Measuring time and frequency The setting on the time base control (above) means that the spot takes 10 ms to move 1 cm horizontally. On the screen, the wave peaks are 4.0 cm apart. So the time between the peaks = $4.0 \times 10 = 40$ ms (0.04 s). This is the period of the wave cycle. So the frequency = 1/period = 1/0.04 = 25 Hz.

F1 Liquid and gas pressure

Density, volume, mass, and weight

The links between the above four quantities are useful when dealing with problems on pressure.

If a material of uniform density ρ has a mass m and occupies a volume V, then

$$\rho = \frac{m}{V} \quad \text{(see B1)}$$

In the diagram in the panel below, the column of liquid has a depth h and a base area A. In this case,

volume of liquid = hA

mass of liquid = ρhA (because $m = \rho V$)

weight of liquid = ρghA (because weight = mg)

Pressure in a liquid

If a liquid exerts a force F on an area A (at right-angles to it), then the pressure p is given by this equation:

$$p = \frac{F}{A} \quad \text{(see B1)}$$

Note:

- If the area A is vanishingly small, the above equation gives the *pressure at a point* in a liquid.
- Pressure is a scalar, *not* a vector (though the force it creates is a vector).
- The unit of pressure is the pascal (Pa). 1 Pa = 1 N m^{-2}.

The liquid on the right has a density ρ. The weight of the liquid column exerts a downward force F on the base area A.

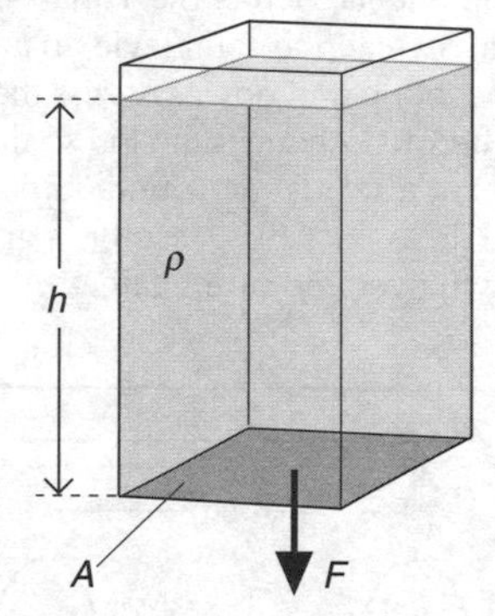

F = weight of column
$= \rho ghA$
(see previous panel)

But $p = F/A$.

So $p = \rho gh$

The above equation gives the pressure p at a depth h in a liquid of density ρ.

Note:

- The pressure acts in *all* directions (see B1). However, the force produced is at right-angles to any area in contact.
- The pressure does *not* depend on the area A. In the liquid below, all points at the same depth h are at the same pressure, ρgh.

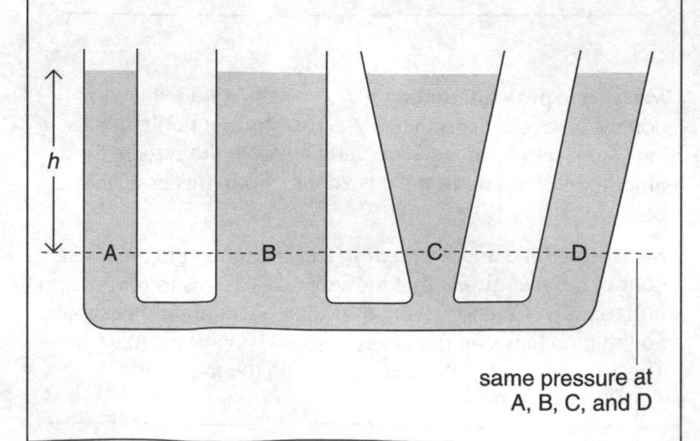

Transmitting pressure

Liquids and gases are called ***fluids*** because they can flow.

According to ***Pascal's principle***, a pressure applied to any part of an enclosed fluid is transmitted to all other parts. This principle is used in ***hydraulic machines*** like the oil-filled jack below.

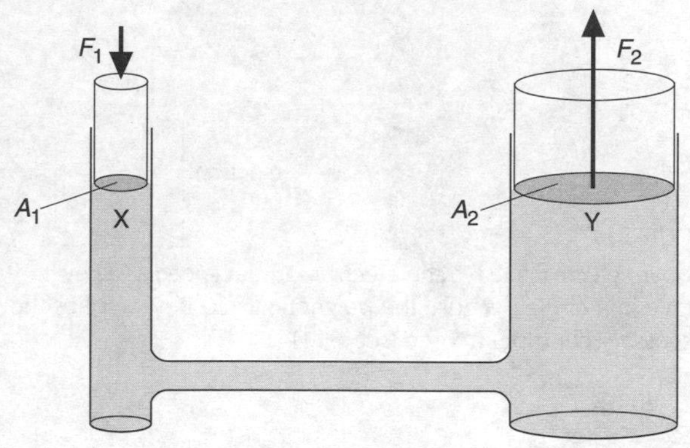

In the above system
pressure at X (due to force F_1 acting on A_1) = F_1/A_1
So, by Pascal's principle pressure at Y = F_1/A_1
But pressure at Y = F_2/A_2

So $$\frac{F_2}{F_1} = \frac{A_2}{A_1}$$

Note:

- The pressures at X and Y would *not* be the same if their levels were different. There would be ρgh to allow for.
- Pressure is only transmitted without reduction if the liquid is static, and not flowing (see B9).

Barometer

Atmospheric pressure can be measured using a simple barometer like the one on the right. The space at the top of the glass tube is a vacuum, so the pressure there is zero. The column of mercury in the tube is supported by the pressure of the atmosphere outside. The greater the pressure, the longer the column.

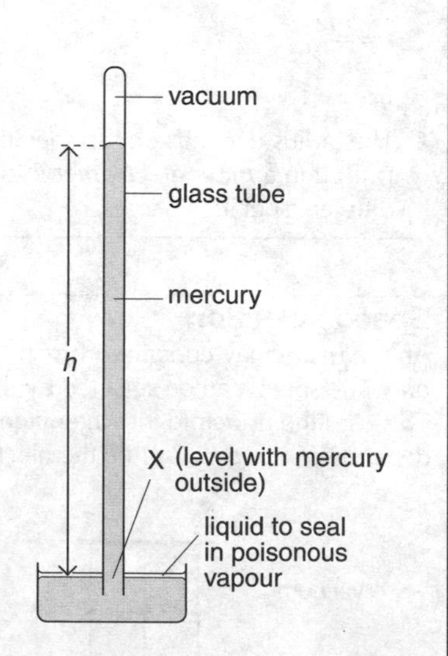

atmospheric pressure = pressure at X
$= \rho gh$

Standard atmospheric pressure (at sea level) will support a column of mercury 0.76 m long. As g is 9.81 N kg^{-1} and the density of mercury is 13.6×10^3 kg m^{-3},

standard atmospheric pressure = $13.6 \times 10^3 \times 9.81 \times 0.76$
$= 1.01 \times 10^5$ Pa

The ***millibar* (mb)** is a unit of pressure used in meteorology.
1 mb = 100 Pa.
So, standard atmospheric pressure = 1.01×10^3 mb.

U-tube manometer

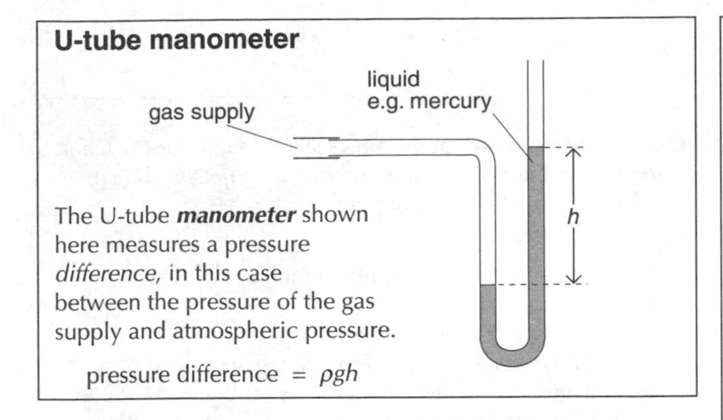

The U-tube ***manometer*** shown here measures a pressure *difference*, in this case between the pressure of the gas supply and atmospheric pressure.

pressure difference $= \rho g h$

Upthrust and Archimedes' principle

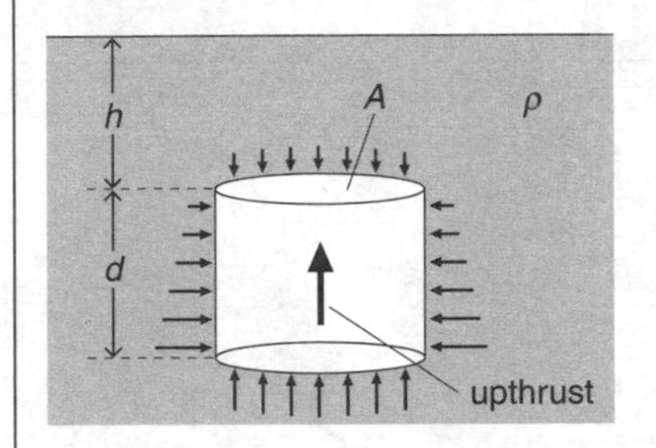

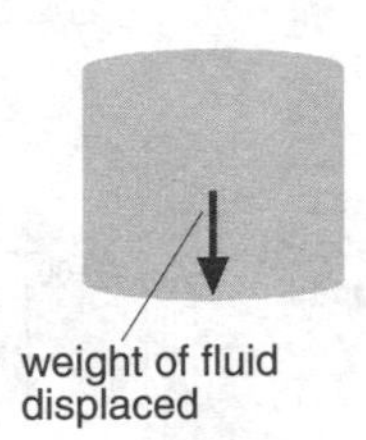

The cylinder above is immersed in a fluid (in this case, a liquid). As a result, fluid is ***displaced***. The mass of fluid displaced is ρdA.

The pressure on the bottom of the cylinder is greater than on the top, so the upward force on the cylinder is greater than the downward force. The *resultant* upward force is the difference between the two. It is called the ***upthrust***.

$$\text{upward force} = \rho g(h + d)A$$

$$\text{downward force} = \rho g h A$$

So $\quad \text{upthrust} = \rho g(h + d)A - \rho g h A = \rho g d A$

But $\quad \rho g d A = \text{weight of fluid displaced}$

So $\quad \text{upthrust} = \text{weight of fluid displaced}$

This is known as ***Archimedes' principle***.

Note:
- The principle applies to all fluids (liquids *and* gases).
- The principle applies to an immersed object of any shape, including one which is only partly immersed.

Floating stability

The upthrust on a boat is the resultant of many forces distributed over the immersed part of its hull. The point at which the upthrust acts is called the ***centre of buoyancy***. It is where the centre of gravity of the displaced fluid would be, and its position changes when the hull rolls.

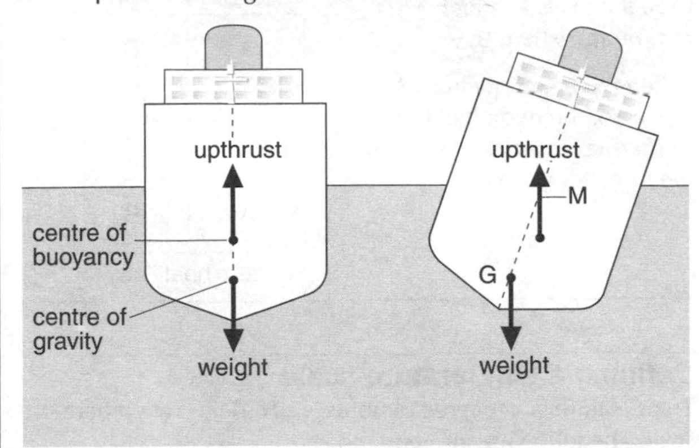

The boat above is in a position of stable equilibrium (see B5). If it starts to roll to one side, the upthrust and the weight form a couple which turns it back again.

The ***metacentre*** (M) is the point where the line of the upthrust meets the centre line of the boat. For stable equilibrium, the metacentre must be *higher* than the centre of gravity (G).

For *maximum* stability the metacentre must be as *high* as possible and the boat loaded so that its centre of gravity is as *low* as possible. (In practice, this is not always desirable because a boat which recovers too rapidly from wave rocking can make passengers sea sick.)

If a boat is loaded so that the metacentre is below the centre of gravity, as on the right, then the equilibrium is unstable. Once the boat has started to roll, the upthrust and the weight form a couple which continues to roll the boat over.

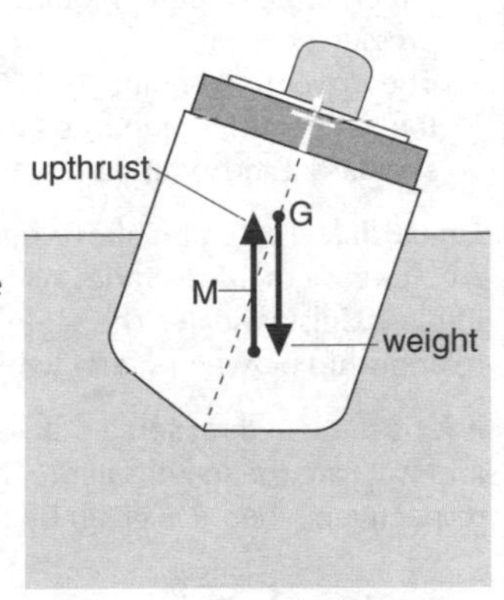

Flotation

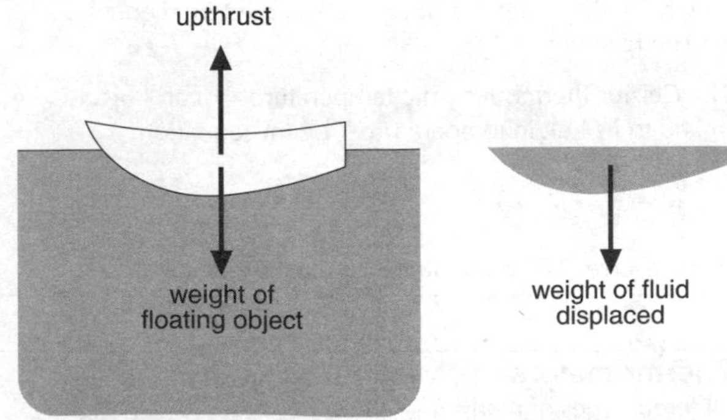

The forces on the boat above are in equilibrium (see B5), so, weight of boat = upthrust on boat. But, by Archimedes' principle, upthrust = weight of fluid displaced. So

weight of floating object = weight of fluid displaced

This is known as the ***principle of flotation***.

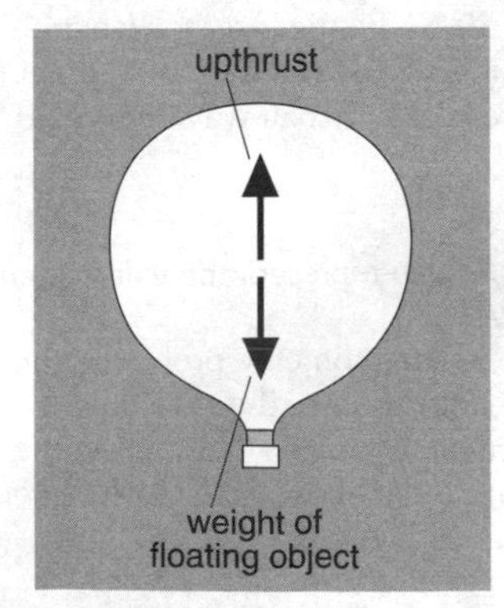

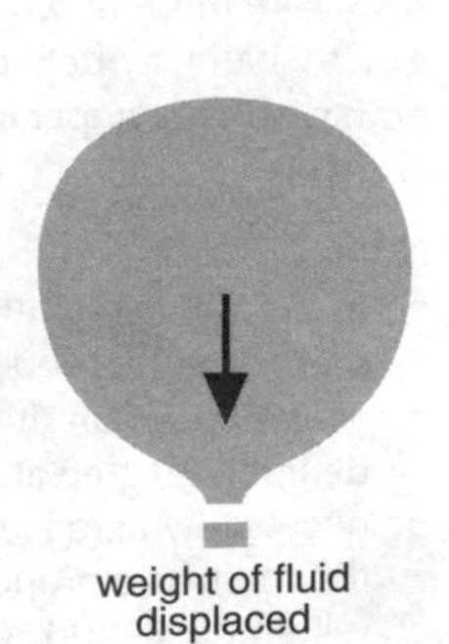

A given volume of hot air weighs less than the same volume of cold air because it has a lower density.

A hot-air balloon like the one above just floats in the cold air around it when the total weight of hot air, fabric, and load is equal to the weight of cold air displaced.

F2 Temperature

Temperature and thermal equilibrium

Objects A and B are in contact. If heat flows from A to B, then A is at a higher ***temperature*** than B.

When the heat flow from A to B is zero, the two objects are in ***thermal equilibrium*** and at the same temperature.

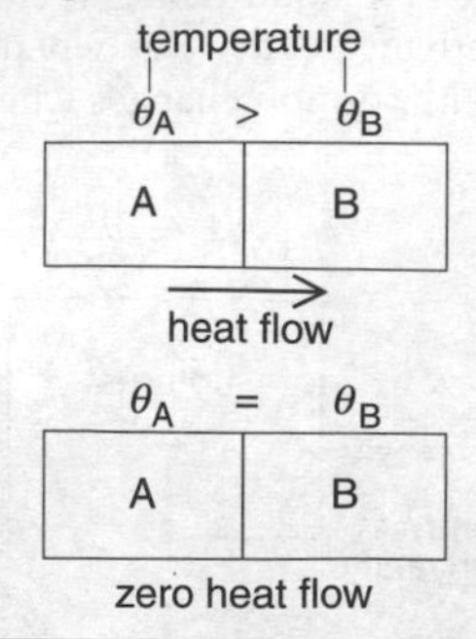

The zeroth law of thermodynamics states that if objects A and B are each in thermal equilibrium with an object C, then they are also in thermal equilibrium with each other.

Note:
- Any objects in thermal equilibrium are at the same temperature.
- Thermodynamics deals with the links between heat and other forms of energy. The zeroth law is so named because the first and second laws of thermodynamics (see F3) had already been stated when the need for a more basic law was realised.

Defining a temperature scale

Temperature is a 'degree of hotness'. To define a temperature scale, the following are required.

Thermometric property This is some property of a material that varies continuously with hotness. For example:
- the length of a column of mercury in a glass capillary tube (the mercury expands when heated, as shown in the diagram on the right),
- the resistance of a coil of platinum wire,
- the pressure of a trapped gas kept at fixed volume.

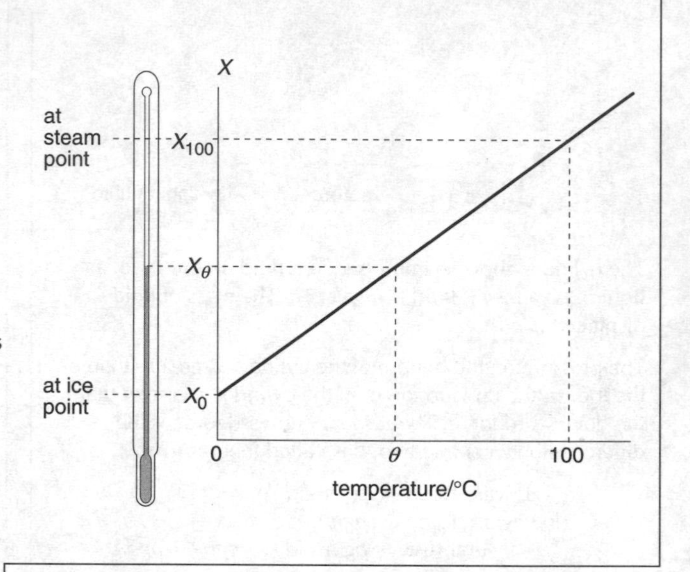

Fixed points These are *defined* reference points against which other temperatures can be judged. For example, on the Celsius scale:
- The *lower* fixed point (0 °C) is the ***ice point.*** This is the temperature of pure, melting ice at standard atmospheric pressure.
- The *upper* fixed point (100 °C) is the ***steam point.*** This is the temperature of the steam above pure boiling water at standard atmospheric pressure.

Linear link The graph above right shows how the length X of a column of mercury varies with temperature on the mercury-in-glass Celsius scale. The scale is *defined* such that the relationship between X and temperature is linear.

If X_θ is the length at some unknown temperature θ, and X_0 and X_{100} are the lengths at the ice and steam points respectively, then θ is given by the following equation:

$$\frac{\theta}{100} = \frac{X_\theta - X_0}{X_{100} - X_0}$$

For example, if $X_0 = 20$ mm, $X_\theta = 30$ mm, and $X_{100} = 40$ mm, then $\theta = 50$ °C. In this case, the length is exactly half way between its values at the ice and steam points, so the temperature is exactly half way between 0 °C and 100 °C.

Note:
- In the above equation, X can represent the value of *any* thermometric property.
- Scales based on different thermometric properties must, by definition, agree at the fixed points. But they do not necessarily agree at any other points. (In practice, the thermometric properties listed above give close agreement within the 0–100 °C range.)

Thermodynamic temperatures

The ***Kelvin scale*** (see F5) is a thermodynamic scale, related to the average kinetic energy per particle (e.g. molecule), and not to any thermometric property. Its definition, in terms of the efficiency of a reversible heat engine, is theoretical and cannot be used practically. However, its scale divisions closely match those of a constant-volume gas thermometer (see next page).

On the Kelvin scale:
- ***Absolute zero*** is 0 kelvin (0 K). This is the temperature at which all substances have minimum internal energy.
- The ice point is 273.15 K.

The Celsius thermodynamic temperature θ of an object is linked to its Kelvin temperature T by this equation:

$$\frac{\theta}{°\mathrm{C}} = \frac{T}{\mathrm{K}} - 273.15$$

In the above, 273 is adequate for most purposes.

Thermometers

Different types of thermometer are described on the right. Some give a direct reading. With others, the temperature must be worked out from other measurements. Before a thermometer can be used, it must be *calibrated* by marking on scale divisions or by preparing a system for converting readings into temperatures. For accurate calibration, a constant-volume gas thermometer is used as a standard.

Type of thermometer	Principle	Range/K	Advantages (+) Disadvantages (–)
mercury-in-glass (see opposite page)	When the temperature rises, mercury expands and moves further up a capillary tube.	234 to 700	+ Portable – Not very accurate – Fragile
constant-volume gas 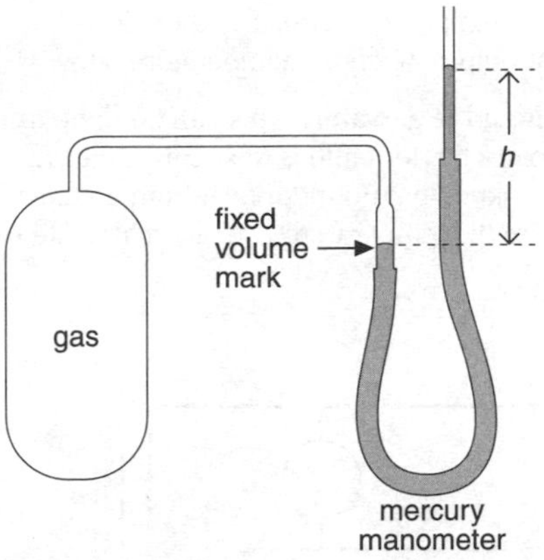	The bulb contains a fixed mass of hydrogen, helium, or nitrogen gas. When the temperature rises, the pressure of the gas rises, and a larger height difference h is needed in the manometer to keep the gas at fixed volume. (The open limb of the manometer can be raised to achieve this.) The pressure of the gas is equal to ρgh plus atmospheric (see F1).	3 to 1750	+ Wide range + Accurate, sensitive + Used as a standard – Not direct reading – Slow response; not suitable for varying temperatures – Cumbersome, fragile
platinum resistance 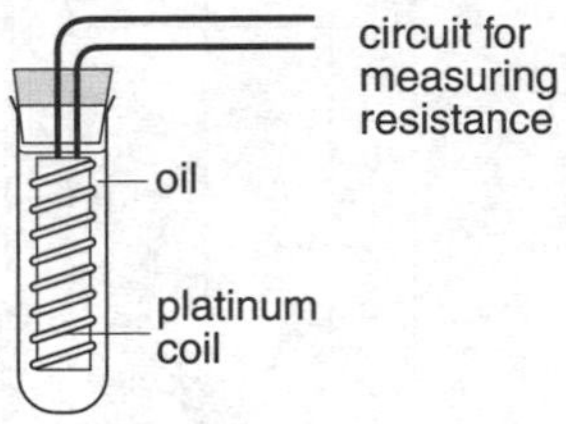	When the temperature rises, the resistance of the platinum coil rises. The resistance is measured accurately with a circuit which makes use of balanced PDs.	25 to 1750	+ Wide range + Accurate, sensitive – Not direct reading – Slow response; not suitable for varying temperatures
thermistor 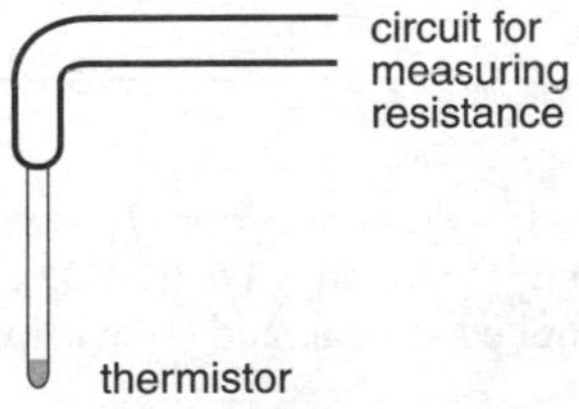	When the temperature rises, the resistance of the thermistor falls. The resistance is measured using a circuit which can include a meter calibrated in degrees.	250 to 450	+ Can be linked to other circuits or computer – Limited range – Not very accurate
thermocouple 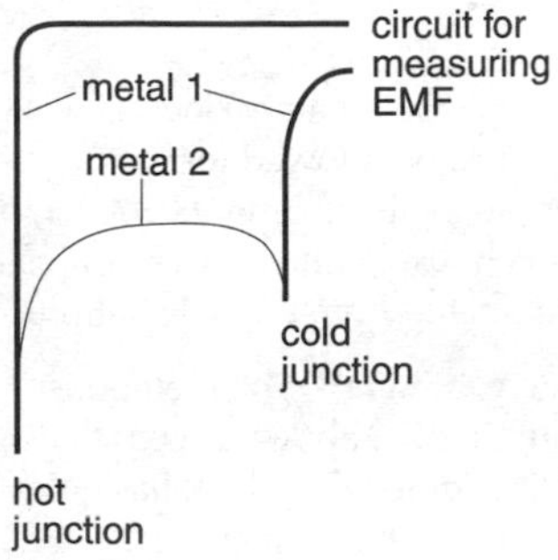	The circuit contains two junctions of dissimilar metals. When one junction is hotter than the other, a small ***thermoelectric*** EMF is generated. The EMF depends on the temperature difference. It is measured using a circuit which can include a meter calibrated in degrees.	80 to 1400	+ Wide range + Small, robust + Quick response + Can be linked to other circuits or computer – Less accurate than constant-volume gas and platinum resistance thermometers
infrared radiation 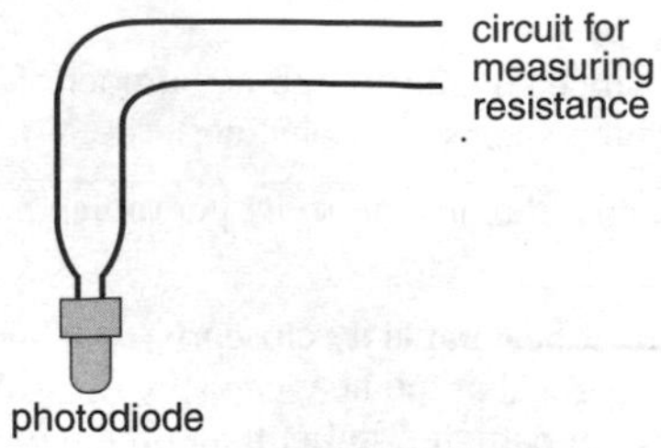	An object gives off more infrared radiation when its temperature rises. This makes the resistance of the photodiode decrease. The resistance is measured using a circuit which can include a meter calibrated in degrees.	225 to 3750	+ No direct contact + Suitable for normal and very high temperatures + Can be linked to other circuits or computer – Less accurate than constant-volume gas and platinum resistance thermometers

F3 Internal energy, heat, and work

Internal energy

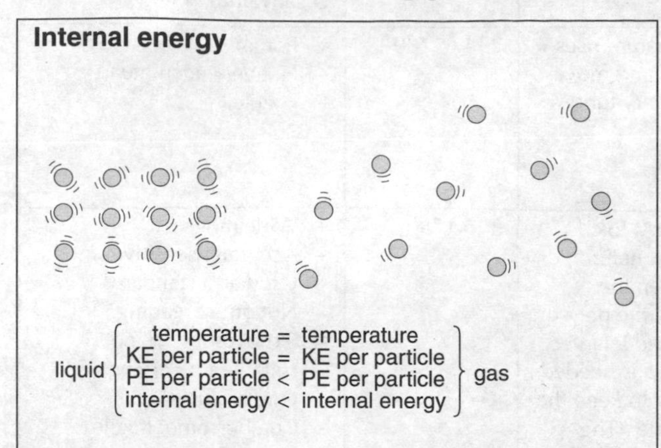

In the liquid, above left, the particles (e.g. molecules) are in motion, so they have kinetic energy (KE). This motion has moved them apart, against the forces of attraction, so they also have potential energy (PE). The total of their KEs and PEs is the ***internal energy*** of the liquid.

Above right, the liquid has become a gas. The temperature is the same as before. So the average KE of each particle due to its linear motion is the same (see F5). However, the average PE is more because of the increased separation of the particles. The gas has more internal energy than the liquid.

The first law of thermodynamics

An object can be given more internal energy:

- by supplying it with heat,
- by doing work on it (i.e. by compressing it) (see B2).

If ΔU is the increase in internal energy when heat Q is supplied to an object *and* work W is done on it, then according to the ***first law of thermodynamics***:

increase in internal energy = heat supplied *to* object + work done *on* object

In symbols $\Delta U = Q + W$

Note:

- The joule is the unit of internal energy, heat, and work.

Heat capacity

heat input Q

ΔT temperature rise

C heat capacity

The ***heat capacity*** of an object is given by this equation:

$$\text{heat capacity} = \frac{\text{heat input}}{\text{temperature rise}} \qquad C = \frac{Q}{\Delta T}$$

For example, if a heat input of 4000 J causes a temperature rise of 2 K, then the heat capacity is 2000 J K^{-1}.

When a solid or liquid is heated, it expands very little and does almost no work. So virtually all the heat supplied is used to increase its internal energy. That follows from the first law of thermodynamics. If $W = 0$, then $\Delta U = Q$.

Specific heat capacity

The heat capacity per unit mass is called the ***specific heat capacity*** (see F5 for typical values). If a substance's specific heat capacity is c, then, for a mass m:

$$Q = mc\Delta T \qquad (1)$$

Water has a high specific heat capacity (4200 J kg^{-1} K^{-1}). This makes it a good 'heat storer'. A relatively large heat input is needed for any given temperature rise, and there is a relatively large heat output when the temperature falls.

Measuring c for a liquid (e.g. water) This can be done using the equipment below. The principle is to supply a measured mass of liquid with a known amount of heat from an electric heating coil, measure the temperature rise, and calculate c using equation (1).

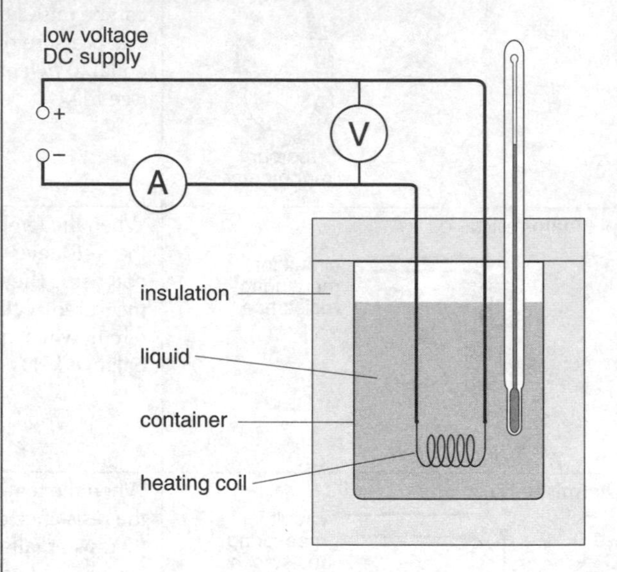

If the PD across the coil is V, and a current I passes for time t, then the electrical energy supplied = VIt (see D2).
If all this energy is supplied as heat, and none is lost:

$$VIt = mc\Delta T$$

Knowing the mass m and temperature rise ΔT of the liquid, its specific heat capacity c can be calculated.

Note:

- When the water heats up, its container does as well. For greater accuracy, this must be allowed for.
- Some heat is lost, despite the insulation. However, there are experiments in which, using different sets of results, heat losses can be eliminated from the calculation.

Measuring c for a solid (e.g. a metal) The method is essentially the same as that shown above, except that a solid block is used instead of the liquid. The block has holes drilled in it for an electric heater and a thermometer.

Molar heat capacity

One mole of any substance contains the same number of particles (6.02×10^{23} atoms, ions, or molecules) (see A1).

The ***molar heat capacity*** is the heat capacity per mole, (rather than per unit mass).

Many solids have a molar heat capacity close to 25 J mol^{-1} K^{-1}. This suggests that the heat capacity of a solid depends on the number of particles, rather than on the mass.

Specific latent heat of vaporization

Energy is needed to change a liquid into a gas, even though there is no change in temperature (see F5).

- Most of the energy is needed as extra internal energy (separating the particles means giving them more PE).
- Some energy is needed to do work in pushing back the atmosphere (because the gas takes up more space).

The ***specific latent heat of vaporization*** of a substance is the heat which must be supplied per unit mass to change a liquid into a gas, without change in temperature.

If Q is the heat supplied, m is the mass, and l_v is the specific latent heat of vaporization, then

$$Q = ml_v \qquad (2)$$

The specific latent heat of vaporization of water is 2.3×10^6 J kg^{-1}. So, to turn 2 kg of water into water vapour (at the same temperature) would require 4.6×10^6 J of heat.

Measuring l_v This can be done using the equipment below. The principle is to supply boiling liquid (e.g. water) with a known amount of heat from an electrical heater, find the mass of vapour formed as a result, and calculate l_v using equation (2). The vapour is cooled, condensed, and collected as a liquid so that its mass can be measured.

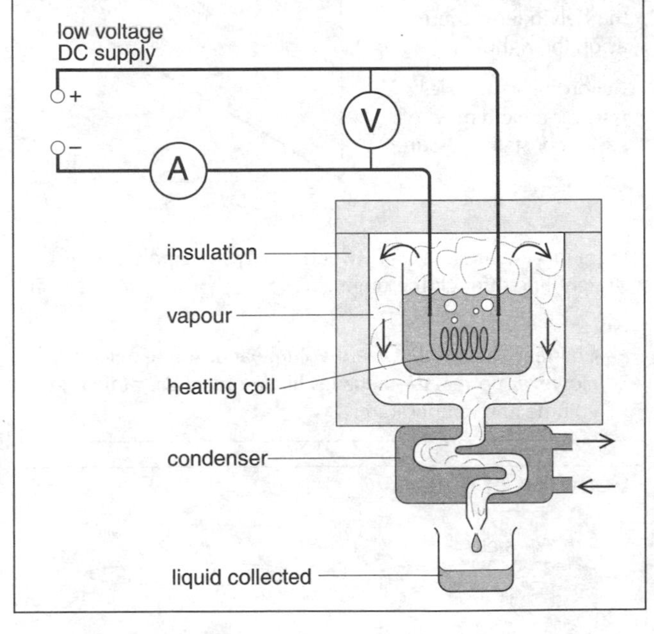

Specific latent heat of fusion

The ***specific latent heat of fusion*** of a substance is the heat which must be supplied per unit mass to change a solid into a liquid, without change in temperature.

If Q is the heat supplied, m is the mass, and l_f is the specific latent heat of fusion, then

$$Q = ml_f$$

The specific latent heat of fusion of water is 3.3×10^5 J kg^{-1} (about $\frac{1}{7}$ of its specific latent heat of vaporization).

Converting work into heat

Work can be completely converted into heat. For example, if a gas is compressed and then left to cool to its original temperature, its internal energy is unchanged, so $\Delta U = 0$. If W is the work done *on* the gas, then from the first law of thermodynamics, $0 = Q + W$, so $-Q = W$.

Q is the heat supplied *to* the gas, so $-Q$ is the heat given out *by* the gas. It is equal to the work done *on* the gas.

Converting heat into work

Petrol, diesel, jet engines, and the boilers-plus-turbines in powers stations are all ***heat engines***. They convert heat into work. But the process can never be 100% efficient. Some heat must always be wasted. This idea is expressed by the ***second law of thermodynamics***. This can be stated in several forms, one of which is as follows:

No continually working heat engine can take in heat and completely convert it into work.

Heat naturally flows from a higher to a lower temperature. So, without a temperature difference, there is no flow of heat. All heat engines take heat from one material at high temperature (e.g. a burning petrol-air mixture) and pass on less heat to a material at a lower temperature (e.g. the atmosphere). The difference is converted into work.

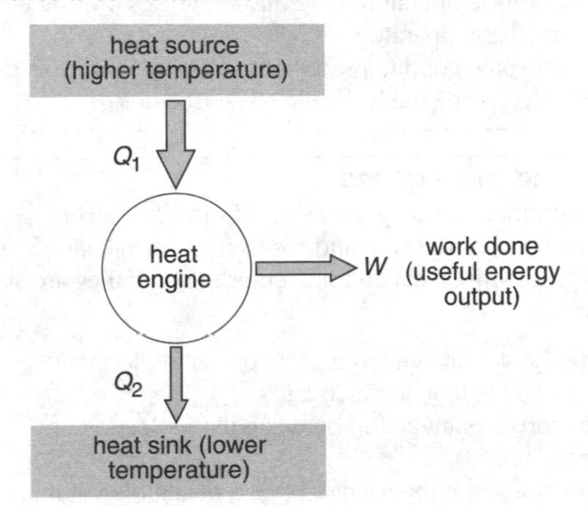

If an engine takes in heat Q_1 from a ***heat source*** and puts heat Q_2 into a ***heat sink***, then the work done $W = Q_1 - Q_2$.

The engine's efficiency is calculated like this (see A2):

$$\text{efficiency} = \frac{\text{useful energy output}}{\text{energy input}} = \frac{W}{Q_1} = \frac{Q_1 - Q_2}{Q_1}$$

An ideal heat engine is one which converts the maximum possible amount of heat into work, so there are no energy losses because of friction. For an engine like this, operating between a source temperature T_1 (in K) and a sink temperature T_2, it can be shown that

$$\text{efficiency} = \frac{T_1 - T_2}{T_1} \qquad (3)$$

For an ideal heat engine, operating between, say, 1000 K (burning fuel) and 300 K (typical atmospheric temperature)

$$\text{efficiency} = \frac{1000 - 300}{1000} = 0.7, \text{ or } 70\%$$

Efficiencies of real engines are much less than this – for example, 30% for a petrol engine. So, in practice, engines waste more heat than they convert into work.

High and low grade energy

Some forms of energy are more useful for doing work than others. For example, in a large electric motor, electrical energy can be converted into work with very high efficiency. Electrical energy is ***high grade energy***.

By comparison, internal energy is ***low grade energy***. The materials around us contain huge amounts of it, but it is unavailable for useful work unless the material is at a higher temperature than its surroundings. Equation (3) shows that the lower the temperature difference, the more unavailable the energy becomes – the more ***degraded*** it is. The waste heat from an engine is very low grade energy.

F4 The behaviour of gases

Boyle's law

Experiments show that, for a fixed mass of gas at constant temperature, the pressure p decreases when the volume V is increased. A graph of p against V for air is shown on the right.

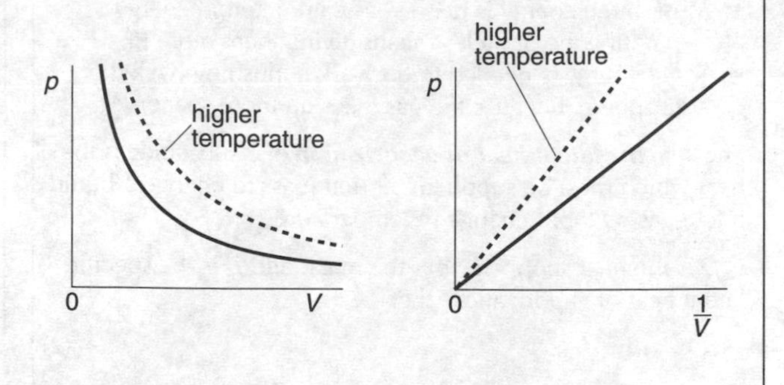

According to ***Boyle's law***, for a fixed mass of gas at constant temperature,

pV = constant

If pV = constant, then $p \propto 1/V$. So a graph of p against $1/V$ is a straight line through the origin.

Note:

- The value of the constant depends on the mass of the gas and on its temperature. The dashed lines show the effects of raising the temperature.
- Under some conditions, the behaviour of real gases departs from that predicted by Boyle's law (see below).

Real and ideal gases

Most common gases are made up of molecules. For convenience, in this unit and the two following, the particles of all gases will be called 'molecules', even if they are single atoms.

An ***ideal gas*** is one which exactly obeys Boyle's law. It can be shown (see F6) that, for such a gas:

- The forces of attraction between the molecules are negligible.
- The molecules themselves have a negligible volume compared with the volume occupied by the gas.

Ideal gases do not exist. However, real gases approximate to ideal gas behaviour at low densities and at temperatures well above their liquefying points.

Ideal gas behaviour is assumed in the rest of this unit.

The pressure law

Before reading this panel, see F2 on temperature.

For a fixed mass of gas at constant volume, the pressure p increases with the Kelvin temperature T, as on the right.

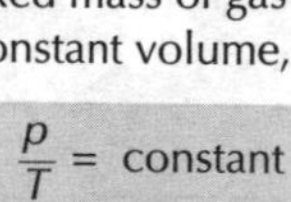

According to the ***pressure law***, for a fixed mass of gas at constant volume,

$\dfrac{p}{T}$ = constant

From this equation, $p \propto T$, which is why the graph is a straight line through the origin.

Note:

- The link between p and T can be used to *define* a temperature scale. However, it can be shown that, for an ideal gas, this scale exactly matches a thermodynamic scale based on the same fixed points (see F6).
- The pressure law predicts that the pressure of any ideal gas should be zero at absolute zero. This concept is used to define the zero point (0 K) on the Kelvin scale and to find its Celsius equivalent (–273.15 °C).

Charles's law

For a fixed mass of gas at constant pressure, the volume V increases with the Kelvin temperature T, as on the right.

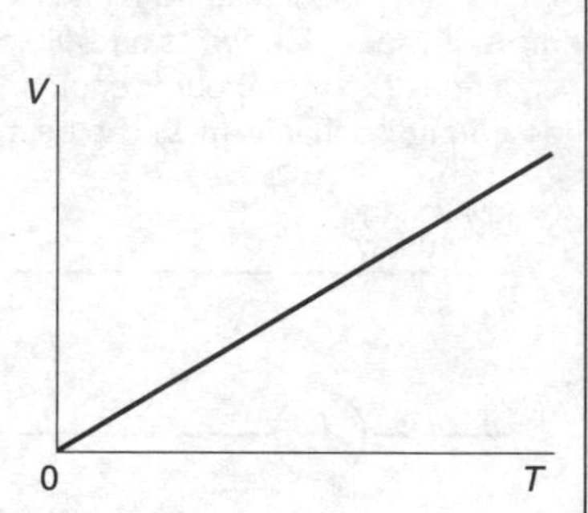

According to ***Charles's law***, for a fixed mass of gas at constant pressure,

$\dfrac{V}{T}$ = constant

From this equation, $V \propto T$, which is why the graph is a straight line through the origin.

Note:

- Charles's law predicts zero volume at absolute zero. However no real gas behaves like an ideal gas at near-zero volume and temperature.

Equation of state

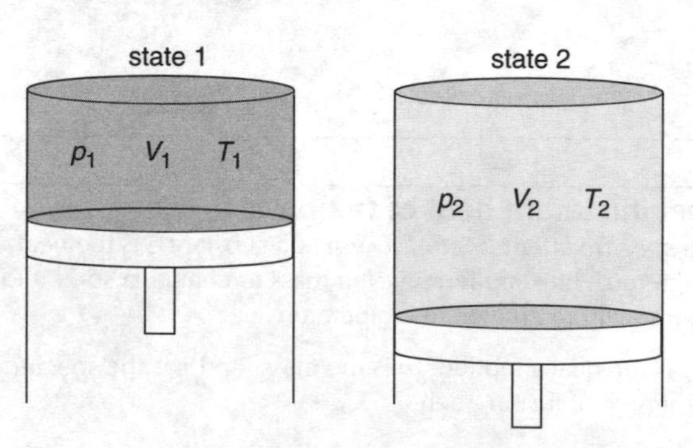

The three gas laws can be combined in a single equation.

If a fixed mass of gas changes from state 1 to state 2, at a different pressure, volume, and Kelvin temperature, as above, then

$$\frac{p_1 V_1}{T_1} = \frac{p_2 V_2}{T_2}$$

This is called the ***equation of state*** for an ideal gas.

Note, in the above equation:

- If $V_1 = V_2$, then $p_1 V_1 = p_2 V_2$. This is Boyle's law.
- If $V_1 = V_2$, then $p_1/T_1 = p_2/T_2$. This is the pressure law.
- If $p_1 = p_2$, then $V_1/T_1 = V_2/T_2$. This is Charles's law.

The ideal gas equation

From the equation of state, pV/T = constant. The constant can have different values depending on the type and mass of gas. However, if the amount of gas being considered is one mole (6.02×10^{23} molecules) (see A1), then the constant is the same for all gases:

for one mole of any gas $\quad pV = RT$

R is called the ***universal molar gas constant***. Its value is 8.31 J mol^{-1} K^{-1}.

for n moles of any gas $\quad pV = nRT$

Note:

- The number of moles $n = m/M$, where m is the mass of the gas and M is its ***molar mass*** (the mass per mole).

Molar masses of some common gases, in kg mol^{-1}	
hydrogen gas	2×10^{-3}
nitrogen gas	28×10^{-3}
oxygen gas	32×10^{-3}

Work done by an expanding gas

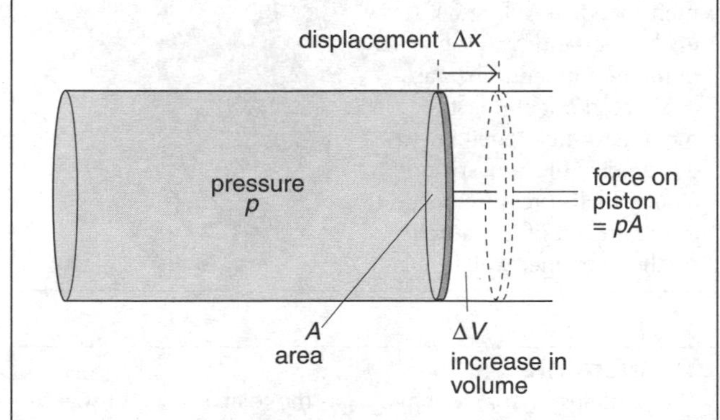

Above, a gas at pressure p exerts a force pA on the piston, and moves it a short distance Δx. If the expansion of the gas is so small that the pressure does not change:

work done by gas = force × displacement = $pA\Delta x$

But $A\Delta x = \Delta V$, the increase in volume.

So $\quad$ work done by gas = $p\Delta V$

Note:

- For convenience, p has been called the pressure. Really, it is the pressure *difference* across the piston (i.e. gas pressure minus atmospheric pressure).
- If the volume of the gas is *decreased* by ΔV, then $p\Delta V$ is the work done *on* the gas.

The graph below left shows the expansion of a gas at constant pressure. The area under the graph gives the work done by the gas ($p\Delta V$). The same principle applies when the pressure is not constant, as shown below right.

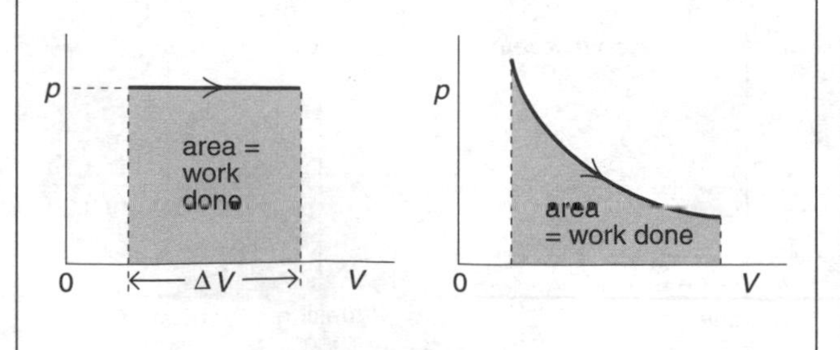

Adiabatic and isothermal expansion

According to the first law of thermodynamics, $\Delta U = Q + W$. The symbols are explained in F3. However, when dealing with an expanding gas, it is useful to know that

Q = heat *taken in* by gas
$-\Delta U$ = *decrease* in internal energy
$-W$ = work done *by* gas

An ***adiabatic*** expansion is one in which no heat is taken in or given out, so $Q = 0$. If during an adiabatic expansion, a gas does work $-W$ it follows that $-W = -\Delta U$ (because $Q = 0$). So there is a decrease in internal energy equal to the work done. As a result, the temperature of the gas falls.

Note:

- Rapid expansions are adiabatic because the gas has negligible time to take in heat from its surroundings.
- Adiabatic *compression* produces a temperature *rise*.

An ***isothermal*** expansion is one in which the temperature is constant. There is no change in internal energy, so $\Delta U = 0$. As before, the gas does work $-W$. However, as $\Delta U = 0$, it follows that $-W = -Q$. So the gas takes in heat from it surroundings equal to the work done.

Note:

- For an isothermal expansion, a gas must stay in thermal equilibrium with its surroundings. In practice, this means a very slow expansion.
- During isothermal *compression*, a gas *gives out* heat.

Indicator diagrams

Pressure–volume graphs are called ***indicator diagrams***. They can be used to show the cycle of changes taking place in an engine. The diagram below shows, in simplified form, what happens in a cylinder of a petrol engine, where there is compression and expansion of a gas as a piston goes up and down.

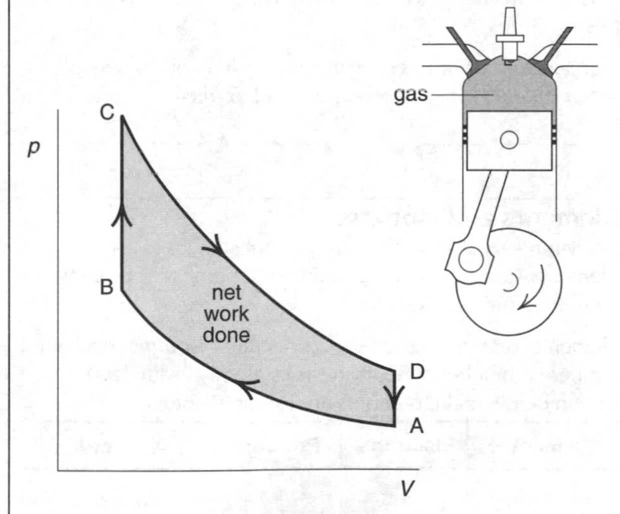

A to B Gas (air–petrol mixture) is compressed adiabatically by the rising piston. This causes a rise in temperature.

B to C Ignited by a spark, the mixture explodes. The further rise in temperature causes a further rise in pressure.

C to D The hot, high-pressure gas pushes down the piston as it expands adiabatically, and the temperature falls.

D to E The warm, waste gas is removed and replaced by cooler, fresh gas mixture, ready for the next cycle.

Note:

- From A to B, work is done *on* the gas. From C to D, work is done *by* the gas. The shaded area therefore represents the net work done during the cycle.

F5 Atoms and molecules in motion

Atoms

All matter is made from ***atoms***. It would take more than a million million atoms to cover this full stop.

An atom has a tiny central ***nucleus*** made of ***protons*** and ***neutrons*** (apart from the simplest atom, hydrogen, whose nucleus is a single proton). Orbiting the nucleus are much lighter particles called ***electrons***.

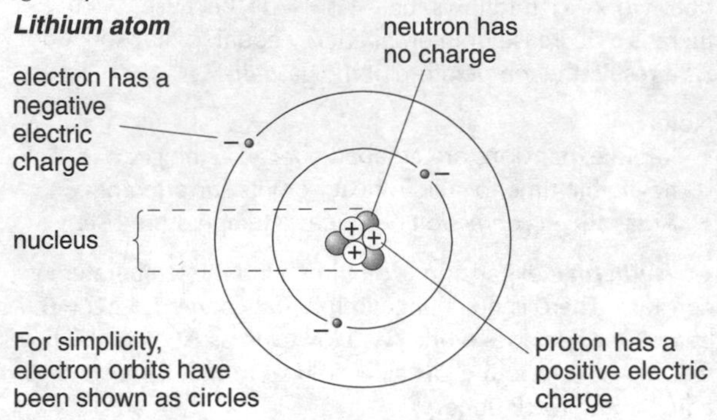

An atom has the same number of electrons as protons, so the amounts of negative and positive charge balance.

Unlike charges (– and +) attract each other. This ***electric force*** holds electrons in orbit around the nucleus.

Like charges (– and –, also + and +) repel each other. However, the particles in the nucleus are held together by a ***strong nuclear force***, which is strong enough to overcome the repulsion between the protons.

Atoms can stick together, in solids for example. The forces that bind them are attractions between opposite charges.

Moving electrons In metals, some of the electrons are only loosely held to their atoms. These ***free electrons*** can drift between the atoms. The electric current in a wire is a flow of free electrons.

If an atom gains or loses electrons, it is left with an overall – or + charge. Charged atoms are called ***ions***.

Elements and isotopes

Everything is made from about 100 substances called ***elements***. Each element has a different number of protons (and therefore electrons) in its atoms.

Elements exist in different versions, called ***isotopes***, each with a different number of neutrons in its atoms. Examples are shown below (italic numbers are for rarer isotopes).

Element	Electrons	Protons	Neutrons
hydrogen	1	1	0 or *1* or *2*
helium	2	2	*1* or 2
lithium	3	3	*3* or 4
carbon	6	6	6 or *7* or *8*
uranium	92	92	*142* or *143* or 146

The total of protons plus neutrons in an atom is called the ***nucleon number***. It is used when naming different isotopes, for example: carbon-12, carbon-13, carbon-14.

Radioactive isotopes These have atoms with unstable nuclei. The nuclei break up, emitting ***nuclear radiation***. The three main types of nuclear radiation are ***alpha*** particles, ***beta*** particles and ***gamma*** waves (see C1).

Solids, liquids and gases

According to the ***kinetic theory***, matter is made up of tiny, randomly moving particles. Each particle may be a single atom, a group of atoms called a ***molecule***, or an ion. The three normal ***phases*** of matter are solid, liquid, and gas.

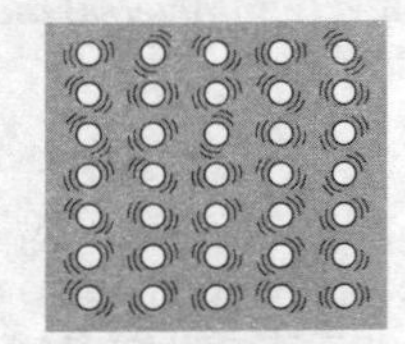

Solid The particles are held close together by strong forces of attraction. They vibrate, but about fixed central positions, so a solid keeps a fixed shape and volume.

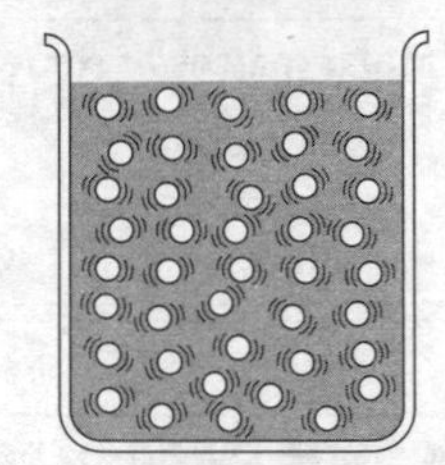

Liquid The particles are held close together. But the vibrations are strong enough to overcome the attractions, so the particles can change positions. A liquid has a fixed volume, but it can flow to fill any shape.

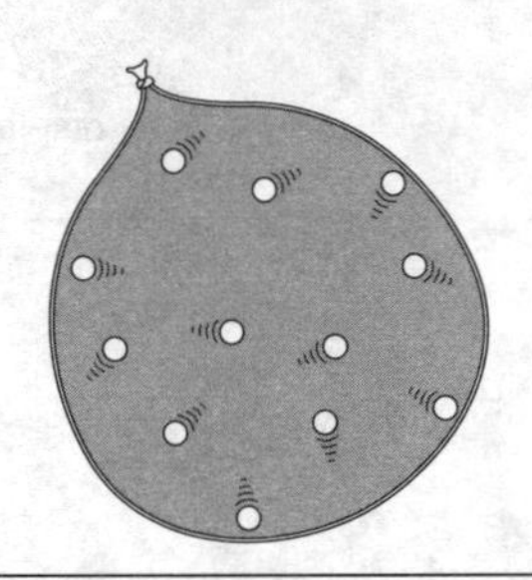

Gas The particles move at high speed, colliding with each other and with the walls of their container. They are too spread out and fast-moving to stick together, so a gas quickly fills any space available. Its pressure is due to the impact of its particles on the container walls.

Temperature

The particles in, for example, a gas move at a range of speeds. However, the higher the temperature, the faster the particles move on average.

If two objects at the same temperature are in contact, there is no flow of heat between them. This is because the average kinetic energy of each particle due to its vibrating or speeding motion is the same in each object, so there is no overall transfer of energy from one object to the other.

Celsius scale On this scale, pure water freezes at 0 °C and boils at 100 °C (under standard atmospheric conditions).

Kelvin scale This has the same sized 'degree' as the Celsius scale, but its 'zero' is ***absolute zero*** (–273 °C), the temperature at which particles have the minimum possible kinetic energy. (The laws governing the behaviour of atoms do not permit zero energy.)

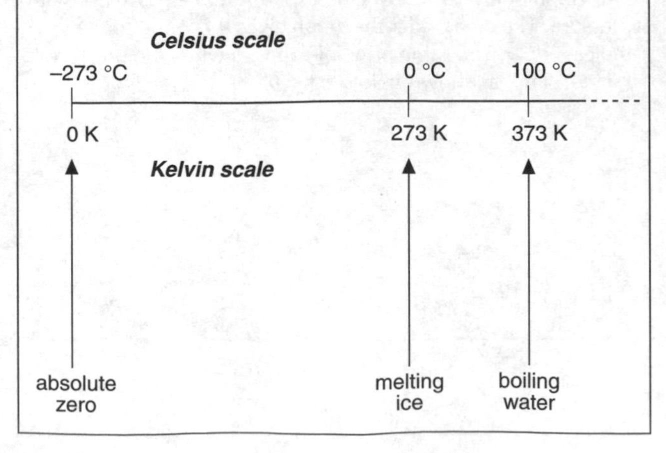

Linking heat and temperature

If, say, a block of copper absorbs heat, its internal energy increases and its temperature rises.

Copper has a ***specific heat capacity*** of 390 J kg^{-1} K^{-1}. This means that 390 J of energy are required to raise the temperature of 1 kg of copper by 1 K.

Specific heat capacities, in J kg^{-1} K^{-1}

copper	390	aluminium	910
iron	470	ice	2100
glass	670	water (liquid)	4200

If a solid of mass m and specific heat capacity c is to increase its temperature by ΔT, then the heat input required is given by the following equation:

heat input = $mc\Delta T$

For example, to raise the temperature of 2 kg of copper by 10 K, the heat input required = 2 × 390 × 10 = 7800 J.

Changing phase

The graph shows what happens when a very cold solid (ice) takes in heat at a steady rate. Melting and boiling are both examples of a change of ***phase***.

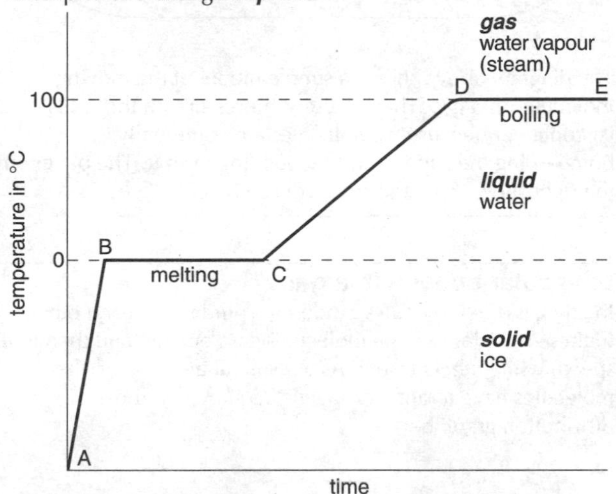

A to B The temperature rises until the ice starts to melt.

B to C Heat is absorbed, but with no rise in temperature. The energy input is being used to overcome the attractions between the particles as the solid changes into a liquid.

C to D The temperature rises until the water starts to boil.

D to E Heat is absorbed, but with no rise in temperature. The energy input is being used to separate the particles as the liquid changes into a gas (water vapour).

A liquid, such as water, starts to turn to gas well below its boiling point. This process is called ***evaporation***. It happens as faster particles escape from the surface.

Boiling is a rapid type of evaporation in which vapour bubbles, forming in the liquid, expand rapidly because their pressure is high enough to overcome atmospheric pressure.

The heat required to change a liquid into a gas (or a solid into a liquid) is called ***latent heat***. When water evaporates on the back of your hand, it takes the latent heat it needs from your hand. That is why there is a cooling effect.

Latent heat is released when a gas changes back into a liquid (or a liquid changes back into a solid).

Heat transfer

Heat can be transferred by ***conduction***, ***convection***, and ***radiation***, as well as by evaporation.

Conduction In all materials, fast-moving particles in one region can gradually pass on energy to neighbouring particles, and hence on to all the particles.

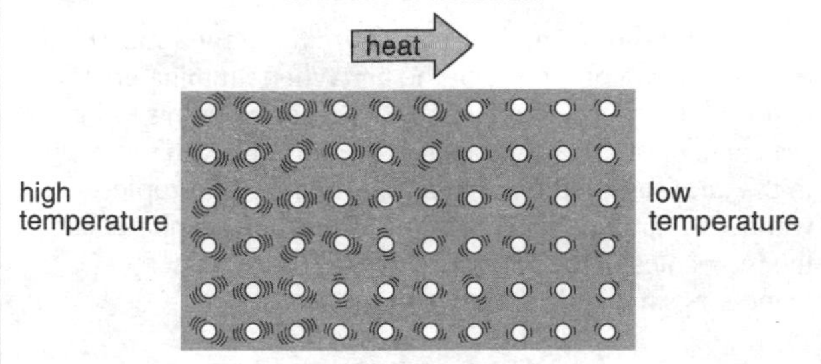

Metals are the best conductors of heat. This is because they have free electrons which can transfer energy rapidly from one part of the material to another. These same electrons also make metals good conductors of electricity.

Non-metal solids and liquids are normally poor conductors of heat because they do not have free electrons. Bad conductors are called ***insulators***. Gases are especially poor conductors: most insulating materials rely on tiny pockets of trapped air for their effect.

Convection Heat is carried by a circulating flow of particles in a liquid or gas.

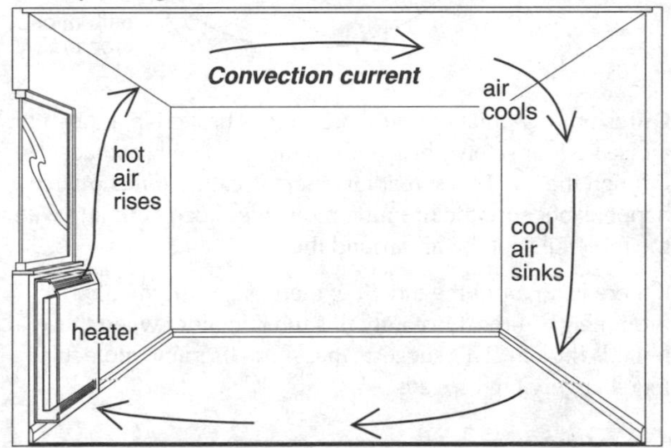

Most room heaters rely on convection. Hot air from the heater expands and floats upwards through the cooler air around it. Cooler air sinks to replace the hot air which has risen. In this way, a ***convection current*** is set up.

Radiation Hot objects radiate energy in the form of electromagnetic waves such as infrared (see C1). The higher the temperature, the more they emit. When this radiation is absorbed by other things, it produces a heating effect. So it is known as ***thermal radiation***.

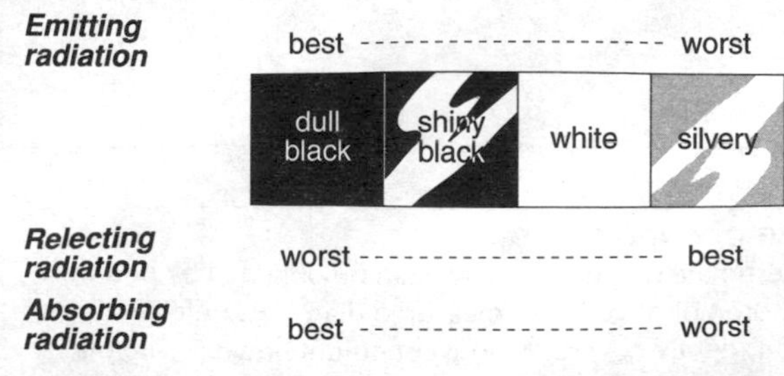

Black surfaces are the best emitters of thermal radiation and also the best absorbers. (They look black because they absorb light.)

Shiny surfaces are poor emitters and also poor absorbers. They reflect most of the radiation that strikes them.

F6 Kinetic theory

Molecules in motion

According to the ***kinetic theory***, matter is made up of randomly-moving particles (e.g. molecules). The following effects provide evidence to support this theory for gases:

Brownian motion Smoke from a burning straw is mainly oil droplets which drift through the air. When illuminated, these oil droplets are just big enough to be seen as points of light, but small enough to be affected by collisions with molecules in the air. Observed through a microscope, the droplets wander in random, zig-zag paths as they are bombarded by the molecules of the air around them. These random wanderings are called ***Brownian motion***.

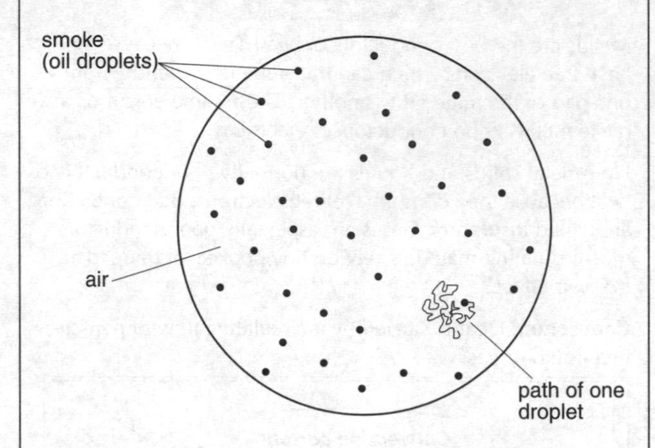

Diffusion If a phial of liquid bromine is broken in a tube of air (sealed for safety), brown bromine gas slowly spreads through the air. This spreading effect is called ***diffusion***. It happens because the bromine molecules keep colliding with the molecules of the air around them.

If there is no air in the tube (i.e. there is a vacuum), the bromine gas almost instantly fills the container when the phial is broken. This suggests that some bromine molecules travel at very high speeds.

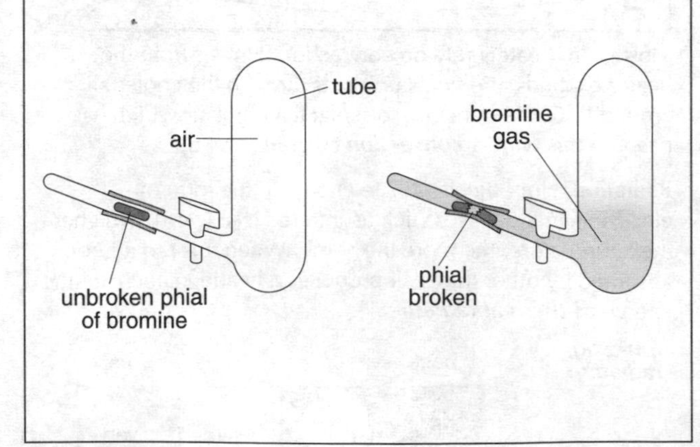

Size of a molecule

An estimate of molecular size can be obtained by putting a tiny drop of olive oil, of measured diameter, onto clean water, lightly covered with lycopodium powder. The oil spreads to form a very thin, circular film whose edge is made visible by the powder. Knowing the volume of oil, and the diameter of the film, the film's thickness can be calculated. Assuming that it is just one molecule thick, one molecule works out to be about 10^{-9} m. (Note: molecules vary considerably in size. Those in olive oil are relatively large.)

Kinetic theory for an ideal gas

The laws governing the behaviour of ideal gases can be deduced mathematically from the kinetic theory, as shown on the next page. In using the theory, the following assumptions are made:

- The motion of the molecules is completely random.
- The forces of attraction between the molecules are negligible.
- The molecules themselves have a negligible volume compared with the volume occupied by the gas.
- The molecules make perfectly elastic collisons (see B7) with each other and with the walls of their container.
- The number of molecules is so large that there are billions of collisions per second.
- Each collision takes a negligible time.
- Between collisions, each molecule has a steady speed.

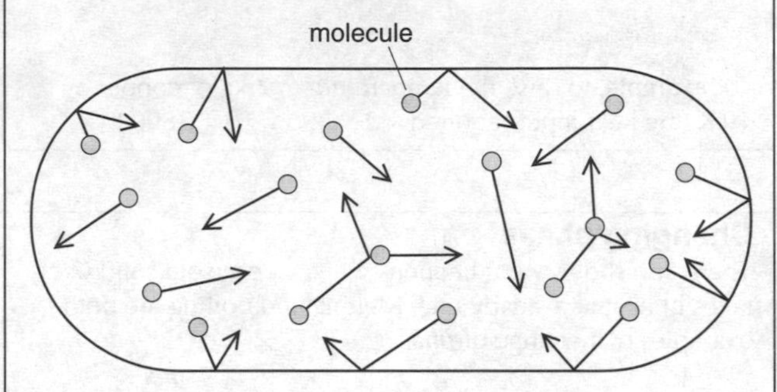

The diagram above shows a simple model of the moving molecules in a gas. The gas exerts a pressure on the walls of its container because its molecules are continually bombarding the surface and rebounding from it. The panel on the right shows how this pressure can be calculated.

Molecular speeds in a gas

In any gas, the molecules randomly collide with each other. In these collisions, some molecules gain energy (and therefore speed) while others lose it. As a result, at any instant, the molecules have a range of speeds, as shown in the distribution graph below.

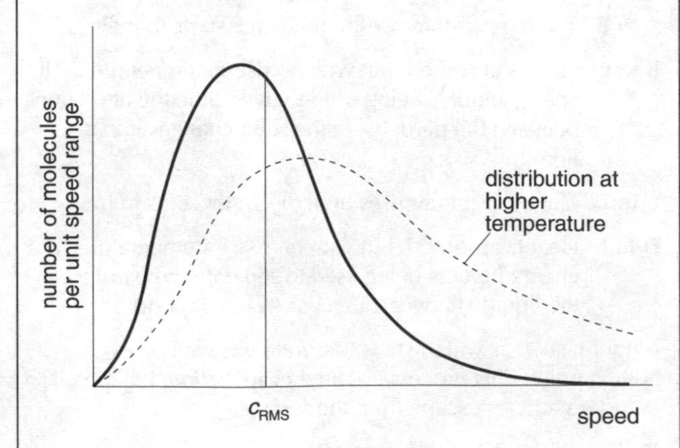

The temperature and the pressure of a gas depend, not on the average speed of the molecules, but on the average of (speed)^2, as explained on the next page. For this reason, it is useful to define the ***mean square speed*** of the molecules. This is the average of (speed)^2 for all the molecules. Its square root is called the ***root mean square speed***, or ***RMS speed***.

RMS speed is represented by the symbol $\boldsymbol{c_{RMS}}$, or alternatively by $\sqrt{\overline{c^2}}$ or $\sqrt{\langle c^2 \rangle}$.

Pressure due to an ideal gas

The pressure of an ideal gas can calculated by considering the motion of its molecules. This has been done below for a gas in a spherical container, though the result applies for any shape. To begin with, it is assumed that all the molecules have the same speed, v.

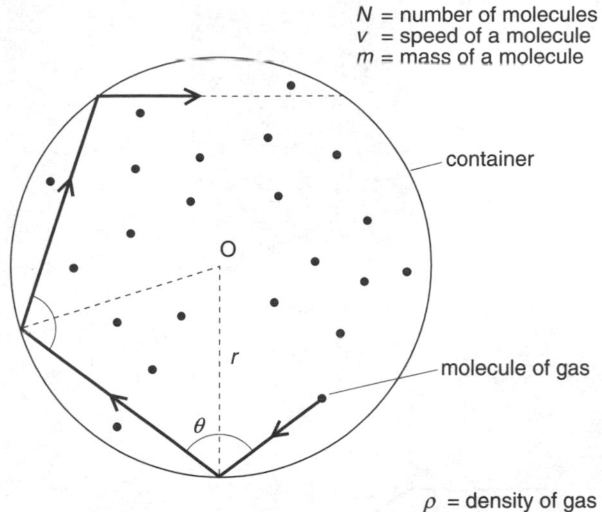

ρ = density of gas

V = volume of gas = $4\pi r^3/3$ (1)

surface area of container = $4\pi r^2$ (2)

Note: m has a different meaning from that used in F4.

The stages in the calculation are as follows:

- The labelled molecule above makes a series of collisions with the container, but always at the same angle θ. Just before each collison it has a component of momentum of $mv \cos \theta$ away from O. This is reversed by the collision. So, for each collision,

  change in momentum (away from O) = $2mv \cos \theta$ (3)

- For the labelled molecule, the distance between collisions is always $2r \cos \theta$. As time = distance/speed,

  time between collisions = $2r \cos \theta / v$ (4)

- The average force exerted by the molecule on the container can found by dividing the change in momentum (3) by the time (4) (see B6).

  This gives force due to one molecule = mv^2/r

  So total force due to all molecules = Nmv^2/r (5)

  The force does not depend on θ. A smaller angle gives a higher momentum change, but less frequent collisions.

- The pressure is found by dividing the force (5) by the area (2).

  This gives $p = Nmv^2/4\pi r^3$.

  Combining this with (1) gives $pV = Nmv^2/3$. (6)

- The pressure depends on v^2. But in reality molecules travel at a range of speeds, so v^2 should be replaced by the mean value of v^2 for all the molecules. This is c_{RMS}^2.

  So, equation (6) should be rewritten as follows:

  $$pV = \frac{1}{3} Nmc_{RMS}^2 \quad (7)$$

  For one mole of gas $$pV = \frac{1}{3} N_A mc_{RMS}^2 \quad (8)$$

  where N_A is the Avogadro constant (6.02×10^{23}) (see A1).

- Nm is the total mass of gas. So Nm/V is its density. Equation (7) can therefore be rewritten

  $$p = \frac{1}{3} \rho c_{RMS}^2$$

Linking kinetic energy and temperature

In a gas, each molecule has kinetic energy because of its linear motion. This is called ***translational*** kinetic energy. Its average value depends on c_{RMS}. For simplicity, average translational kinetic energy will just be called KE.

So KE per molecule = $\frac{1}{2} mc_{RMS}^2$ (9)

This can be linked with the Kelvin temperature as follows.

- According to the ideal gas equation, for one mole of an ideal gas $pV = RT$ (see F4). This equation is used to define the ideal gas scale of temperature.
- Combining $pV = RT$ with (8) gives

  $$\frac{1}{3} N_A mc_{RMS}^2 = RT \quad (10)$$

- Combining the above equation with (9) gives

  $$\text{KE per molecule} = \frac{3}{2}\frac{R}{N_A} T$$

- R/N_A is the gas constant per molecule. Known as the ***Boltzmann constant***, k, its value is 1.38×10^{-23} J K^{-1}. Using k, the above equation can be rewritten:

  $$\text{KE per molecule} = \frac{3}{2} kT$$

Note:

- The KE per molecule is proportional to T. That is why the Kelvin thermodynamic scale (based on energy – see F2 and F4) coincides with the ideal gas scale.

Deducing Boyle's law

According to equation (7) $pV = \frac{1}{3} Nmc_{RMS}^2$

On the right-hand side of this equation:

- If the mass of gas is fixed, N is constant.
- If the temperature is steady, the KE per molecule (9) is constant, so mc_{RMS}^2 is constant.

Therefore, it follows that, for a fixed mass of gas at steady temperature, pV is constant. This is Boyle's law (see F4).

Deducing Avogadro's law

According to ***Avogadro's law***, equal volumes of ideal gases under the same conditions of temperature and pressure contain equal numbers of molecules.

This law can be deduced from equation (7). If two gases are at the same temperature, then mc_{RMS}^2 is the same for each. If they are at the same pressure and volume, then pV is the same for each. So, from equation (7), N must also be the same for each.

Calculating c_{RMS}

$N_A m$ is the molar mass M of a gas (see F4). So equation (10) can be rewritten

$$\frac{1}{3} Mc_{RMS}^2 = RT$$

Rearranged, this gives $c_{RMS} = \sqrt{3RT/M}$.

With R, T, and M known, c_{RMS} can be calculated. For example, for nitrogen (the main gas in air) at a room temperature of 300 K, c_{RMS} works out at 517 m s^{-1}.

F7 Heat transfer

Processes and particles

Heat (thermal energy) can be transferred by ***evaporation***, ***conduction***, ***convection***, and ***radiation***. Basic descriptions of these processes are given in F5. For simplicity, the explanations which follow refer to the motion of molecules. In reality, the particles of a substance may be molecules, atoms, or ions.

Evaporation

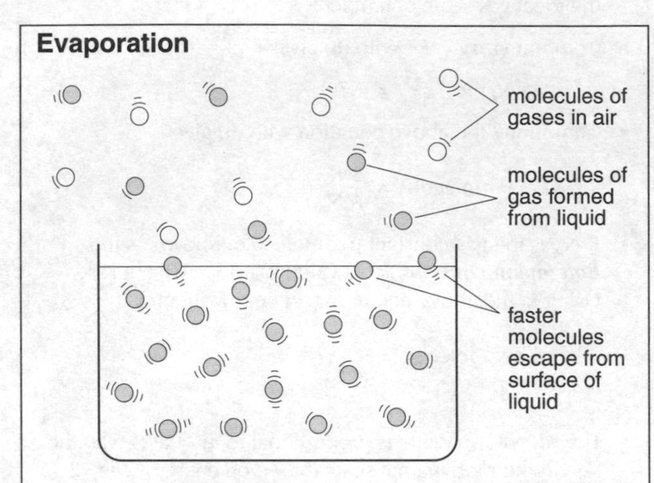

When a liquid evaporates, molecules escape from its surface and move about freely as a gas.

In a liquid, the vibrating molecules keep colliding with each other, some gaining kinetic energy and others losing it. At the surface, some of the faster, upward-moving molecules have enough kinetic energy to overcome the attractions from other molecules and escape from the liquid. With these faster molecules gone, the average KE of those left behind is reduced i.e. the temperature of the liquid falls. That is why evaporation has a cooling effect.

The rate of evaporation (and therefore the rate at which heat is lost from a liquid) is increased if:

- the surface area is increased (more of the faster molecules are near the surface),
- the temperature is increased (more of the molecules have enough kinetic energy to escape),
- the pressure is reduced (escaping molecules are less likely to rebound from other molecules back into the liquid),
- there is a draught across the surface (escaping molecules are removed before they can rebound),
- gas is bubbled through the liquid.

Thermal conduction: the processes

In gases Fast-moving molecules pass on kinetic energy to slower-moving ones when they collide with them. In this way, heat is slowly conducted through the gas.

In non-metal solids and liquids The molecules are ***coupled*** to each other by the forces between. So the molecules with most vibrational energy pass on some of this to those with less energy. However, this process of heat conduction is slow compared with that described next.

In metals Metals contain ***free electrons*** (see D2) which are in thermal equilibrium with the surrounding atoms. These electrons travel at high speeds, and transfer energy quickly from one part of the metal to another. That is why metals are such good conductors of heat. (They also conduct some heat by the transfer of vibrational energy.)

Thermal conductivity

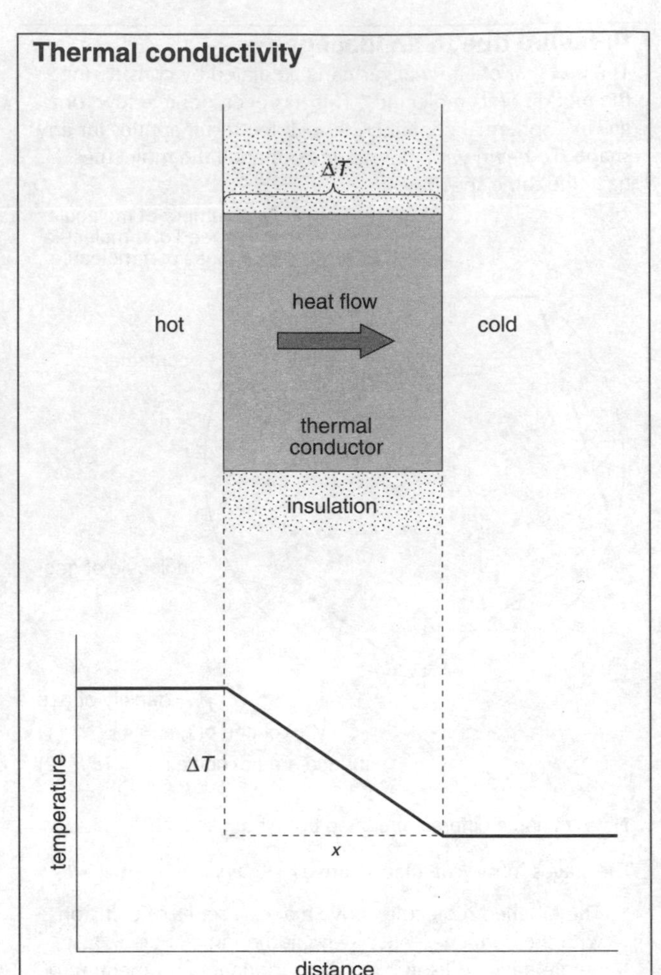

Above, there is a temperature difference ΔT across a block of material of thickness x. As a result, heat ΔQ flows through the material in time Δt. The flow is in the direction of *decreasing* temperature.

The graph shows how the temperature falls, through the block. $\Delta T/x$ is the temperature gradient. It is negative.

$\Delta Q/\Delta t$ is the rate of flow of heat. It is proportional to:

- the temperature gradient (a larger temperature difference or a thinner block give a greater heat flow),
- the cross-sectional area A (a larger area gives a greater heat flow).

The above principles can be expressed as an equation:

$$\text{rate of flow of heat} = -k \times \text{area} \times \text{temperature gradient} \qquad \frac{\Delta Q}{\Delta t} = -kA\frac{\Delta T}{x} \quad (1)$$

k is called the ***thermal conductivity*** of the material. Some typical values of k are given below.

Note:

- In the above equations, the minus sign indicates that the heat flow is in the direction of *decreasing* temperature.
- Rate of flow of heat is the same as power. Its unit is the watt (W).
- As no heat escapes from the sides of the block, the rate of flow of heat is the same throughout.
- k is *defined* by the above equation. Good conductors have high k values. Good insulators have low k values.

Thermal conductivities, in W m^{-1}K^{-1}	
copper	400
aluminium	238
air (at normal temperature and pressure)	0.03

Conductivity equations compared

The equations dealing with heat flow and charge flow are of a similar form (see also D2).

For a thermal conductor

rate of flow of heat = constant × temperature difference

For an electrical conductor

rate of flow of charge = constant × potential difference
(current)

Thermal conduction through layers

On the right, a layer of brick is covered with insulating foam to reduce the heat flow. Knowing T_1 and T_2, x_1 and x_2, and k_1 and k_2, the temperature T at the boundary between the two materials can be found, and also the rate of flow of heat.

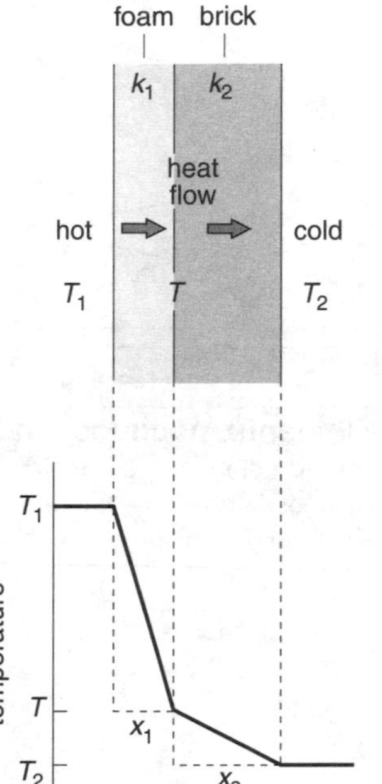

As no heat escapes from the sides, the rate of flow of heat, $\Delta Q/\Delta t$, must be the same through both layers. So, from (1),

$$\frac{\Delta Q}{\Delta t} = k_1 A \frac{(T_1 - T)}{x_1}$$

$$= k_2 A \frac{(T - T_2)}{x_2}$$

T can be found using the right-hand parts of the above equations, and rearranging. With T known, either of these parts gives the value of $\Delta Q/\Delta t$.

Note:

- In the above example $k_2 > k_1$. For the *same* rate of flow of heat, the material with the *lower* k must have the *higher* temperature gradient

U-values

Heating engineers use ***U-values***, rather than k values, when calculating heat losses through walls, windows, and roofs. A U-value is defined by the following equation:

rate of flow of heat = U-value × area × temperature difference

Using the symbols in the panel on the left,

$$\frac{\Delta Q}{\Delta t} = \text{U-value} \times A\,\Delta T \qquad (2)$$

From (1) and (2), it follows that, for a material of thermal conductivity k, the U-value = k/x. So, unlike k, the U-value depends on thickness. For good insulation, a low U-value is needed. The requirements for this are a low k and a high thickness. Here are some typical U-values:

U-values in W m^{-2} K^{-1}	
single brick wall	3.6
double brick wall with air space	1.7
window, single glass layer	5.7
double-glazed window	2.7

Convection

Room heaters (including so-called 'radiators') and refrigerators lose most of their heat by convection. A hot surface heats the air next to it. The hot air rises, to be replaced by cooler air which then heats up, and so on. This is called ***natural convection***. The surface will lose heat more quickly of air is blown across it. This is known as ***forced convection***.

Convection is caused by an upthrust (see F1). A region of cold air just floats in the cold air around it because it displaces its own weight. However, when heated, it expands. It weighs the same as before, but now displaces more cold air, so the increased upthrust pushes it upwards. (The same principle applies to other gases and to liquids.)

Thermal radiation

Vibrating and spinning molecules in one object give off electromagnetic radiation whose energy can be absorbed by molecules in another object so that they speed up. This radiation is called ***thermal radiation***. From most warm or hot objects, it is mainly infrared (see C1).

Some surfaces are better absorbers of thermal radiation than others (see F5). A perfect absorber (i.e. one which reflects no radiation) is called a ***black body***. It is also the best possible emitter of thermal radiation. The Sun, odd though it may sound, is effectively a black body radiator.

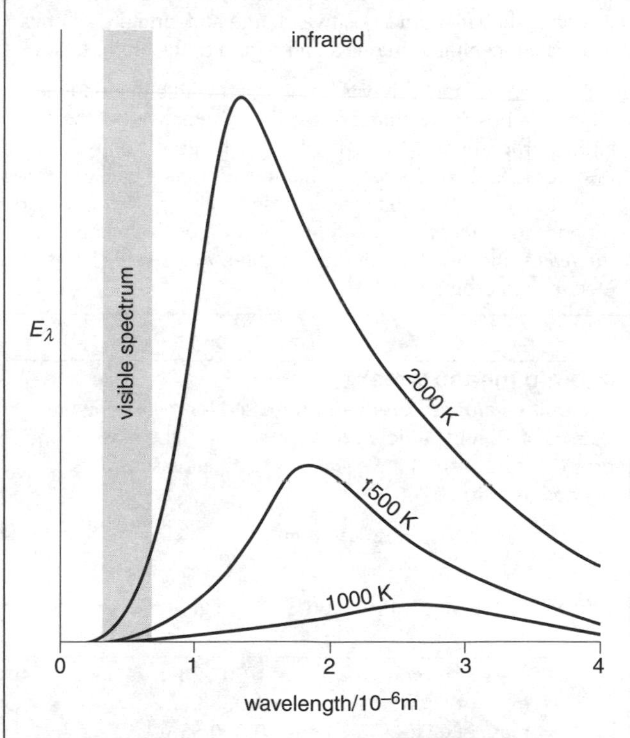

The graph above shows how radiated energy is distributed across different wavelengths for a black body radiator at various temperatures. The E_λ axis represents the relative energy output per second per unit range of wavelength.

Note:

- As the temperature increases, the total energy output per second increases (in proportion to T^4).
- As the temperature increases, the peak wavelength becomes less. By 1000 K some of the radiation has reached the red end of visible spectrum, so the object is glowing red hot. When the peak is within the visible spectrum, the object is glowing white hot.

For a radiator which is not a black body, the lines of the graph are of a similar form, but the peaks are lower.

G1 The nuclear atom

Evidence for a nucleus

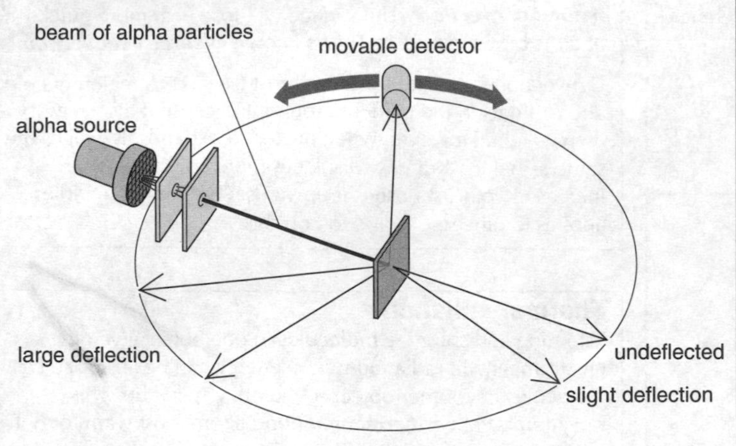

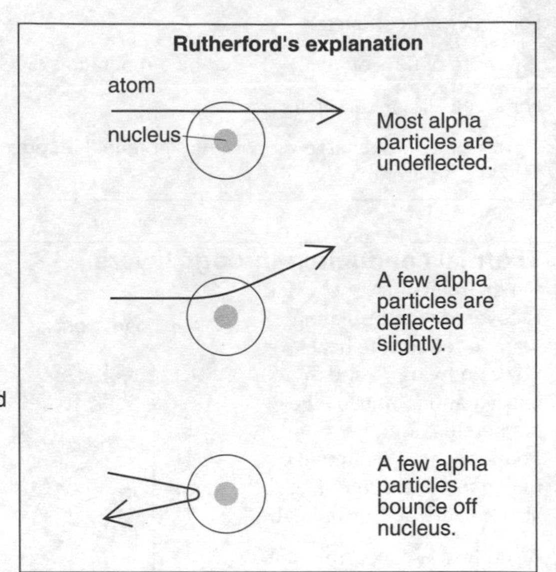

A neutral atom contains equal amounts of positive (+) and negative (–) charge (see F5). The above experiment, supervised by Ernest Rutherford in 1911, first provided evidence that an atom's positive charge and virtually all of its mass is concentrated in one small region of the atom.

A thin piece of gold foil was bombarded with alpha particles, which are positively charged (see next page). Most of the alpha particles passed straight through the gold atoms. But a few were repelled so strongly that they bounced back or were deflected through large angles. These results led Rutherford to propose this model of the atom: a heavy, positively charged ***nucleus*** at the centre, with much lighter, negatively charged electrons in orbit around it.

Atomic measurements

The ***unified atomic mass unit*** **(u)** is used for measuring the masses of atomic particles. It is very close to the mass of one proton (or neutron). However, for practical reasons, it is defined as follows:

$$1\text{ u} = \frac{\text{mass of carbon-12 atom}}{12}$$

Converting into kg, 1 u = 1.66×10^{-27} kg.

mass of proton	1.007 28 u
mass of neutron	1.008 67 u
mass of electron	0.000 55 u
charge on proton	$+1.60 \times 10^{-19}$ C
charge on electron	-1.60×10^{-19} C
diameter of an atom	$\sim 10^{-10}$ m
diameter of a nucleus	$\sim 10^{-14}$ m

Note:

- The proton and neutron have approximately the same mass – about 1800 times that of the electron.
- ~ means 'of the order of' i.e. 'within a factor ten of'.
- The diameter of an atom is $\sim 10^4$ times that of its nucleus. (Atom size varies from element to element.)
- Confusingly, the symbol *e* may be used to represent the charge on an electron (–) or a proton (+). In this unit, the charge on an electron will be called $-e$.

Elements, nuclides, and isotopes

For most elements, a sample contains a mixture of different versions. These have the same number of protons (and electrons) but different numbers of neutrons (see F5).

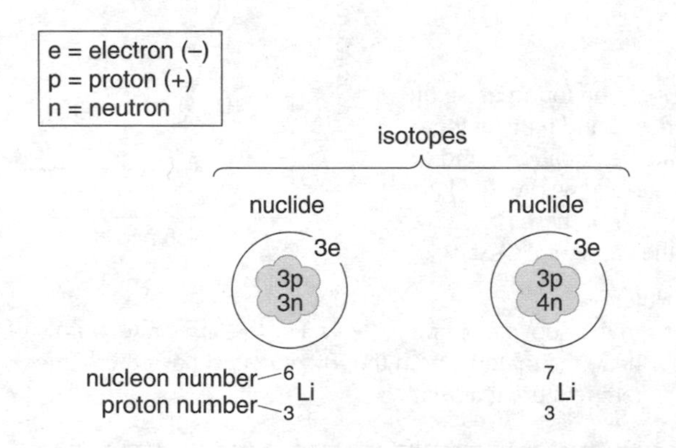

Nuclide This is any particular version of an atom. Above are simple models of the two naturally occurring nuclides of lithium, along with the symbols used to represent them.

Nucleon number *A* As protons and neutrons are called ***nucleons***, this is the total number of protons plus neutrons in the nucleus. It was once called the ***mass number***.

Proton number *Z* This is the number of protons in the nucleus (and therefore the number of electrons in a neutral atom). It was once called the ***atomic number***.

Isotopes These are atoms with the same proton number but different nucleon numbers. They have the same electron arrangement and, therefore, the same chemical properties.

The following statements illustrate the meanings of the terms *element, nuclide,* and *isotope*.

- Lithium is an element.
- Lithium-6 is a nuclide; lithium-7 is a nuclide.
- Lithium-6 and lithium-7 are isotopes.

Note:

- A nuclide is commonly referred to as 'an isotope', though strictly speaking, this is incorrect.

Stability of the nucleus

If the number of neutrons ($A - Z$) in the nucleus is plotted against the number of protons (Z) for all known nuclides, the general form of the graph is like this:

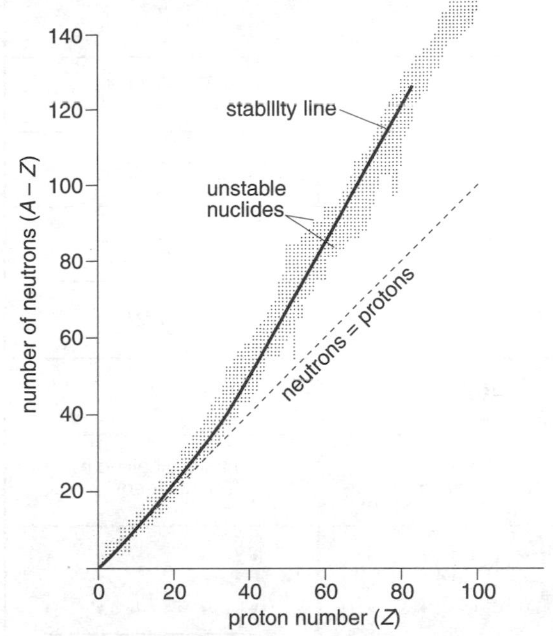

Stable nuclides These have nuclei which are stable. They occur along the solid line.

- Light nuclides (those at the lower end of the graph) have about equal numbers of protons and neutrons.
- The heaviest nuclides have about 50% more neutrons than protons.
- The most stable nuclides tend to have an even number of protons and an even number of neutrons. This is because each group of 2 protons and 2 neutrons in the nucleus makes an especially stable combination.

Unstable nuclides These occur either side of the solid line. They have unstable nuclei which, in time, *disintegrate* (break up) – usually by emitting an ***alpha*** particle or ***beta*** particle and maybe ***gamma*** radiation as well (see also G2). The disintegration is called ***radioactive decay***.

Element	Symbol	Z
hydrogen	H	1
helium	He	2
lithium	Li	3
beryllium	Be	4
boron	B	5
carbon	C	6
nitrogen	N	7
oxygen	O	8
fluorine	F	9

Element	Symbol	Z
neon	N	10
sodium	Na	11
iron	Fe	26
cobalt	Co	27
nickel	Ni	28
protactinium	Pa	91
uranium	U	92

Alpha (α) decay

An alpha (α) particle consists of 2 protons and 2 neutrons, so it has a charge of $+2e$. It is identical to a nucleus of helium-4 and may be represented by either of these symbols:

$^{4}_{2}\alpha$ $\quad$ $^{4}_{2}\text{He}$

If an atom emits an α particle, its proton number is decreased, so it becomes the atom of a different element. For example, an atom of radium-226 emits an α particle to become an atom of radon-222, as shown by this equation:

$$^{226}_{88}\text{Ra} \rightarrow {}^{222}_{86}\text{Rn} + {}^{4}_{2}\alpha$$

radium-226 $\quad$ radon-222 $\quad$ α particle

Note:

- Radon-222 and the α particle are the ***decay products***.
- The nucleon numbers on both sides of the equation balance (226 = 222 + 4) because the total number of protons and neutrons is conserved (unchanged).
- The proton numbers balance (88 = 86 + 2) because the total amount of positive charge is conserved.
- Alpha decay tends to occur in heavy nuclides which are below the stability line (see above graph) because it produces a nuclide which is closer to the line.

Beta (β) decay

β⁻ decay This is the most common form of beta decay. The main emitted particle is an electron. It has a charge of $-1e$ and may be represented by either of these symbols:

$^{0}_{-1}\beta$ $\quad$ $^{0}_{-1}\text{e}$

Note:

- The beta particle is not a nucleon, so it is assigned a nucleon number of 0.
- The 'proton number' is –1 because the beta particle has an equal but opposite charge to that of a proton.

During β^- decay, a neutron is converted into a proton, an electron, and an almost undetectable particle with no charge and near-zero mass called an antineutrino. The electron is emitted, along with the antineutrino. For example, an atom of boron-12 emits a β^- particle to become an atom of carbon-12, as described by this equation:

$$^{12}_{5}\text{B} \rightarrow {}^{12}_{6}\text{C} + {}^{0}_{-1}\text{e} + {}^{0}_{0}\bar{\nu}$$

boron-12 $\quad$ carbon-12 $\quad$ β^- particle $\quad$ antineutrino

Note:

- The nucleon numbers balance on both sides of the equation. So do the proton numbers.
- β^- decay tends to occur in nuclides above the stability line (see graph at top of page).

β⁺ decay Here, the main emitted particle is a ***positron***, with the same mass as an electron, but a charge of $+1e$. It is the ***antiparticle*** of an electron. For example, an atom of nitrogen-12 emits a β^+ particle to become an atom of carbon-12, as described by the following equation:

$$^{12}_{7}\text{N} \rightarrow {}^{12}_{6}\text{C} + {}^{0}_{+1}\text{e} + {}^{0}_{0}\nu$$

boron-12 $\quad$ carbon-12 $\quad$ β^+ particle $\quad$ neutrino

Nuclear reactions

One element changing into another is called a ***transmutation***. It can occur when atoms are bombarded by other particles. For example, if a high-energy α particle strikes and is absorbed by a nucleus of nitrogen-14, the new nucleus immediately decays to form a nucleus of oxygen-17 and a proton. This is an example of a ***nuclear reaction***. It can be described by the following equation:

$$^{14}_{7}\text{N} + {}^{4}_{2}\alpha \rightarrow {}^{17}_{8}\text{O} + {}^{1}_{1}\text{p}$$

nitrogen-14 $\quad$ α particle $\quad$ oxygen-17 $\quad$ proton

Note:

- The nucleon numbers balance on both sides of the equation. So do the proton numbers.

G2 Radiation and decay

Properties of alpha, beta, and gamma radiation

Type of radiation	α	β	γ
nature	2 p + 2 n	e	electromagnetic (see C1 and C2)
charge	$+2e$	$-1e$	no charge
speed (typical) (c = speed of light)	$0.1c$	up to $0.9c$	c
energy (typical)	10 MeV	0.03 to 3 MeV	1 MeV
ionizing effect: ion pairs per mm in air	$\sim 10^5$	$\sim 10^3$	~ 1
penetration (typical)	stopped by: 50 mm air 0.5 mm paper	stopped by: 5 mm aluminium	intensity halved by 100 mm lead
effect of magnetic field (B out of paper) *not to scale*	+	slow fast –	(undeflected)

Note:
- α, β, and γ radiations all cause ***ionization*** – they remove electrons from atoms (or molecules) in their path. The removed electron (–) and the charged atom or molecule (+) remaining are called an ***ion pair***. The ionized material can conduct electricity.
- α particles interact the most with atoms in their path, so they are the most ionizing and the least penetrating.
- Unlike α particles, β particles are emitted from their source at a range of speeds.
- For γ radiation emitted from a point source in air, intensity $\propto$ 1/(distance from source)^2 (see C2).
- γ radiation is not stopped by an absorber, but its intensity is reduced.
- α and β particles are deflected by magnetic fields, as predicted by Fleming's left-hand rule (see E5 and E8). They are also deflected by electric fields.

Detecting alpha, beta, and gamma radiation

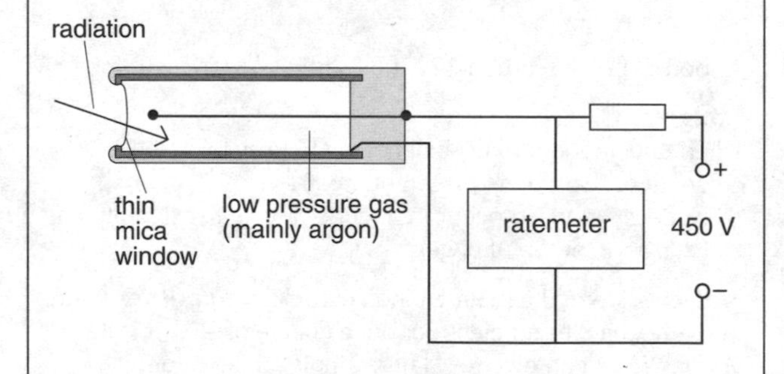

Detectors In the ***G–M tube*** (Geiger–Müller tube) above, a high voltage is maintained across the gas. When, say, a beta particle enters through the thin window, it ionizes the gas and makes it conduct. This causes a pulse of current in the circuit. The ***ratemeter*** registers the ***count rate*** (average number of pulses per second).

In a ***solid state detector***, incoming radiation ionizes the semiconductor material in a diode (see D1 and D2).

Telling the radiations apart To tell one type of radiation from another, absorbing materials of different thicknesses can be placed between the source and the detector. For example, a thick aluminium plate will stop α and β particles, but not γ radiation.

Background radiation In radioactivity experiments, allowance must be made for background radiation. This is low-level radiation whose sources include radioactive materials naturally present in rocks and soil, and cosmic radiation (high-energy particles from space).

Activity

The ***activity*** of a radioactive source is the number of disintegrations occurring within it per unit time.

The SI unit of activity is the ***becquerel*** (**Bq**):

$$1 \text{ becquerel} = 1 \text{ disintegration s}^{-1} \quad (1 \text{ Bq} = 1 \text{ s}^{-1})$$

The activity of a typical laboratory source is $\sim 10^4$ Bq.

Each disintegration produces an α or β particle and, in many cases, γ radiation as well. The γ radiation is emitted as a 'packet' of wave energy called a ***photon*** (see G4). Particles and photons cause pulses in a detector, so the count rate is a measure of the activity of the source.

Note:
- The activity of a source is unaffected by chemical changes or physical conditions such as temperature. However, it does decrease with time (see next page).

The decay law

Unstable nuclei disintegrate spontaneously and at random. However, the more undecayed nuclei there are, the more frequently disintegrations are likely to occur. For any particular radioactive nuclide, on average

activity ∝ number of undecayed nuclei

If N is the number of undecayed atoms after time t, then the activity is the rate of change of N with t. So, in calculus notation, the activity is $-dN/dt$. (The minus sign indicates that a *decrease* in N gives *positive* activity.)

With a suitable constant, λ, the above proportion can be rewritten as an equation:

$$-\frac{dN}{dt} = \lambda N \qquad (1)$$

λ is called the ***radioactive decay constant***. Each radioactive nuclide has its own characteristic value. (Note that the symbol λ is also used for wavelength.)

By applying calculus to equation (1), a link between N and t can be obtained:

$$N = N_0 e^{-\lambda t} \qquad (e = 2.718) \qquad (2)$$

where N_0 is the initial number of undecayed nuclei.

This is known as the ***radioactive decay law***.

A graph of N against t has the form shown above right. The graph is an *exponential* curve (see also E2).

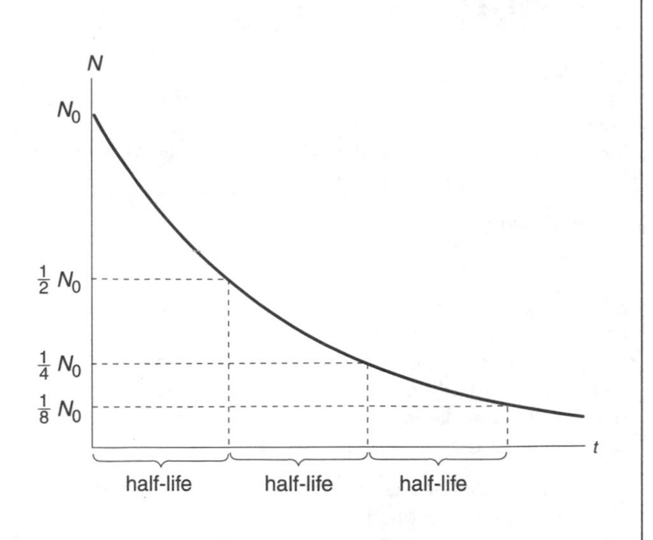

Half-life

There are two alternative definitions for this term.

The ***half-life*** of a radioactive nuclide is:

A the average time taken for the number of undecayed nuclei to halve in value,

B the average time taken for the activity to halve.

Version A is illustrated in the graph above. One feature of the exponential curve is that the half-life is the same from whichever point you start.

In equation (2), the half life, $t_{\frac{1}{2}}$, is the value of t for which $N = N_0/2$. Substituting this in the equation, taking logs, and rearranging gives

$$t_{\frac{1}{2}} = \frac{0.693}{\lambda} \qquad (0.693 = \ln 2)$$

Combining equations (1) and (2) gives $-dN/dt = \lambda N_0 e^{-\lambda t}$. So the activity ($-dN/dt$) also decreases exponentially with time, and a graph of *activity* against *time* has the same general form as the graph above. Version B of the half-life definition follows from this.

Half-lives for some nuclides

potassium-40	1.3×10^9 years
plutonium-239	24 400 years
carbon-14	5730 years
strontium-90	28 years
magnesium-28	21 hours
radon-224	55 seconds

Uses of radioisotopes

Elements are a mixture of isotopes. The radioactive ones are called ***radioisotopes***. Some can be produced artificially by transmutations in a nuclear reactor. They have many uses.

Tracers Radioisotopes can be detected in very small (and safe) quantities. This means that they can be used as tracers – their movements can be tracked. Examples include:

- tracking a plant's uptake of fertilizer from roots to leaves by adding a tracer to the soil water,
- detecting leaks in underground pipes by adding a tracer to the fluid in the pipe.

Testing for cracks γ rays have the same properties as short-wavelength X-rays, so they can be used to photograph metals to reveal cracks. A γ source is compact and does not need an electrical power source like an X-ray tube.

Cancer treatment γ rays can penetrate deep into the body and kill living cells. So a highly concentrated beam from a cobalt-60 source can be used to kill cancer cells in a tumour. Treatment like this is called ***radiotherapy***.

Carbon dating Living organisms are partly made from carbon which is recycled through their bodies and the atmosphere as they obtain food and respire. A tiny proportion is radioactive carbon-14 (half-life 5730 years). This is continually forming in the upper atmosphere as nitrogen-14 is bombarded by cosmic radiation. When an organism dies, no new carbon is taken in, so the proportion of carbon-14 is gradually reduced by radioactive decay. By measuring the activity, the age of the remains can be estimated to within 100 years. This method can be used to date organic materials such as wood and cloth.

Dating rocks When rocks are formed, some radioisotopes become trapped. As decay continues, the proportion of radioisotope (e.g. potassium-40) decreases, while that of its decay product (e.g. argon-40) increases. The age of the rock can be estimated from the proportions.

Smoke detectors These contain a tiny α source which ionizes the air in a small chamber so that it conducts a current. Smoke particles entering the chamber attract ions and reduce the current. This is sensed by a circuit which triggers the alarm.

G3 Nuclear energy

Energy and mass

One conclusion from Einstein's ***theory of relativity*** is that energy has mass. If an object gains energy, it gains mass. If it loses energy, it loses mass. The change of energy ΔE is linked to the change of mass Δm by this equation:

$$\Delta E = \Delta mc^2$$

where c is the speed of light: 3×10^8 m s^{-1}

c^2 is so high that energy gained or lost by everyday objects produces no detectable change in their mass. However, the energy changes in nuclear reactions produce mass changes which are measurable. For example, when a fast α particle is stopped, its mass decreases by about 0.2%. The mass of an object when it is at rest is called its ***rest mass***.

With nuclear particles, energy is often measured in MeV (the electronvolt, eV, is defined in E8):

$$1 \text{ MeV} = 1.60 \times 10^{-13} \text{ J}$$

From data on mass changes, scientists can calculate the energy changes taking place. With nuclear particles, mass is usually measured in u (see G1). By converting 1 u into kg and applying $\Delta E = \Delta mc^2$, it is possible to show that

1 u	is equivalent to	931 MeV
change in mass		change in energy

Mass defect

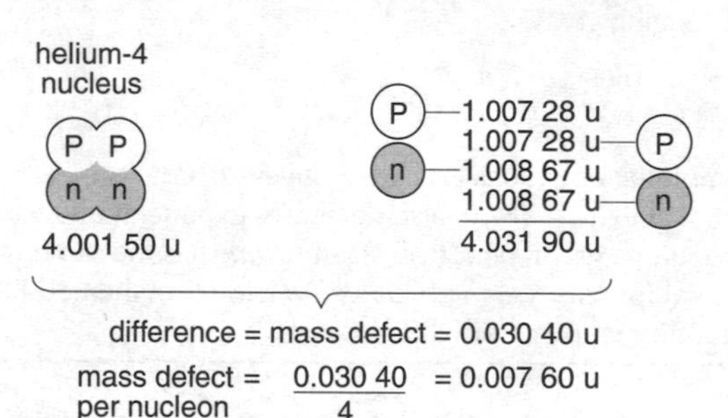

A helium-4 nucleus is made up of 4 nucleons (2 protons and 2 neutrons). The calculation above shows that the nucleus has less mass than its four nucleons would have as free particles. The nucleus has a ***mass defect*** of 0.030 40 u.

The reason for the mass defect is as follows. In the nucleus, the nucleons are bound together by a strong nuclear force. As work must be done to separate them, they must have less potential energy when bound than they would have as free particles. Therefore, they must have less mass.

All nuclides have a mass defect (apart from hydrogen-1 whose nucleus is a single proton). For example:

	Mass defect	Mass defect per nucleon
hydrogen-2	0.002 40 u	0.001 20 u
iron-56	0.528 75 u	0.009 44 u
lead-208	1.757 84 u	0.008 45 u
uranium-238	1.935 38 u	0.008 13 u

Binding energy

The ***binding energy*** of a nucleus is the energy equivalent of its mass defect. So it is the energy needed to split the nucleus into separate nucleons. For example, a helium-4 nucleus has a mass defect of 0.030 40 u. As 1 u is equivalent to 931 MeV, 0.030 40 u is equivalent to 28.3 MeV. So the binding energy of the nucleus is 28.3 MeV.

Note:

- The term 'binding energy' is rather misleading. 'Unbinding energy' would be better. 28.3 MeV is the energy needed to 'unbind' the nucleons in helium-4.

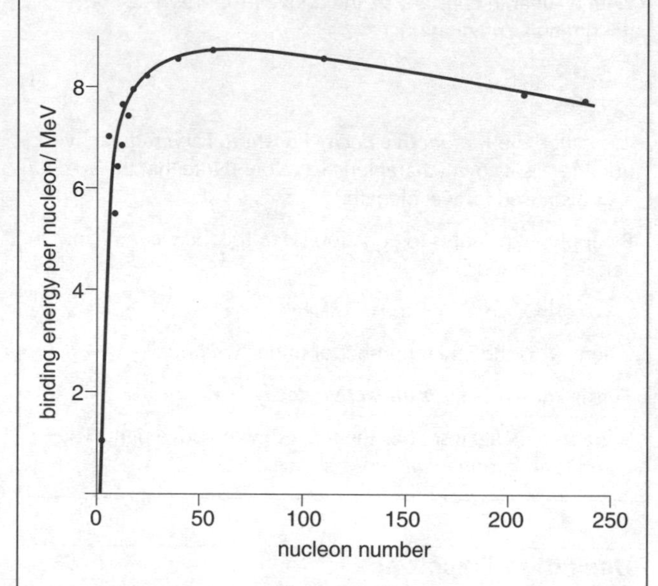

The stability of a nucleus depends on the ***binding energy per nucleon***. The graph above shows how this varies with nucleon number. The line gives the general trend; points for some individual nuclides have also been included.

Note:

- Nuclei near the 'hump' of the graph are the most stable, because they need most 'unbinding energy' per nucleon.
- A graph of *mass defect* against *nucleon number* has the same general form as the graph above.

If nucleons become rearranged so that they have a *higher* binding energy per nucleon, there is an *output* of energy.

Radioactive decay Unstable nuclei decay to form more more stable products, so energy is released. In α decay, for example, this is mostly as the kinetic energy of an α particle. When the α particle collides with atoms, it loses KE and they speed up. So radioactive decay produces heat.

Nuclear reactions The ***fission*** and ***fusion*** reactions on the next page give out energy. During fission, heavy nuclei split to form nuclei nearer the 'hump' of the graph. During fusion, light, nuclei *fuse* (join) to form heavier ones.

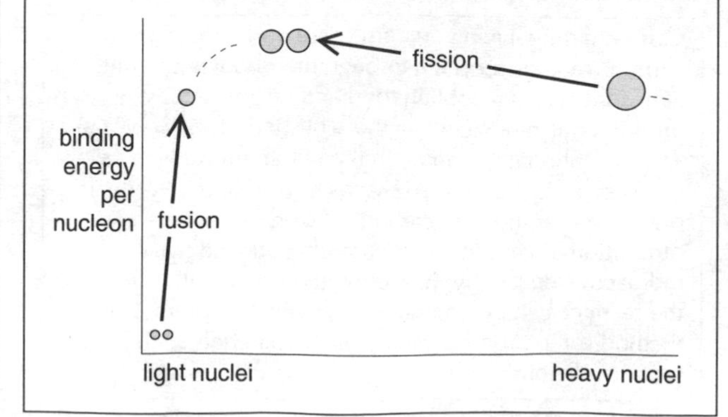

Nuclear fission

During ***nuclear fission***, a heavy nucleus (e.g. of uranium or plutonium) splits to form two nuclei of roughly the same mass, plus several neutrons. Rarely, fission happens spontaneously. More usually, it occurs when a neutron hits and is captured by the nucleus. For example, here is a typical fission reaction for uranium-235:

$$^{235}_{92}U + ^{1}_{0}n \rightarrow ^{144}_{56}Ba + ^{90}_{36}Kr + 2^{1}_{0}n$$

The reaction releases energy, mostly as KE of the heavier decay products (see also B7). So fission is a source of heat.

Note:
- The energy released per atom by fission (about 200 MeV) is about 50 million times greater than that per atom from a chemical reaction such as burning.

Chain reaction The fission reaction above is started by one neutron. It gives off neutrons which may cause further fission and so on in a ***chain reaction***:

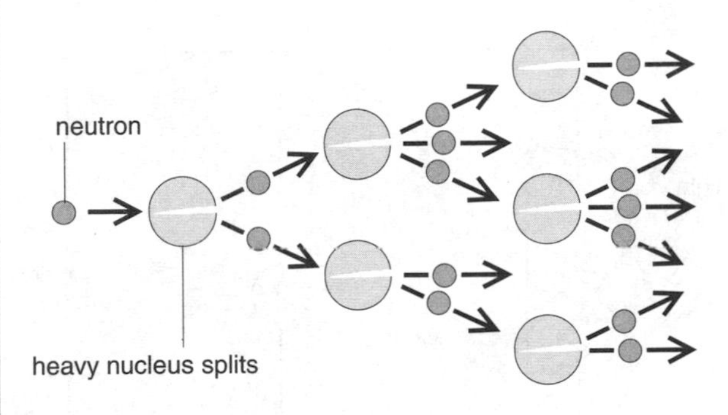

Uncontrolled chain reactions are used in nuclear weapons. Controlled chain reactions take place in ***nuclear reactors*** (see right) and release energy at a steady rate. The most commonly used fissionable material is uranium-235.

To maintain a chain reaction, a minimum of one neutron from each fission must cause further fission. However, to achieve this, these problems must be overcome:
- If the fission material is less than a certain ***critical size***, too many neutrons escape without hitting nuclei.
- The fission of uranium-235 produces medium-speed neutrons. But slow neutrons are better at causing fission.
- Less than 1% of natural uranium is uranium-235. Over 99% is uranium-238, which absorbs medium-speed neutrons without fission taking place.

Thermal reactors

In a nuclear power station, the heat source is usually a ***thermal reactor***. (Otherwise, the layout is as for a fuel-burning station – see D4.) In the reactor, there is a steady release of heat as fission of uranium-235 takes place. It is known as a *thermal* reactor because the neutrons are slowed to speeds associated with thermal motion.

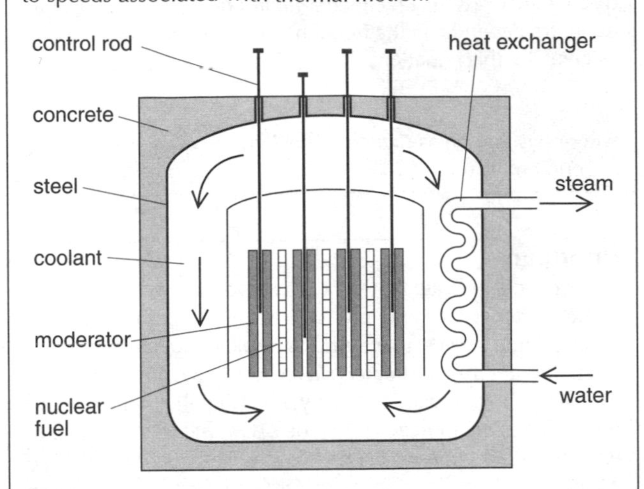

Nuclear fuel This is uranium dioxide in which the natural uranium has been enriched with extra uranium-235. 1 kg of this fuel gives as much energy as about 25 tonnes of coal.

Moderator This is a material which slows down the medium-speed neutrons produced by fission. Some reactors use graphite as a moderator. Others use water.

Control rods These are raised or lowered to control the rate of fission. They contain boron, which absorbs neutrons.

Coolant (e.g. water or carbon dioxide gas) This carries heat from the reactor to the heat exchanger.

Note:
- Many useful radioisotopes are made by bombarding stable isotopes with neutrons in the core of a reactor.

Safety issues Nuclear radiation can damage or kill living cells, so reactors have thick concrete shielding to absorb it. However, any radioactive gas or dust which escapes into the atmosphere is especially dangerous because it may be taken into the bodies of living things via food or water. 'Spent' fuel from reactors contains highly active decay products. Not all have short half-lives, and some will require safe, sealed storage for thousands of years.

Nuclear fusion

Reactors using ***nuclear fusion*** are many years away. Current research is based on the fusion of hydrogen-2 (called ***deuterium***) and hydrogen-3 (called ***tritium***):

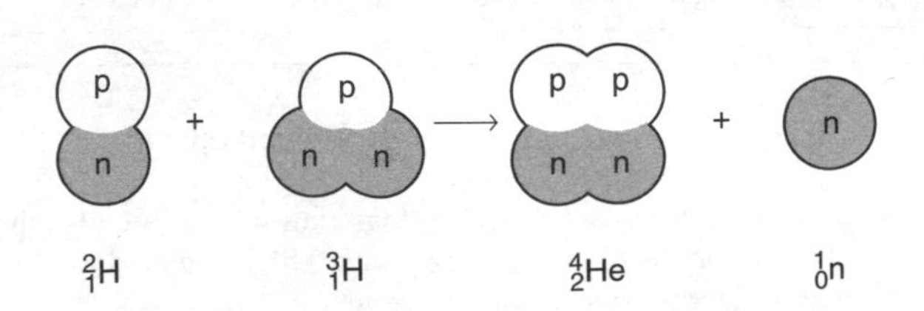

Although the energy release per fusion is less than 10% of that from a fission reaction, fusion is the better energy source if the processes are compared *per kg* of material.

Fusion is much more difficult to achieve than fission because the hydrogen nuclei repel each other.
- For the nuclei to collide at a high enough speed for fusion, the gas has to be heated to 10^8 K or more.
- No ordinary container can hold such a hot material and keep it compressed. Scientists are experimenting with magnetic fields to trap the nuclei.

The advantages of a fusion reactor will be:
- Fuels will be readily obtainable. For example, deuterium can be extracted from sea-water.
- The main waste product, helium, is not radioactive.
- Fusion reactors have built-in safety. If the system fails, fusion stops.

The Sun gets its energy from the fusion of hydrogen, though using a different reaction from that on the left. Its huge size and gravity maintain the conditions needed.

G4 Quantum theory

Quantum energy

To explain certain features of thermal radiation (see F7), Planck (in 1900) put forward the theory that energy cannot be divided into smaller and smaller amounts. It is only emitted in discrete 'packets', each called a ***quantum***. The energy E of a quantum depends on the frequency f of the radiating source, as given by this equation:

$$E = hf$$

where h is known as ***Planck's constant***. Its value, found by experiment, is 6.63×10^{-34} J s.

For electromagnetic radiation, $c = f\lambda$ (see C2), so the equation on the left can be rewritten as $E = hc/\lambda$, where c is the speed of light and λ the wavelength.

Note:
- The shorter the wavelength (and therefore the higher the frequency), the greater the energy of each quantum.
- A quantum is an extremely small amount of energy.

quantum of red light: energy = 2 eV
quantum of violet light: energy = 4 eV
(1 eV = 1.60×10^{-19} J – see E8)

Photons

Some effects indicate that light is a wave motion. Examples include interference and diffraction (see C3). But there are others which suggest that light has particle-like properties. These include the ***photoelectric effect*** below. Einstein (in 1905) was able to explain this by assuming that light (or other electromagnetic radiation) is made up of 'packets' of wave energy, called ***photons***. Each photon is one quantum of energy.

The photoelectric effect

When some substances are illuminated by light (or shorter wavelengths), electrons are emitted from their surface. This is called the ***photoelectric effect***. The electrons are emitted with a range of kinetic energies, up to a maximum.

Experiments show that:
- Increasing the intensity of the light increases the number of electrons emitted per second.
- For light beneath a certain ***threshold frequency***, f_0, no electrons are emitted, even in very intense light.
- Above f_0, the maximum KE of the electrons increases with frequency, but is not affected by intensity. Even very dim light gives some electrons with high KE.

The wave theory cannot explain the threshold frequency, or how low-amplitude waves can cause high-KE electrons.

Einstein's quantum explanation Each photon delivers a quantum of energy, hf, which is absorbed by an electron. Energy Φ is needed to free the electron from the surface. If hf is more than this, the remainder is available to the electron as KE (though most electrons lose some KE before emission because they interact with other atoms). So

hf	=	Φ	+	$\frac{1}{2}m_e v_{max}^2$	(1)
energy delivered by photon		energy needed to free electron from surface		KE of electron (with no further energy losses)	

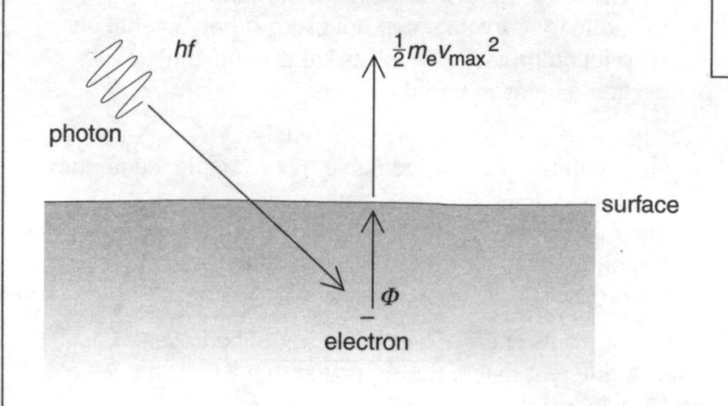

Investigating the photoelectric effect

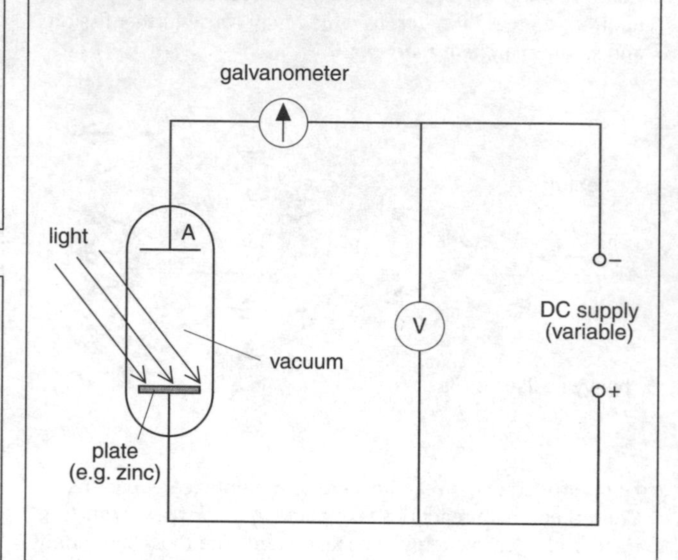

The principle of an experiment to investigate the photo electric effect is shown above. The material being investigated (e.g. zinc) is illuminated with light of known frequency f. Emitted electrons reach plate A, so the galvanometer detects a current in the circuit. The maximum KE of the emitted electrons is found by applying just enough *opposing* voltage, V_s, to *stop* them reaching A, so that the galvanometer reading falls to zero.

V_s is called the ***stopping voltage***. At this voltage

$$eV_s = \tfrac{1}{2}m_e v_{max}^2 \quad \text{(see E8)}$$

So, if equation (2) below is correct,

$$eV_s = hf - hf_0$$

Therefore, if V_s is measured for light of different frequencies, a graph of V_s against f should be of the form shown above.

Note:
- The number of electrons emitted is proportional to the number of photons absorbed.
- Φ is called the ***work function***. Materials with a low Φ emit electrons in visible light. Those with a higher Φ require the higher-energy photons of ultraviolet.
- If $hf < \Phi$, no electrons are emitted.
- The energy of a photon at the threshold frequency = hf_0 = Φ. So, equation (1) can be rearranged and rewritten:

$$\tfrac{1}{2}m_e v_{max}^2 = hf - hf_0 \quad (2)$$

Spectral lines

A ***spectrum*** (see C1) contains a mixture of wavelengths, but not always in a continuous range. For example, if there is an electric discharge through hydrogen at low pressure, the gas emits particular wavelengths only, so the spectrum is made up of lines (visible colours, ultraviolet and infrared), some of which are shown below:

Hydrogen

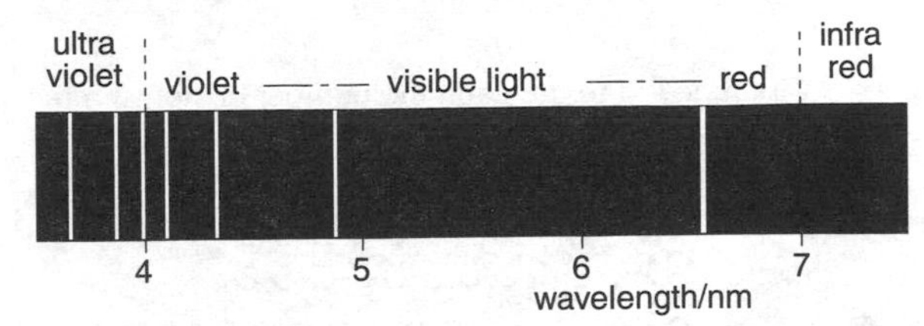

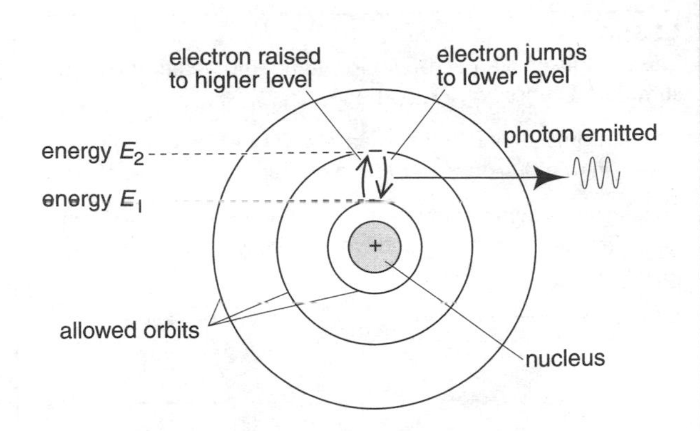

Bohr's quantum explanation (1913) In an atom, the electrons can move around the nucleus in certain allowed orbits only (top right). An electron has a different amount of energy (KE + PE) in each orbit. It may be raised to a higher orbit, for example, by colliding with an electron from another atom. When it jumps back to a lower orbit, it loses energy ($E_2 - E_1$) which is emitted as a photon. So

$$hf = E_2 - E_1$$

An electron jump is called a ***transition***.

- The greater the energy change ($E_2 - E_1$) of the transition, the higher the frequency f of the photon.
- Each possible transition gives a different spectral line.

Bohr's allowed-orbit analysis only works for the simplest atom, hydrogen. It has now been replaced by a mathematical, ***wave mechanics*** model of the atom in which allowed orbits are replaced by allowed ***energy levels***. However, the above equation still applies.

A line spectrum is a feature of any gas in which individual atoms do not interact. If atoms exert forces on each other, many more energy levels are created. Tightly-packed atoms or molecules which are vibrating, rotating, or colliding with each other have so many possible energy states that the spectrum is a continuous range of colours.

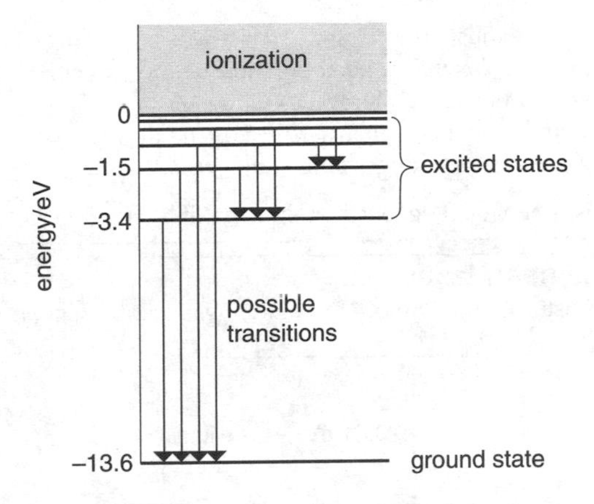

The main energy levels and transitions for hydrogen (with isolated atoms) are shown above.

Note:

- If an atom is in its ***ground state***, no electron has an unoccupied energy level beneath it.
- If an atom is in an ***excited state***, an electron has been raised to a higher energy level, so there is an unoccupied level beneath it.
- If an atom is in an ***ionized state***, an electron has been raised above the highest energy level (i.e. it has escaped). From the energy scale on the above chart, the minimum energy required to ionize a hydrogen atom is 13.6 eV.

Emission and absorption spectra

If light is radiated directly from its source, its spectrum is called an ***emission spectrum***. Examples include the line spectrum above and the continuous spectrum of the Sun.

The Sun's emission spectrum is crossed by many faint, dark lines. These are an ***absorption spectrum***. They occur because some wavelengths emitted by the Sun's core are absorbed by cooler gases (e.g. hydrogen) in its outer layers.

Some of the lines in the absorption spectrum of hydrogen are shown below. When the Sun's radiation passes through the gas, the atoms *absorb* photons whose energies match those in their emission spectrum. They then re-emit photons of these energies, but in all directions, so the intensity in the forward direction is reduced for those wavelengths.

Hydrogen (absorption)

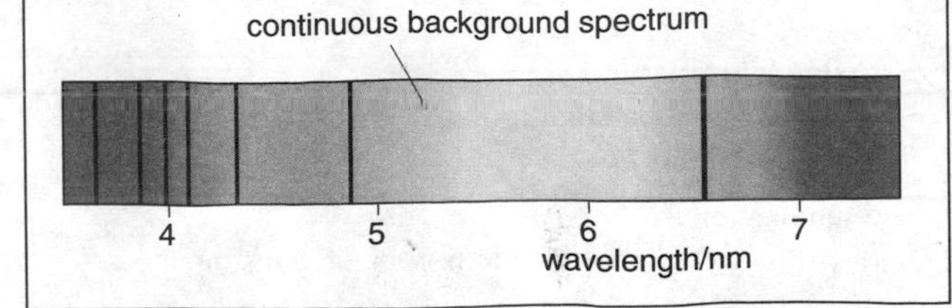

Wave–particle duality

Light waves have particle-like properties. De Broglie (in 1922) suggested that the converse might also be true: matter particles, such as electrons, might have wave-like properties. There might be ***wave–particle duality***.

According to de Broglie, if a particle of momentum p is associated with a ***matter wave*** of wavelength λ, then

$$\lambda = \frac{h}{p}$$

If a beam of electrons is passed through a thin layer of graphite, the electrons form a diffraction pattern. This suggests that the rows of atoms are acting rather like a diffraction grating (see C3). Measurements indicate that the electron wavelength is 10^{-10} m, as predicted by the de Broglie equation. This is much shorter than light wavelengths. For more on ***electron diffraction***, see G5.

G5 Applications of quantum theory

Lasers

The term ***laser*** is an acronym for '**l**ight **a**mplification by the **s**timulated **e**mission of **r**adiation'.

When atoms are excited there are normally fewer excited atoms than atoms in lower energy states.

The diagrams below show the relative numbers of excited and unexcited atoms normally and when there is population inversion.

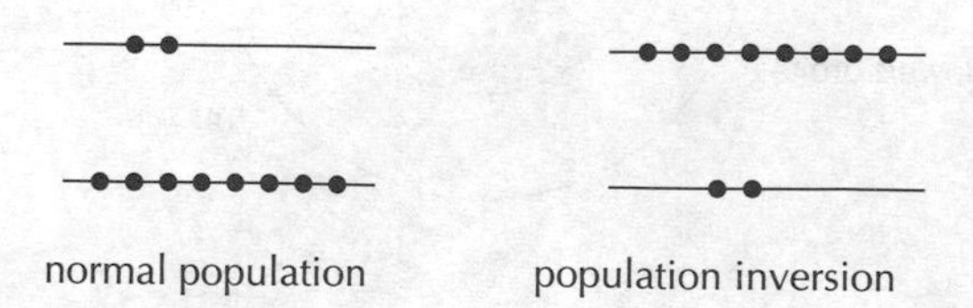

In a laser there must be

- a ***population inversion***; i.e. there must be more excited atoms than unexcited atoms,
- an excited state that is ***metastable***; i.e. one in which the atoms remain for a longer time than is usual.

How this is achieved depends on the type of laser.

Helium–neon laser

This consists of a mixture of 15% helium gas and 85% neon.

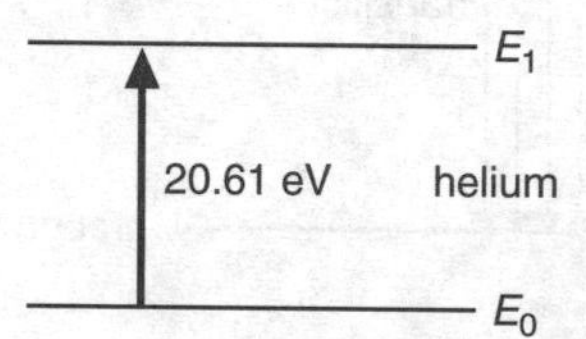

Some helium atoms are excited into the metastable state E_1 by electrical discharge through the gas.

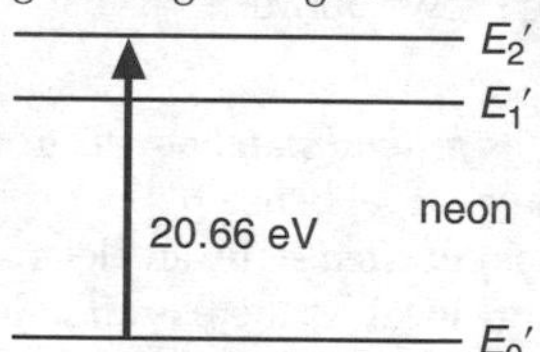

Helium atoms collide with neon atoms, exciting them to the metastable state E_2'. The helium provides this energy by dropping into the ground state (providing 20.61 eV) and by losing kinetic energy (0.05 eV).
There is population inversion between this and the E_2' level.

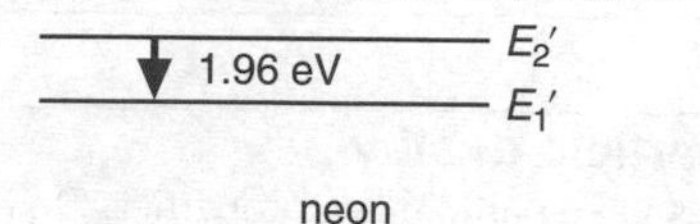

When some of the excited atoms fall into the E_1' level the photons collide with other excited atoms and stimulate them to fall. The emitted photons cause further atoms to emit photons. These reflect back and forth from mirrors in the laser tube as shown above right. The result is an intense light in which all the photons have the same frequency and phase and move in the same direction.

Laser safety

Because of their intense localized energy great care has to be take to avoid injury when using lasers. Even with low-energy beams the eyes are particularly vulnerable. It is necessary to avoid inadvertently looking at a beam reflected from a mirror or other good reflector. The use of special spectacles is advisable.

Schematic representation of laser operation

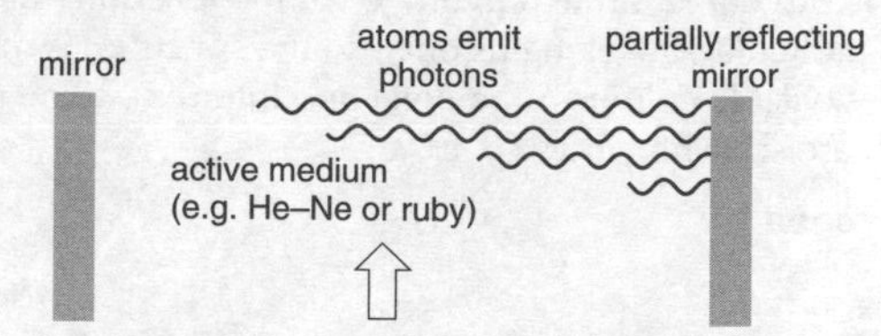

Energy is pumped in to maintain the number of atoms in the excited metastable state.

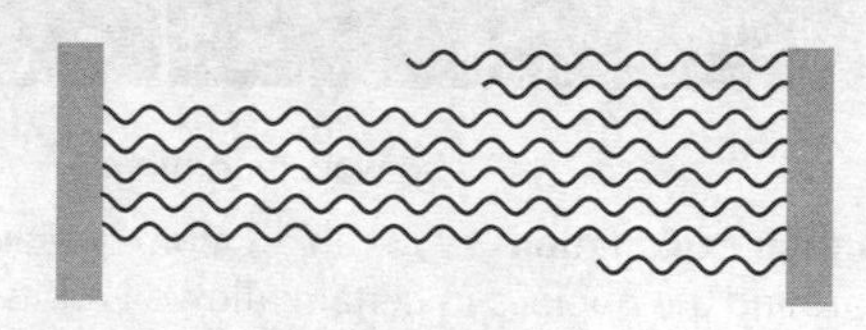

Photons bounce back and forth, increasing in number by stimulating emission of further photons.

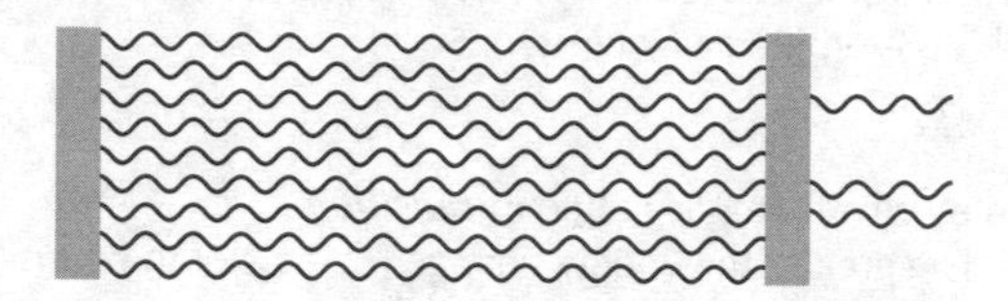

Some (about 1%) of the photons pass through the partial reflecting mirror. Energy is continually pumped in and the laser light can be emitted continuously.

Use of lasers in CDs

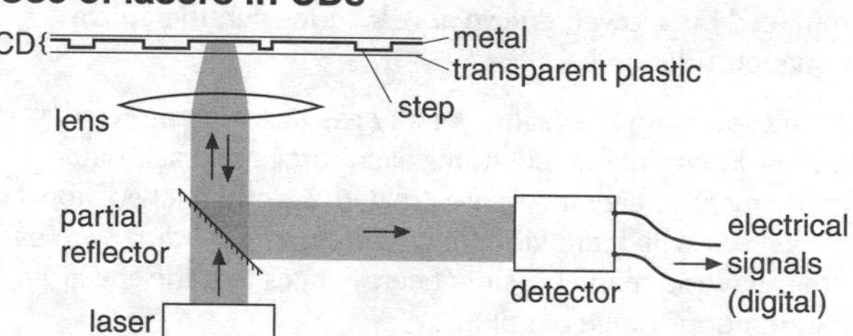

Within a CD, there is a metal layer with a spiral track of tiny steps (bumps) on it. These and the spaces between represent the 0s and 1s of the digitized signal. Light from a laser is focused onto the track and reflected. But where there is a step, the reflected light is cancelled because of interference effects (see C3). The result is a series of light pulses which the detector converts into electrical signals for processing.

Other uses of lasers

The laser beam is a very narrow beam of intense energy and as a result it has many and varied uses. The following is a list of some of the uses of lasers:

In medicine

- to destroy tissue in a localized area
- to break up kidney stones
- to repair broken tissue (e.g. detached retina)
- to restore sight impaired by a cateract
- to remove decay in teeth (white teeth reflect the laser energy but darker decayed areas do not)

In industry

- to drill fine holes in hard material
- to produce very accurate surveys
- to produce holograms

In communications

- to produce the light beam in fibre optic transmission

Electron diffraction

Electrons have wave-like properties. This is an example of ***wave–particle duality***.

According to de Broglie's equation, electrons with a momentum p (= mv) have a wavelength λ associated with them given by

$$\lambda = \frac{h}{p}$$ where h is the Planck constant (6.63×10^{-34} J s)

The interpretation of the ***wave amplitude*** at any point is that it is related to the ***probability*** of finding the particle at that point. The greater the amplitude the more chance there is of detecting the particle at the point.

In an electron diffraction tube electrons are accelerated and hit a thin film of graphite. The graphite behaves in a similar way to a diffraction grating. The waves are diffracted by the graphite and produce an interference pattern. The pattern observed with an electron diffraction tube is two bright concentric rings. These are produced by two different spacings of atomic layers in the graphite structure.

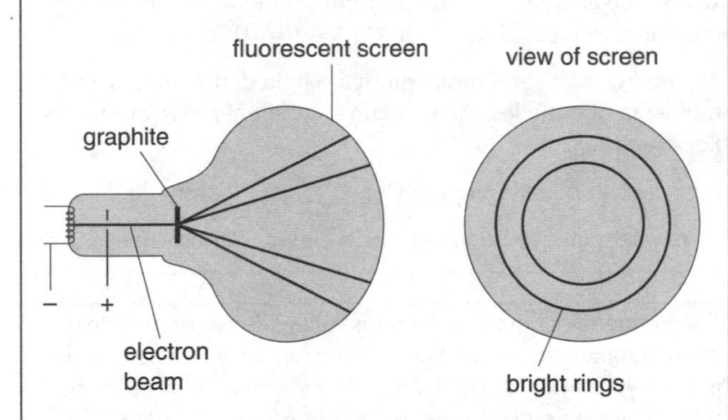

At a point where the electron waves interfere constructively the amplitude is a maximum. This means that there is a high probability of an electron arriving at that point. Lots of electrons arrive there and the screen coated with a phosphor (such as zinc sulphide) glows brightly.
At a point where the amplitude is zero, there is no chance of an electron arriving there so no light is emitted.

Typical de Broglie wavelengths of electrons

An electron accelerated through 2000 V has an associated wavelength of 2.7×10^{-11} m and a kinetic energy of 3.2×10^{-16} J.

The short wavelength is the reason why electron diffraction cannot be observed using ordinary diffraction gratings. The spacing of the 'slits' needs to be very small. Because atoms have diameters of the order of 10^{-10} m the layers of atoms in the graphite behave like a diffraction grating with a small enough spacing.

Electron microscope

The short wavelength of matter waves associated with electrons is put to use in electron microscopes. Smaller wavelengths result in better resolution (see C5). This means that greater magnification can be achieved and therefore more detail is seen in small objects.

Electron stationary waves in atoms

Matter waves can set up stationary waves in a similar way to those set up in a stretched string. This formed the basis of Erwin Schrödinger's use of wave mechanics to predict the energy levels in a hydrogen atom.

This is a simple treatment that outlines the principle of the process in one dimension as follows. It treats the electron as if it were trapped in a box of size equal to the atomic radius.

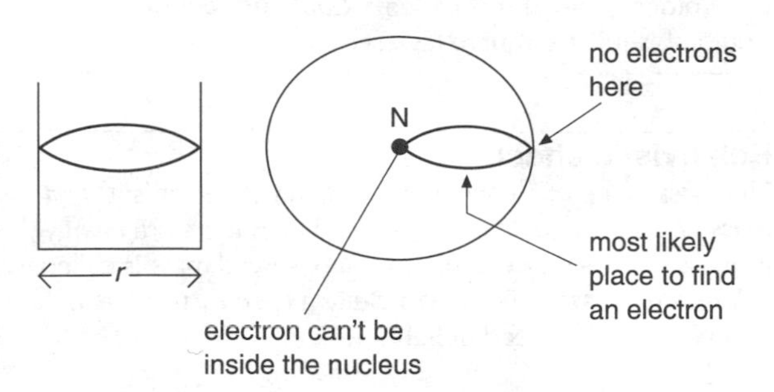

Consider the following 'facts':

- The electron in an atom cannot be in the nucleus or outside the edge so at the nucleus and the edge the amplitude of the wave is zero.
- The electron wave is confined between these points, reflecting backwards and forwards to produce the standing wave.
- The nucleus and the edge of the atom are the nodes of the stationary wave.
- The distance between the nucleus and the edge of the atom is about 1×10^{-10} m.

The possible stationary waves for the electron are as follows:

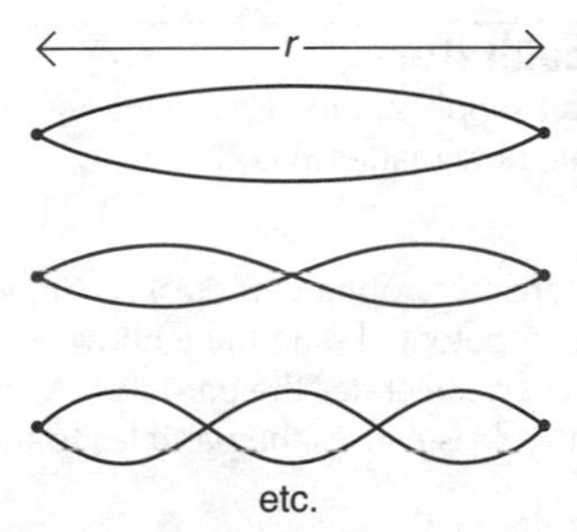

The single loop represents the ground (unexcited) state of the electron. The model suggests that the electron is most likely to be found half way between the nucleus and the edge of the atom.

Energy levels in atoms: excited states

This simple model cannot accurately predict energy levels as no account is taken of the changes in potential energy of the electron at different distances from the nucleus. However, the following shows how one can begin to see why there are discrete energies for the electron.

For one loop wavelength $= 2r = \frac{2r}{1}$

For two loops wavelength $= r = \frac{2r}{2}$

For three loops wavelength $= \frac{2r}{3}$

etc.

Generally wavelength $= \frac{2r}{n}$

where n is the number of loops in the standing wave.

The possible momenta of the electron $= \frac{nh}{2r}\left(\text{i.e. } \frac{h}{\lambda}\right)$

G6 Particle physics – 1

Probing the nucleus

To investigate the nucleus, scientists have broken it into bits using beams of high-energy particles (such as protons), from ***particle accelerators*** (see below).

Protons must be accelerated to very high speeds to penetrate the nucleus. They have to overcome the electric repulsion of the protons there. This is called a ***Coulomb repulsion*** (from Coulomb's inverse square law: see E1).

Being uncharged, neutrons can penetrate the nucleus more easily. But they cannot be directed and controlled by electric and magnetic fields.

Ordinary matter is made up of protons, neutrons, and electrons. But collision experiments with accelerators have produced hundreds of other 'elementary' particles as well. These are described in G8.

Relativistic effects

The mass of a particle when at rest to an observer is its ***rest mass***, m_0. As the particle gains speed (and therefore energy), its mass increases (see G3). Its total observed mass is called its ***relativistic mass***, m. For a particle at a speed v relative to the observer (c is the speed of light),

$$m = \frac{m_0}{\sqrt{1 - v^2/c^2}} \qquad (1)$$

At the speeds achieved in accelerators, the mass increase can be significant. For example, for an electron travelling at 90% of c, m is more than twice m_0.

Energy and mass According to Einstein (see G3), a mass m has an ***energy equivalent*** E as given by this equation:

$$E = mc^2 \qquad (2)$$

For a particle at rest, $E_0 = m_0c^2$. This is its ***rest energy***.

Energy and momentum If a particle's momentum is p, then $p = mv$. So, from equation (2), it follows that

$$p = Ev/c^2 \qquad (3)$$

Using equations (1), (2), and (3), it can be shown that

$$E^2 = m_0{}^2c^4 + p^2c^2$$

Note:

- If E is much greater than E_0, then $E \approx pc$.

Units The energies of particles from accelerators are often measured in GeV: 1 GeV = 1000 MeV = 10^9 eV.

As energy, mass, and momentum are linked, the masses and momenta of particles can be expressed in energy-related units. For example,

mass can be measured in GeV/c^2 (from equation 2)

momentum can be measured in GeV/c (from equation 3).

Particle accelerators

Accelerators can supply charged particles with the energy needed to create new matter in collisions.

Cyclotron

In a cyclotron ions (e.g. alpha particles) are injected at a point near the centre. A potential difference between the 'dee'-shaped electrodes accelerates the particles. A magnetic field vertically into the dees causes the particles to move in a circular path.

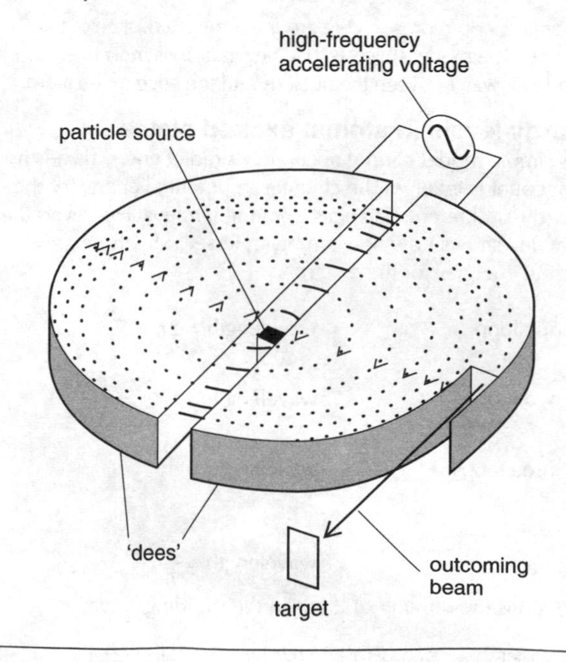

The frequency of the accelerating voltage is adjusted so that the time taken for the particle to travel in a semicircle in a dee is equal to half the period. The time for one revolution of the particle is then equal to the period of the accelerating PD. This condition results in the particle being accelerated every time it crosses from one dee to the other. After each acceleration the particle moves in an orbit of slightly larger radius. When it reaches the outer edge of the cyclotron the particle beam is extracted and used in experiments. The frequency of the orbit remains constant as the particle accelerates (see below) so that there is no need to adjust the accelerating frequency or the magnetic field.

Cyclotron frequency

Whilst travelling in the dees,

$$Bqv = mv^2/r$$

where B is the magnetic flux density, q is the charge on an ion, m is the mass of an ion, v is the velocity of an ion, and r is the radius of the path.

It follows that $v = Bqr/m$

The velocity is also given by $v = 2\pi r/T = 2\pi rf$

so that $$f = \frac{Bq}{2\pi m}$$

This is the cyclotron frequency. Notice that the frequency does not depend on r or the velocity of the particle.

Particles and antiparticles

Most types of particle have a corresponding ***antiparticle*** (see also G1). This has the same rest mass, but at least one property which is opposite to that of the particle. Here are some examples.

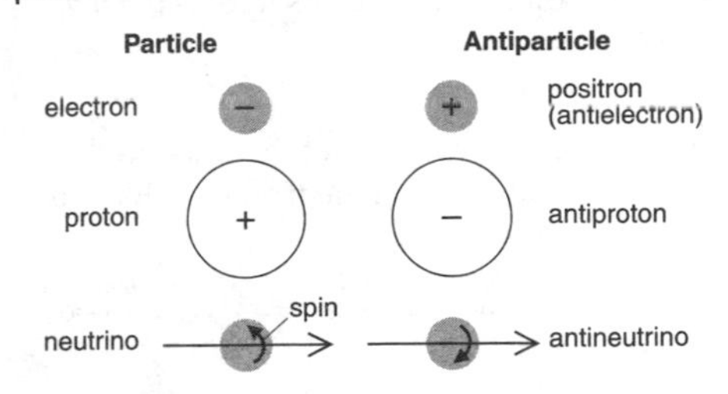

When a particle and its antiparticle meet, in most cases, they ***annihilate*** each other and their mass is converted into energy as given by $E = mc^2$. For example, the annihilation of an electron and positron may produce a pair of gamma photons.

Note:

- There are far more particles than antiparticles in the Universe, so annihilation is extremely rare.

Creating matter Energy can also be converted into mass. For example, if a gamma photon has at least 1.02 MeV of energy, it may, when passing close to a nucleus, convert into an electron-positron pair (total rest mass 1.02 MeV/c^2). In high-energy collisions, heavier particles (and antiparticles) may materialize from the energy supplied.

Linear accelerator (up to 20 GeV)

Charged particles (e.g. electrons or protons) in a vacuum pipe are accelerated through a series of electrodes by an alternating voltage. The frequency is carefully chosen so that, as each electrode goes alternately + and –, particles leaving one electrode are always pulled towards the next. The beam of particles is directed at a target or into a ***synchrotron***.

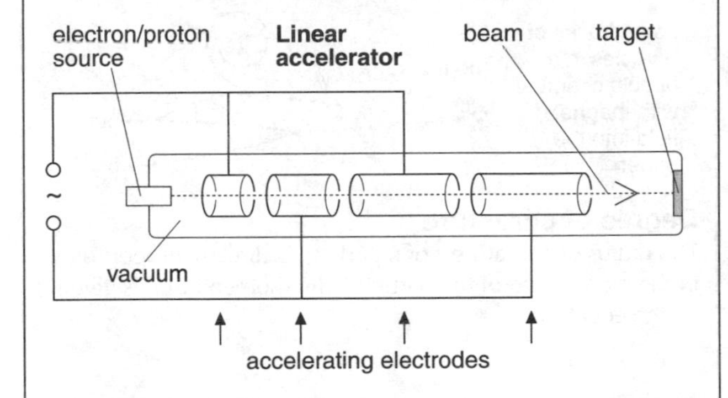

Synchrotron (1000 GeV or more)

In effect, this is a linear accelerator, bent into a ring so that the charged particles can be given more energy each time they go round. Electromagnets keep the particles in a curved path. As the speed increases, the magnetic field strength is increased to compensate for the extra mass.

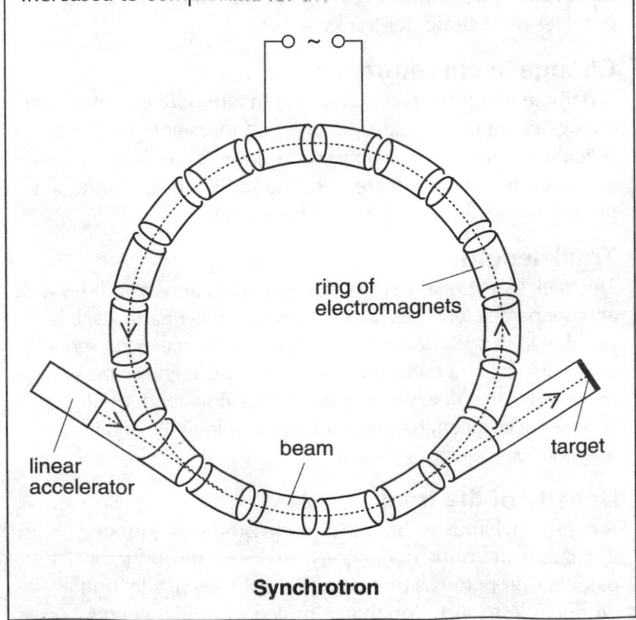

Synchrotron

Detectors

In experiments using accelerators, detectors are needed to reveal the paths of the particles produced.

Bubble chamber This is filled with liquid hydrogen whose pressure is suddenly reduced so that it is ready to vaporize. Charged particles entering the chamber ionize the hydrogen. This triggers vaporization, so that a trail of bubbles is formed along the track of each particle.

Drift chamber This is a gas-filled chamber containing, typically, thousands of parallel wires. Incoming particles cause a trail of ionization in the gas. Their track is worked out electronically by timing how long it takes ionization electrons to drift to the nearest sense wires. A computer processes the signals and displays the tracks graphically.

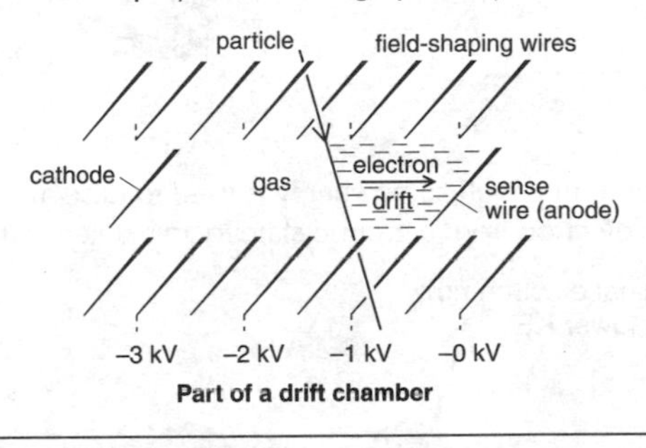

Part of a drift chamber

Colliding beams

One way of conducting nuclear experiments is to allow an accelerated beam to collide with a stationary target.

To obtain the higher-energy collisions that are needed to explore the structure of matter in greater detail two beams of particles are made to collide head on.
This doubles the energy involved in the collision.

The accelerator first accelerates a group of one type of particle (e.g. electrons) to a high energy. This is then stored in a storage ring while a group of particles of another type (e.g. positrons) is accelerated. The beams are then steered so that they interact.

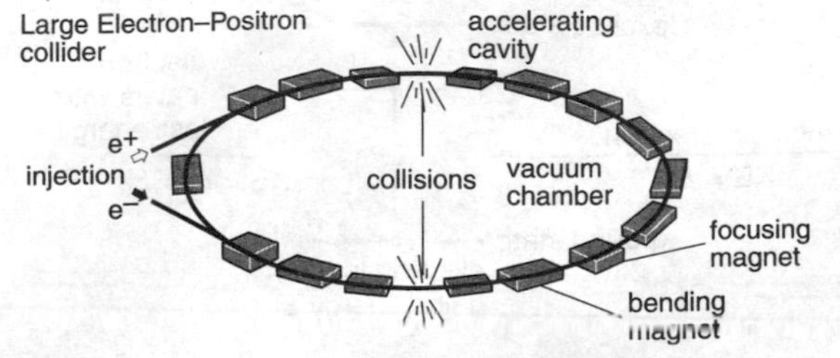

At CERN the energy of each beam is about 90 GeV giving total collision energy of 180 GeV. This enables the production of new particles as the energy appears as mass (from $E = mc^2$).

G7 Particle physics – 2

Electron collisions with atoms

Electrons can collide with atoms elastically or inelastically.

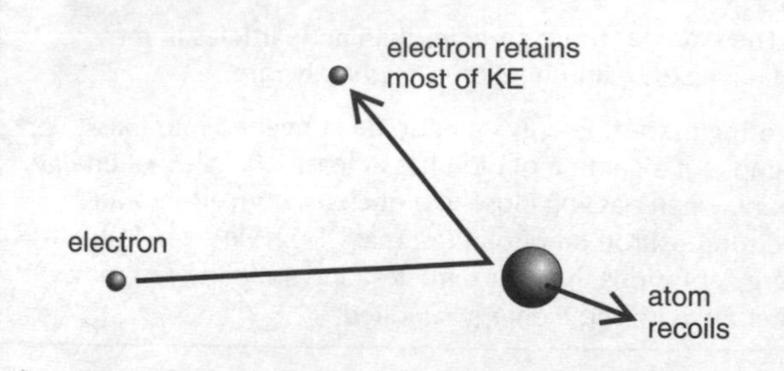

In the elastic collisions (as in all such collisions) kinetic energy is conserved. Since the electron is so much less massive than any atom the electron effectively bounces off the atom, giving up little of its energy to the recoiling atom.

Kinetic energy may be lost in inelastic collisions because of

- ionization, in which some of the energy is used to remove an electron from the atom,

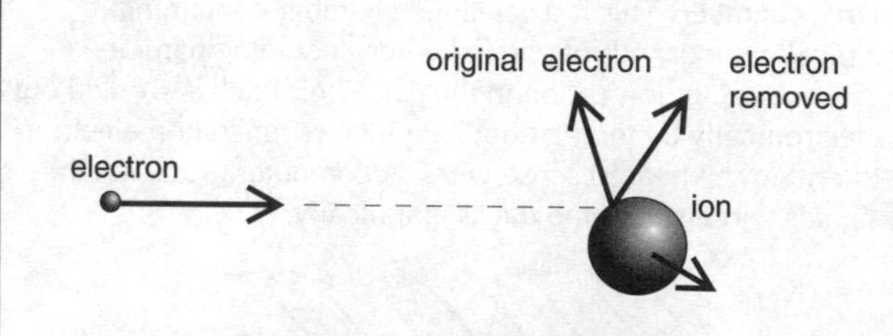

- excitation, in which some energy is used to raise an atomic electron from a ground state to an excited state.

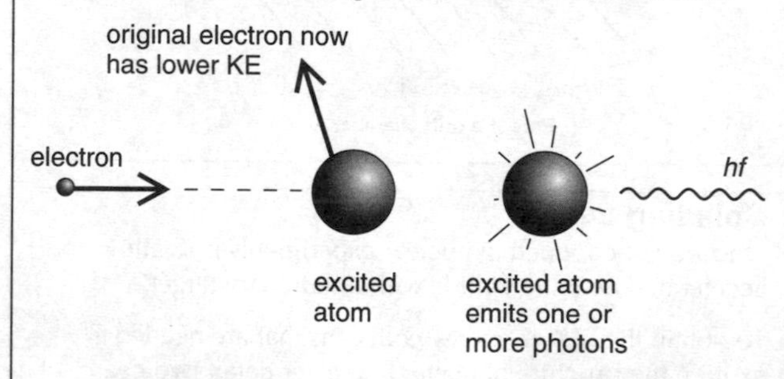

By examining the energy of an electron before and after a collision it is possible to determine the energy levels that an electron can occupy in a particular atom. The loss in energy of an electron in one collision corresponds to an energy difference between the ground state and an excited state.

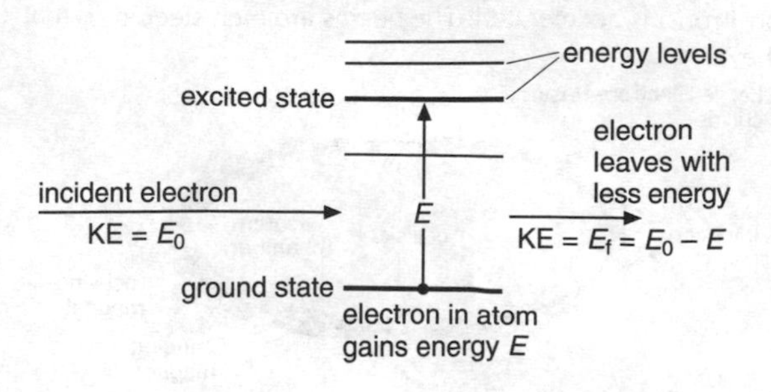

The fact that colliding electrons lose energy in well defined amounts is further evidence for the existence of electron energy levels in atoms.

Identifying particles

A particle's properties can be identified from

- the direction of curvature of its track in a magnetic field
- the radius of curvature of the track
- the change in radius of the path as it loses energy
- the length of the track
- the density of the track.

Although uncharged particles do not leave tracks their properties can be deduced from the tracks of charged particles they interact with.

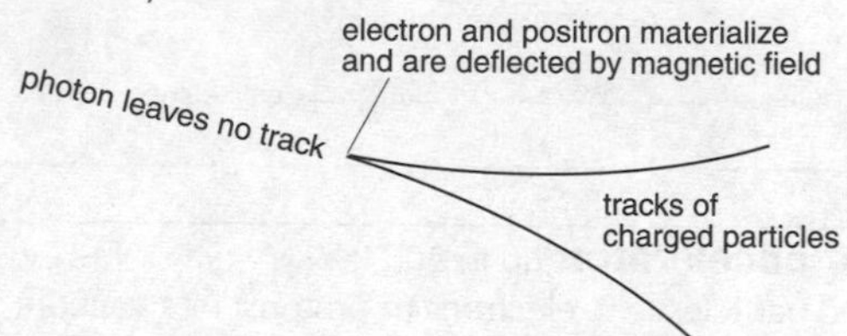

Direction of curvature

When a charged particle moves perpendicularly to a magnetic field it moves in a circular path. The direction of curvature of a track depends on the field direction and the sign of the charge on the particle and can be deduced using the 'left-hand rule'.

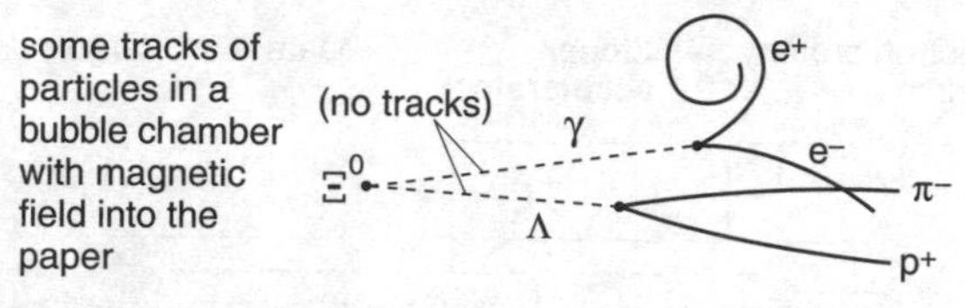

Degree of curvature

The radius of curvature r of a particle is directly proportional to the momentum of the particle. The momentum p is given by the equation

$$p = Bqr$$

where B is the magnetic flux density.

Most subatomic particles have a charge of $+1.6 \times 10^{-19}$ or -1.6×10^{-19} C so $p \propto r$.

For a given particle, kinetic energy = $p^2/2m$, so energetic particles therefore have high momentum and high radius of curvature of their tracks. This means that more energetic particles have straighter tracks.

Change in curvature

A particle that loses energy quickly in a bubble chamber will change curvature quickly. An electron loses energy very quickly and its path is seen to spiral inwards as the radius of curvature of the path decreases. The radius of curvature of a proton, however, would hardly change at all.

Track length

The track length of a particle ends either because it collides with another particle or because it decays into other particles. It is possible to deduce from conservation laws whether subsequent tracks are due to a collision or a decay. The length of the track of a particle that decays provides information about its lifetime. Longer tracks mean that the particle has a long life and is therefore more stable.

Density of the track

A dense track means that the particle produces a lot of ionization and some deductions can be made about the particle's properties. For example a fast-moving electron produces less ionization than a proton of similar energy. However, a fast-moving proton would also produce very little ionization, so although it is possible to draw some conclusions using the density of the track other evidence is needed.

The strong nuclear force

In the nucleus, the nucleons (neutrons and protons) are bound together by the strong nuclear force. The strong force:

- is strong enough to overcome the Coulomb repulsion between protons, otherwise they would fly apart,
- has a short range, $\sim 10^{-15}$m, and does not extend beyond neighbouring nucleons,
- becomes a repulsion at very short range, otherwise the nucleus would collapse.

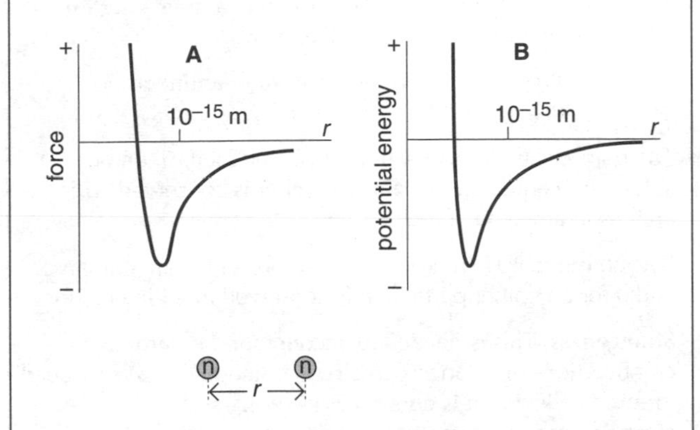

Graph A shows how the strong force varies with separation, for two neutrons. (An *attractive* force is *negative*.)

Graph B shows how the potential energy of the neutrons varies with separation. Minimum potential energy corresponds with the position of zero force in graph A.

Nuclear radii

Results of experiments show that the radius of a nucleus R is proportional to $A^{1/3}$ where A is the nucleon number.

Hence $R = R_0 A^{1/3}$

where R_0 is a constant having a value of 1.2×10^{-15} m.

Nuclear density

The mass of a nucleus $= A \times$ the mass of a nucleon
The volume of a nucleus $= \frac{4}{3}\pi(1.2 \times 10^{-15} A)^3$
The mass of a nucleon is 1.7×10^{-27} kg
The density of a nucleus is therefore 2.3×10^{17} kg m^{-3}

This is enormous when compared with the density of water (1000 kg m^{-3}). The difference is due to the fact that most of the volume of atoms is empty space. Only in neutron stars where the atoms are stripped of their electrons is it possible that such large densities exist on a large scale.

Escaping from the nucleus: α decay

The graph below shows how the potential energy of an α particle varies along a line through the centre of a nucleus. Outside the nucleus, there is Coulomb repulsion (giving positive PE). Inside the nucleus, this is overcome by the strong force (giving negative PE). The result is a potential energy 'well' with a ***Coulomb barrier*** around it. Within this well, there are different energy levels.

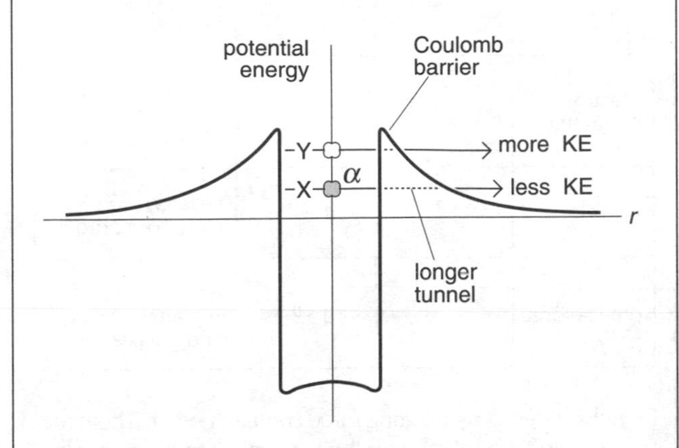

Above, an α particle has formed at X. It seems to be trapped by the Coulomb barrier. But because of the ***uncertainty principle*** (see G8) there is a chance that it may briefly 'borrow' enough energy to tunnel through the barrier and escape. This is called ***quantum mechanical tunnelling***.

An α particle at Y will escape with more KE – and is likely to escape sooner because the tunnel is shorter and easier to pass through. That is why α particles with the *highest* KEs come from nuclides with the *shortest* half-lives.

Measurements of masses in particle physics

A commonly used unit of mass in particle physics is the unit MeV/c^2 or GeV/c^2. This avoids the necessity to continually convert particle energies to mass or vice versa when considering collisions and decays.

$$1 \text{ MeV} \equiv 1.6 \times 10^{-13} \text{ J}$$
$$1.6 \times 10^{-13} \text{ J} = 1.6 \times 10^{-13}/(3.0 \times 10^{8})^2$$
$$\equiv 1.78 \times 10^{-30} \text{ kg}$$

Using this unit the mass of a proton is 960 MeV/c^2 and that of an electron is 0.51 MeV/c^2.

G8 Particle physics – 3

Fundamental forces

Force	Range/m	Relative strength	Effects, e.g.
strong	$\sim 10^{-15}$	1	Holding nucleons in nucleus
electro-magnetic	∞	$\sim 10^{-2}$	Holding electrons in atoms; holding atoms together
weak	$\sim 10^{-17}$	$\sim 10^{-5}$	β decay; decay of unstable hadrons
gravitational	∞	$\sim 10^{-39}$	Holding matter in planets, stars, and galaxies

Particles *interact* by exerting forces on each other. There are four known types of force in the Universe (see chart above). As electric and magnetic forces are closely related, they are regarded as different varieties of one force, the electromagnetic. ***Grand unified theories (GUTs)*** seek to link the strong, weak, and electromagnetic forces. Gravitational force has yet to be linked with the others. It is insignificant on an atomic scale.

Conservation laws

There are conservation laws for *momentum* and *total energy*. However, as mass and energy are equivalent, the total energy must include the rest energy (see G6).

Particles have various ***quantum numbers*** assigned to them. These are needed to represent other quantities which may be conserved during interactions. For example:

Charge In any interaction, this is conserved: it balances on both sides of the equation (see G1 for examples).

Lepton number This is +1 for a lepton, –1 for an antilepton, and 0 for any other particle. For example, a 'free' neutron decays, after about 15 minutes, like this:

neutron	→	proton	+	electron	+	antineutrino
(0)		(0)		(+1)		(–1)

The numbers (in brackets) have the same total, 0, on both sides of the equation, so lepton number is conserved. This applies in any type of interaction.

Baryon number This is +1 for a baryon, –1 for an antibaryon, and 0 for any other particle. It is conserved in all interactions.

Strangeness This is needed to account for the particular combinations of 'strange particles' (certain hadrons) produced in some collisons. It is conserved in strong and electromagnetic interactions, but not in all weak ones.

Charm relates to the likelihood of certain hadron decays.
Spin relates to a particle's angular momentum.
Topness and **bottomness** are further quantum numbers.

Classifying particles

Ordinary matter is made up of protons, neutrons, and electrons. However, in high-energy collisions, many other particles can be created. Most are very short-lived.

Matter particles can be divided into two main groups:

Hadrons (see right) These feel the strong force. They can be subdivided into ***baryons*** (which include protons, neutrons, and heavier particles that these), and ***mesons*** (which are generally lighter than protons).

Leptons (see below) These do not feel the strong force. They have no size and, in most cases, low or no mass. There are three generations of leptons, but only the first (the electron and its neutrino) occurs in ordinary matter.

Generation	Leptons spin $\frac{1}{2}$	
1	electron e^- (charge $-e$)	electron-neutrino ν_e (charge 0)
2	muon μ^- (charge $-e$)	muon-neutrino ν_μ (charge 0)
3	tau τ^- (charge $-e$)	tau-neutrino ν_τ (charge 0)

- All leptons and most hadrons have corresponding antiparticles.
- The neutrino, ν, produced by beta⁻ decay is the electron-antineutrino $\bar{\nu}_e$.

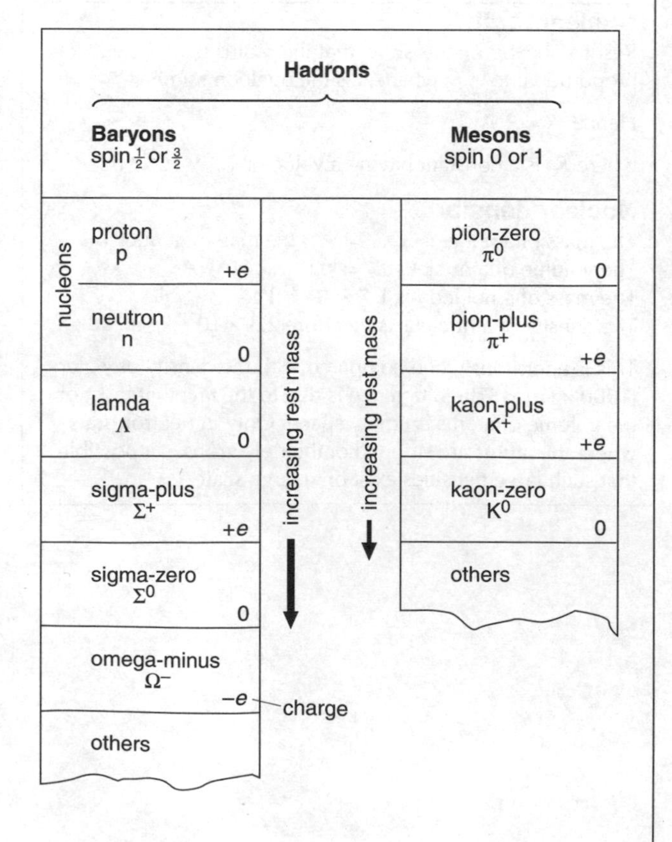

Quarks

The properties and quantum numbers of hadrons can be accounted for by assuming that each particle is a combination of others, called ***quarks***. These have a fractional charge of $+\frac{2}{3}e$ or $-\frac{1}{3}e$. ***Symmetry theory*** predicts that there should be three generations of quarks to match the three generations of leptons. (For each type of quark, there is also a corresponding antiquark.)

Generation	Quarks spin $\frac{1}{2}$		Charge	Baryon number	Strangeness	Charm	Topness	Bottomness
1	up	u	$+\frac{2}{3}e$	$\frac{1}{3}$	0	0	0	0
	down	d	$-\frac{1}{3}e$	$\frac{1}{3}$	0	0	0	0
2	strange	s	$-\frac{1}{3}e$	$\frac{1}{3}$	−1	0	0	0
	charmed	c	$+\frac{2}{3}e$	$\frac{1}{3}$	0	1	0	0
3	top	t	$+\frac{2}{3}e$	$\frac{1}{3}$	0	0	1	0
	bottom	b	$-\frac{1}{3}e$	$\frac{1}{3}$	0	0	0	−1

Note:
- Ordinary matter contains only the first generation of quarks. Very high energies are needed to make hadrons of other quark generations. These hadrons quickly decay into first generation particles.
- Individual quarks have never been detected.

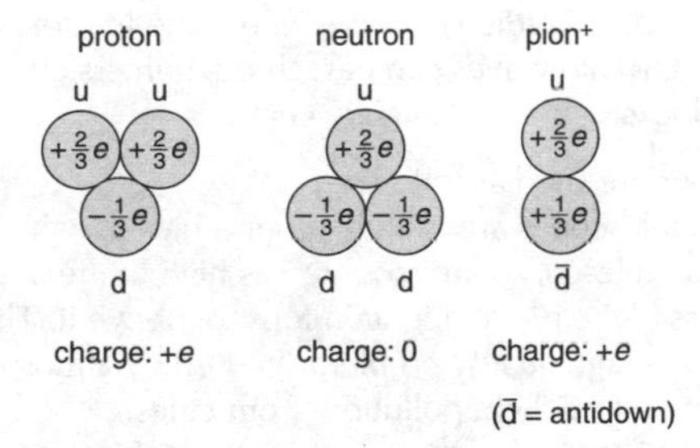

All known hadrons can be constructed from different quark (and antiquark) combinations. Examples are given above.
- Baryons are each made up of three quarks.
- Mesons are each made up of a quark and an antiquark.

Kaon

The kaon K^+ has a strangeness of +1. The kaon consists of an u quark and an antistrange particle (charge $+\frac{1}{3}e$, strangeness +1). Using the table, this gives the K^+ the following quantum numbers:

charge +1
strangeness +1

Energy and the uncertainty principle

According to the ***uncertainty principle***, a particle's momentum and position cannot both have precise values. There is a level of uncertainty about them. One consequence of this is that the law of conservation of energy can, briefly, be *violated* (disobeyed). A particle can have more energy than it 'ought' to, by an amount ΔE, provided that this is paid back in a time Δt, where $\Delta E.\Delta t \approx h$. This has important consequences for the behaviour of particles, including quantum mechanical tunnelling in G7.

Force carriers

Like other particles, nucleons need not be in contact to exert forces on each other. To explain how the strong force is 'carried' from one nucleon (e.g. neutron) to another, the idea of ***exchange particles*** is used:

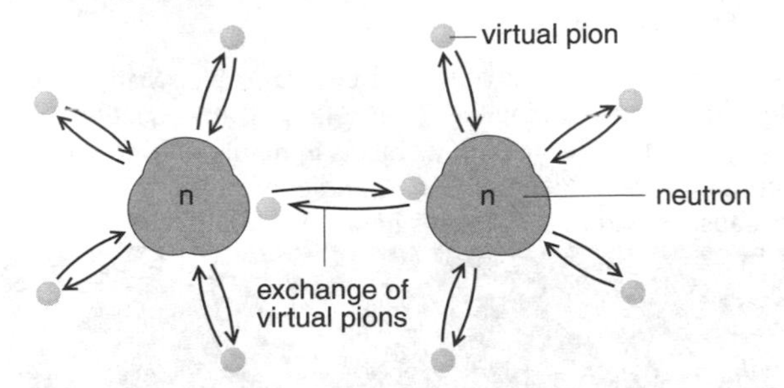

Each nucleon is continually emitting and reabsorbing ***virtual pions***, which surround it in a swarm. When close, two nucleons may exchange a pion. The momentum transfer produces the effect of a force (attractive or repulsive).

Note:
- The emitting nucleons lose no mass, so virtual pions are only allowed their brief existence by the uncertainty principle. To create 'real' pions, the missing mass must be supplied by the energy of a collision.

All the fundamental forces are believed to be carried by exchange particles. For example, electrons repel each other by exchanging ***virtual photons***. This process can be represented by a ***Feynman diagram*** as below. For 'real' photons to exist, energy must be supplied.

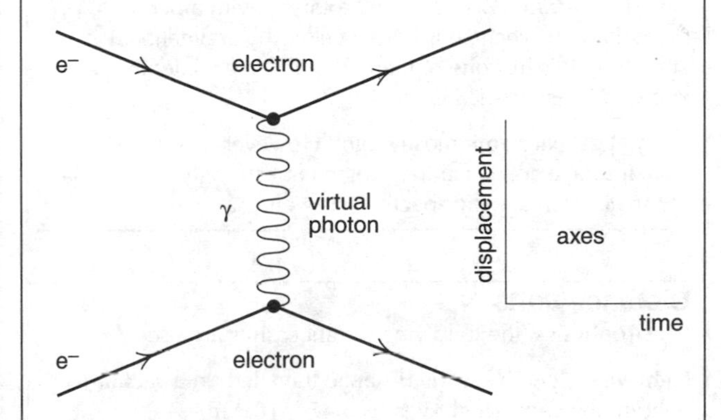

The particles that carry the fundamental forces are known as ***gauge bosons***. They are listed in the chart below.

Note:
- Quarks are bound together by ***gluons***. As nucleons and pions are made of quarks, the gluon would seem to be the basic force carrier for all strong interactions.
- The existence of the graviton is speculation only.

Force	Gauge bosons		
strong	gluon		
electromagnetic	photon		
weak	W^+	W^-	Z^0
gravitational	graviton		

H1 Astrophysics – 1

Solar System, stars, and galaxies

The Earth is one of many ***planets*** in orbit around the Sun. The Sun, planets, and other objects in orbit are together known as the ***Solar System***.

Most of the planets move in near-circular orbits. Many have smaller ***moons*** orbiting them (see E4 for orbital equations and laws). ***Comets*** are small, icy objects in highly elliptical orbits around the Sun. Planets, moons, and comets are only visible because they reflect the Sun's light.

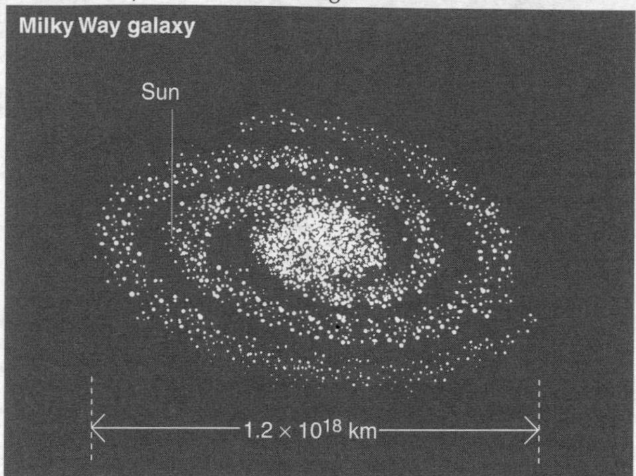

The Sun is one star in a huge star system called a ***galaxy***. Our galaxy contains about 10^{11} stars, as well as interstellar matter (thinly-spread gas and dust between the stars). Our galaxy, called the ***Milky Way***, is slowly rotating, with a period of more than 10^8 years. It is held together by gravitational attraction. It is just one of many billions of galaxies in the known ***Universe*** (see H3).

Normal galaxies emit mostly light. However, about 10% of galaxies have active centres which emit strongly in other parts of the electromagnetic spectrum as well.

Distance units

In astrophysics, the following distance units are used.

Light-year (ly) This is the distance travelled (in a vacuum) by light in one year. 1 light-year = 9.47×10^{15} m.

Astronomical unit (AU) This is the mean radius of the Earth's orbit around the Sun. 1 AU = 1.50×10^{11} m.

Parsec (pc) This is the distance at which the mean radius of the Earth's orbit has an angular displacement of one arc second (1/3600 degree). (See also *Parallax* in H2).

$$1\,\text{pc} = 3.26\,\text{ly} = 2.06 \times 10^5\,\text{AU} = 3.09 \times 10^{16}\,\text{m}$$

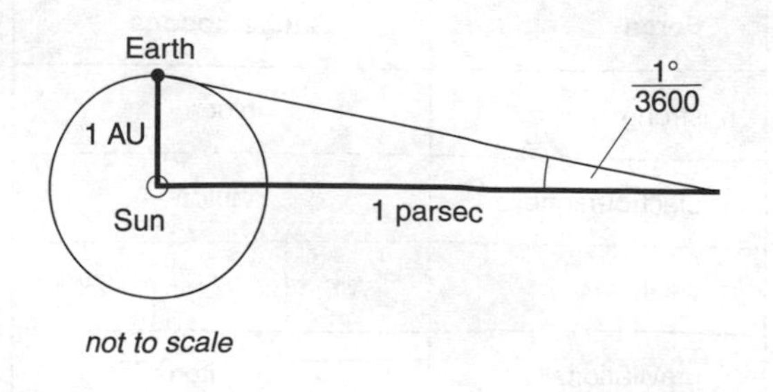

diameter of Earth = 1.3×10^4 km
diameter of Sun = 1.4×10^6 km
radius of Earth's orbit = 1.5×10^8 km = 1 AU

diameter of Solar System = 50 AU
distance to nearest star (*Proxima Centauri*) = 2.7×10^5 AU = 4.2 ly = 1.3 pc

diameter of galaxy (*Milky Way*) = 1.3×10^5 ly = 40 kpc
distance to neighbouring galaxy (*Andromeda*) = 2.2×10^6 ly = 0.7 Mpc

Collectors and detecting radiation

Information about planets, stars, and galaxies is obtained by analysing the electromagnetic radiation they emit. Depending on the source, this can range from radio waves to γ radiation. Some form of telescope is used as the ***collector***. This is linked to a ***detector***.

Electromagnetic radiation	Telescope: collector, e.g.	detector, e.g.
radio	concave metal dish	antenna
infrared	concave mirror	solid state detector
light ultraviolet	concave mirror (see C5)	photographic plate or CCD
X-rays γ-rays	concave metal dish	solid state detector

Charge-coupled device (CCD) This is used in many optical (e.g. light) telescopes instead of a photographic plate. It is more sensitive, and the image on it can be processed electronically. It consists of an array of tiny photodiodes, each contributing one piece *(a **pixel**)* to the whole picture. Signals from the photodiodes are amplified and processed by a computer for display on a screen.

Siting telescopes Most incoming radiation is blocked by the Earth's atmosphere. That which does pass through includes light, parts of the radio spectrum, and some infrared and ultraviolet. Radio waves can pass through interstellar dust, which blocks light from galactic centres.

Telescopes are sited as follows:
- Radio telescopes are usually ground-based.
- Optical telescopes are mounted as high in the atmosphere as possible (e.g. on mountain tops) or above it. This is to reduce image quality problems caused by atmospheric refraction and 'light pollution' from cities.
- Infrared, ultraviolet, X, and γ radiation telescopes are placed in high-altitude balloons or orbiting satellites. Satellite-based instruments include:
  HST, the *Hubble Space Telescope* (optical),
  COBE, the *Cosmic Background Explorer* (microwave),
  IRAS, the *Infrared Astronomical Satellite*.

Telescope design

Optical telescopes These are described in C5.

Radio telescopes Most have a large concave dish to reflect incoming radio waves towards the antenna. Microwaves need a smooth metal reflector. With longer wavelengths, wire mesh can be used, provided the mesh size is less than about $\lambda/20$. A computer-generated 'radio image' is built up by ***scanning*** the source line by line.

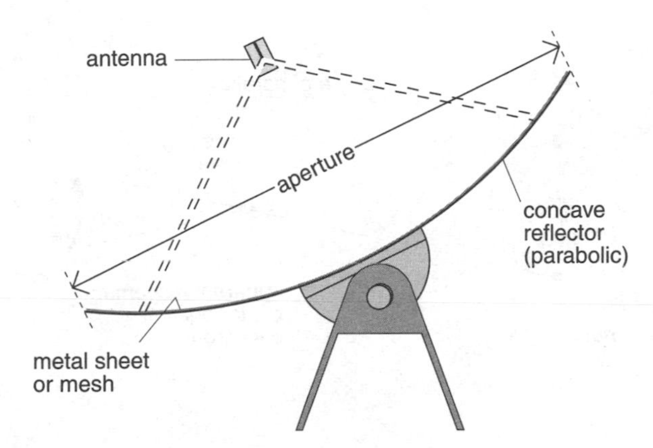

Radio telescopes have to deal with much longer wavelengths than optical instruments, so they need much wider apertures to give the same resolving power (see C5). Effective apertures of many kilometres can be achieved by linking several telescopes electronically.

Luminosity and magnitude

The ***luminosity*** of a star is the rate at which it radiates energy. The unit of luminosity is the watt (W).

The observed brightness of a star depends on its luminosity *and* on its distance from Earth. A very luminous star can appear dim if it is far enough away, because the intensity of its radiation obeys an inverse square law (C2).

The ***apparent magnitude***, *m*, of a star is a measure of its observed brightness. On the scale of *m* values, 0, 1, 2, 3 etc represents an order of *decreasing* observed brightness.

- By definition, a star for which $m = 1$ appears 100 times brighter than a star for which $m = 6$.
- The brightest stars in the sky have negative values of *m*.
- A star's apparent magnitude can be deduced by analysing its image on a photographic plate or CCD.

The ***absolute magnitude***, *M*, of a star is the apparent magnitude it would have if it were 10 parsecs away.

- Absolute magnitude is directly related to luminosity and does not depend on the star's distance from the Earth.

Using the inverse square law for intensity (see C2), it can be shown that a star's distance *d* (in pc), apparent magnitude *m*, and absolute magnitude *M* are linked like this:

$$m - M = 5 \log \frac{d}{10} \qquad (1)$$

For methods of estimating distances to stars, see H2.

Star	Apparent magnitude (*m*)	Absolute magnitude (*M*)	Distance/ pc
Sirius	–1.5	1.4	2.7
Rigel	0.1	–7.2	290
Deneb	1.3	–7.2	490
Proxima C	11.1	15.5	1.3

Spectral analysis

Information about a star's temperature, composition, and motion can be found by analysing its spectrum.

Thermal radiation in F7 includes a graph which shows how the peak wavelength changes with temperature for a black body radiator (e.g. a star). According to ***Wien's law***,

$$\lambda_{max} T = 2.90 \times 10^{-3}\ \text{m K}$$

If λ_{max} is found from a star's spectrum, the surface temperature *T* can be calculated.

Emission and absorption spectra in G4 explains how the composition of the Sun's outer layers can be deduced from its absorption spectrum. This also applies to other stars. For more on stellar (star) spectra, see H2.

The Doppler effect

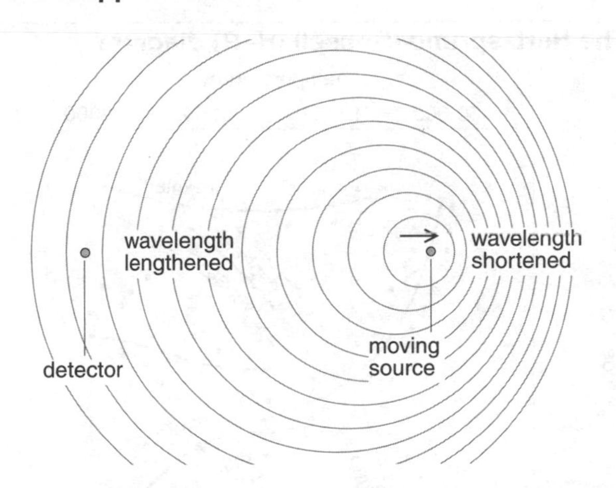

If a wave source is receding (moving away) from a detector, the waves reaching the detector are more spaced out, so their measured wavelength is increased and their frequency reduced. This is an example of the ***Doppler effect***. It causes the change of pitch which you hear when an ambulance rushes past with its siren sounding.

Star motion can be fast enough to cause a detectable Doppler shift in light waves. If a star is moving *away* from the Earth, its spectral lines are shifted towards the *red* end of the spectrum. If *v* is the relative velocity of recession, and *v* is small compared with the speed of light, *c*:

$$\frac{\Delta f}{f} = -\frac{v}{c} \quad \text{and} \quad \frac{\Delta\lambda}{\lambda} = \frac{v}{c}$$

f and λ are the emitted frequency and wavelength. Δf and $\Delta\lambda$ are the observed changes – both defined as *increases*.

- The Sun's rotation causes a broadening of its spectral lines, because light from the receding side is red-shifted while that from the approaching side is blue-shifted. This effect can be used to work out the speed of rotation.

Radar astronomy

The distance of a planet (or moon), and its speed of rotation, can be found by directing radar (microwave) pulses at its surface and analysing the pulses reflected back.

The distance is found by measuring the time interval between the outgoing and returning pulses.

The speed of rotation is found by measuring the Doppler shifts which occur when pulses are reflected from the receding and approaching sides of the planet.

H2 Astrophysics – 2

Classifying stars

Stars can be classified according to their spectra. The main spectral classes are: O, B, A, F, G, K, M. This represents an order from high to low temperature (see diagram below).

Here are details of some of the spectral classes:

- O-type stars are the hottest and appear blue-white. Helium lines are prominent in their absorption spectra.
- A-type stars appear white. Hydrogen lines are prominent in their absorption spectra.
- G-type stars, like the Sun, appear yellow-white. There are many metallic lines in their absorption spectra.
- M-type stars are the coolest, and appear red. Banding in their absorption spectra indicates the presence of molecules.

The Hertzsprung–Russell (H–R) diagram

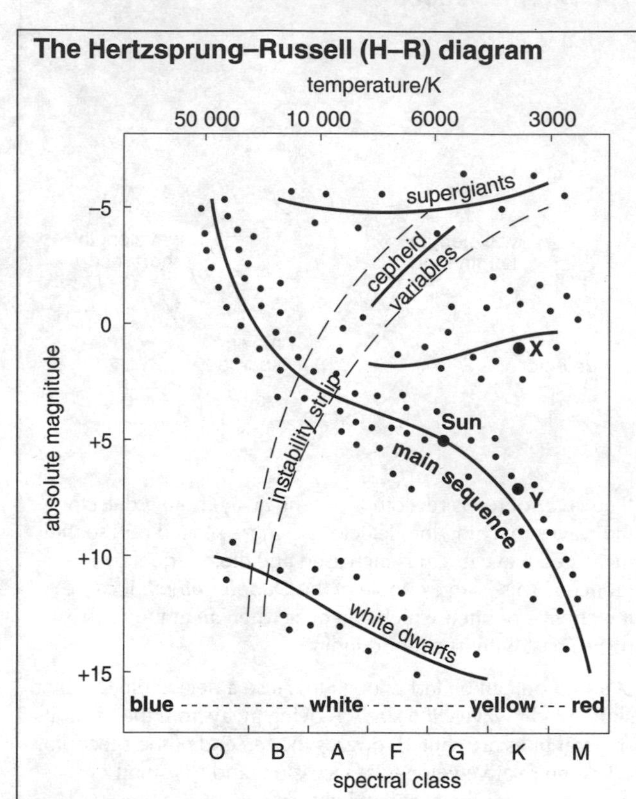

This is a diagram in which the absolute magnitudes of stars are plotted against their spectral classes. A simplified version is shown above.

Note:

- Star X is the same spectral class as star Y, but higher up the diagram, i.e. X is at the same temperature as Y, but radiates more light. So X is larger than Y.
- Terms such as ***giant*** and ***dwarf*** indicate star size.
- The points on an H–R diagram occur in zones. Most stars, including the Sun, belong to the ***main sequence***.

Estimating distances to stars and galaxies

These are the main methods used:

Parallax As the Earth orbits the Sun, nearby stars appear to move agains the background of very distant stars. The nearer the star, the greater its apparent movement. This effect is called ***parallax***.

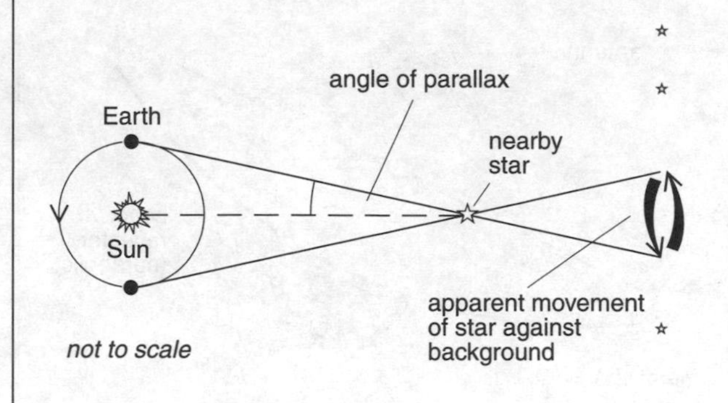

By measuring the angle of parallax, the distance can be calculated using trigonometry. The method is suitable for distances up to ~100 pc. (The pc – parsec, see H1 – was defined to aid calculations involving parallax.)

Inverse square law The distance of a star or galaxy is worked out by comparing its apparent and absolute magnitudes, and using equation (1) in H1. The method is suitable for very distant objects. However, it requires 'standard' sources, of known absolute magnitude, e.g.

- stars whose positions on the H–R diagram can be worked out by spectral analysis
- cepheid variables (see below). These can be observed in other galaxies as well as our own.

The Hubble law Galaxies have a red shift which is proportional to their distance from the Earth (see H3). This can be used to estimate the distance.

Birth of a star

Stars form in huge clouds of gas (mainly hydrogen) and dust called ***nebulae.*** The Sun formed in a nebula about 5×10^9 years ago. The process took about 5×10^7 years:

Gravity pulled more and more nebular matter into a concentrated clump called a ***protostar***. The loss of gravitational PE caused a rise in core temperature which triggered the fusion of hydrogen and the release of energy (see G3). Thermal activity stopped further gravitational collapse. The Sun had become a main sequence star. (Its planets had formed in an orbiting disc of nebular matter.)

Cepheid variables

Most of the stars in the ***instability strip*** (see the H–R diagram above) are ***cepheid variables.*** These show a regular variation in brightness, as in the example in the graph on the right.

The period (of brightness variation) of a cepheid variable is directly related to its absolute magnitude. The more luminous the star, the longer the period. By measuring the period, the absolute magnitude can be found and, from this the distance, using equation (1) in H1.

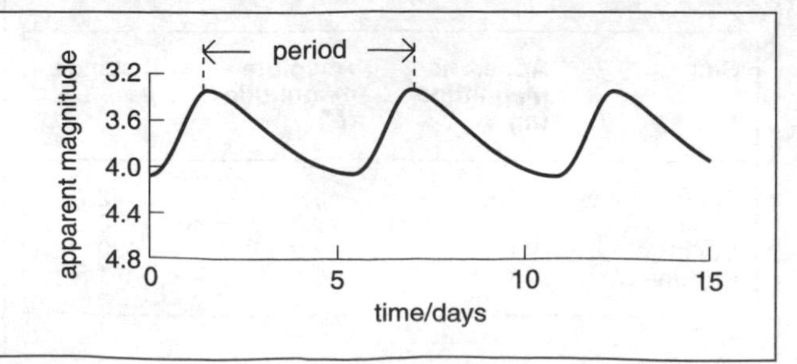

Life and death of a star

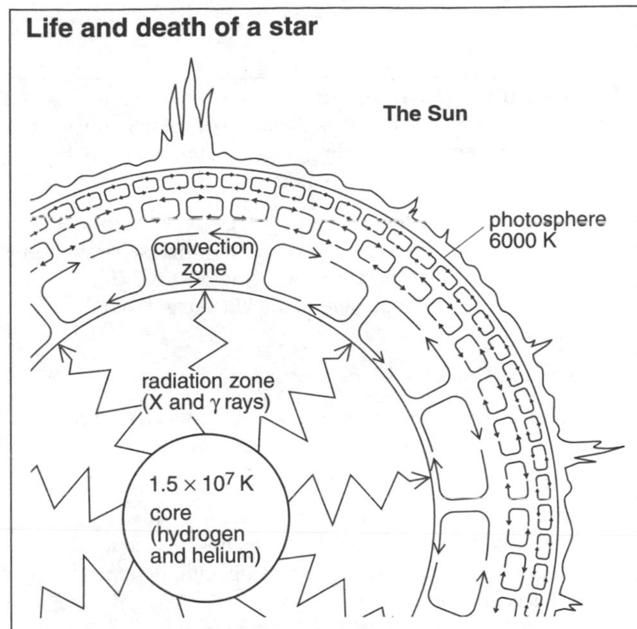

The Sun gets most of its energy from the ***proton–proton chain***, a multi-stage fusion process which converts hydrogen-1 into helium-4. Hotter, more massive stars use the ***CNO cycle***. This also changes hydrogen-1 into helium-4, but involves carbon, nitrogen, and oxygen nuclei.

The Sun is about half way through its life on the main sequence (about 10^{10} years). Hotter, more massive stars consume hydrogen more quickly and have shorter main sequence lives.

When all its hydrogen has been converted into helium, the Sun will take the path shown on the H–R diagram below.

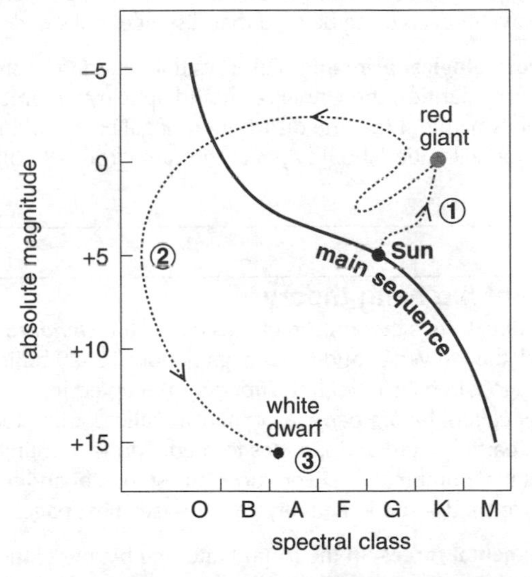

1 The core collapses. The Sun becomes a ***red giant*** as its outer layers expand and cool (and engulf the Earth). With the core temperature rising to over 10^8 K, energy is released by the fusion of helium into carbon.
2 After further changes, the outer layers expand and drift off into space. The core and inner layers become a ***white dwarf*** whose core is so dense that the normal atomic structure breaks down. The electrons form a ***degenerate electron gas*** whose pressure stops further collapse.
3 Fusion ceases. The white dwarf cools and fades for ever.

Note:
- Stars less massive than the Sun end up as white dwarfs, without going through the giant stages.
- Massive stars become giants or supergiants, then end up as ***neutron stars*** or ***black holes*** (see right).

More stellar objects

Supernovae When a massive star enters its giant phase, its core becomes so hot that carbon is fused into heavier elements. If the star exceeds about 8 solar (Sun's) masses, iron is produced. As this is at the top of the binding energy curve (see G3), fusion no longer supplies energy. The core collapses, causing a shock wave which blows away the star's outer layers in a huge explosion called a ***supernova***. For a few days, this is millions of times brighter than a star. Elements ejected from supernovae eventually 'seed' the nebulae in which new stars and planets will form.

Neutron stars If the core of a supernova exceeds about 1.4 solar masses, the degenerate electron gas cannot resist gravitational collapse. Electrons and protons are pushed together to form neutrons. The result is a neutron star – essentially a giant nucleus about 10–30 km across.

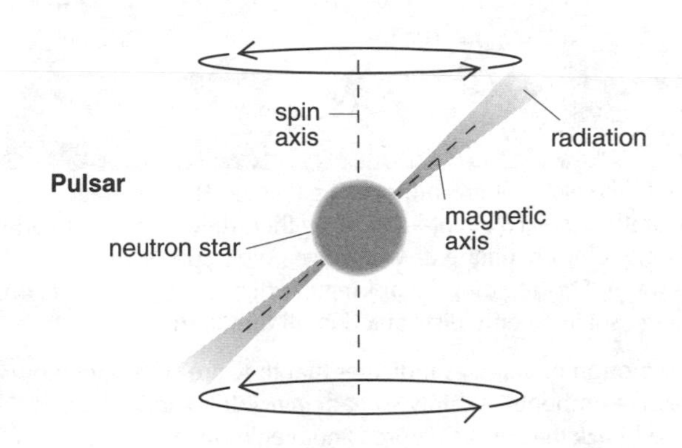

Pulsars These emit radio, light, or X-ray pulses at up to 500 times per second. They are believed to be rapidly spinning neutron stars. Pulses are detected because the star sends out two narrow radiation beams which rotate with it, rather like the beams of light from a lighthouse.

Black holes If the core of a supernova exceeds about 2.5 solar masses, even the neutrons formed cannot resist gravitational collapse. The core shrinks to become a black hole from which no particles or radiation can escape.

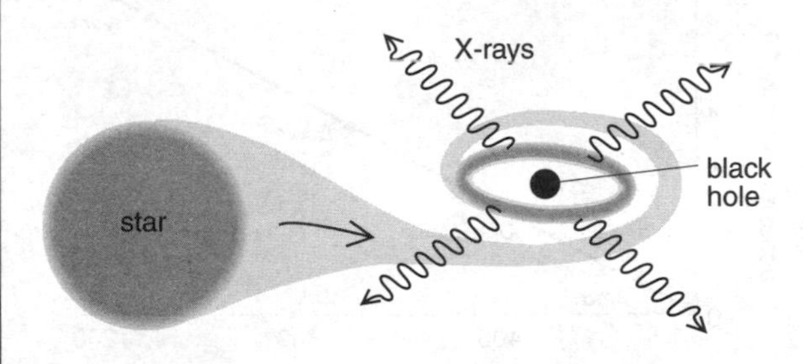

X-ray binary system

Binary stars These are two stars which rotate about a common centre of mass. If they are close, gravity may pull material from one to the other. If one is a neutron star or black hole, material falling into it will give off X-rays.

Quasars These have red shifts which suggest that they are the most distant objects in the Universe (see *Hubble's law* in H3). If they really are distant, they radiate as much energy as some galaxies, but have only the volume of a solar system. Each may be the active centre of a galaxy where nebular matter surrounds a supermassive black hole.

Note:
- Some scientists argue that quasars are closer, less luminous objects whose red shifts have some other cause.

H3 Cosmology

The structure of the Universe

The study of the Universe, its origins, and evolution is called ***cosmology***.

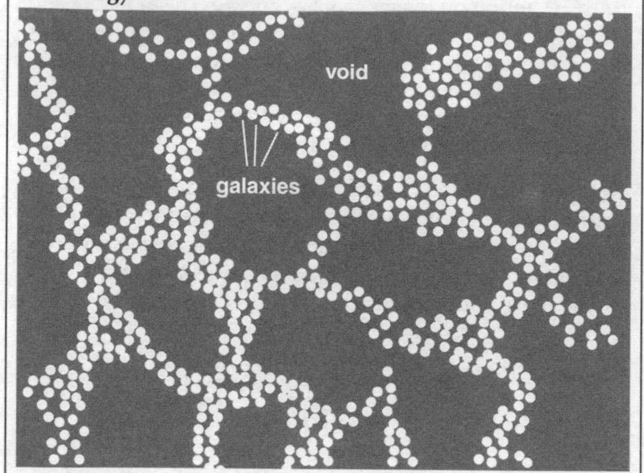

The Universe contains billions of galaxies. Their average separation is ~ 10^6 light-years. Together, they form a network of long, clumpy filaments with huge voids (spaces) in between. Despite their local irregularities, the galaxies are, on a large scale, evenly distributed in all directions.

The motion of galaxies indicates that they are surrounded by massive amounts of thinly spread, invisible material. This is called ***dark matter***. Its nature is not yet known.

Hubble's law

Measurements of Doppler red shifts (see H1) indicate that, in general, the galaxies are receding from each other. The further away the galaxy, the greater is its red shift and, therefore, the greater its recession velocity.

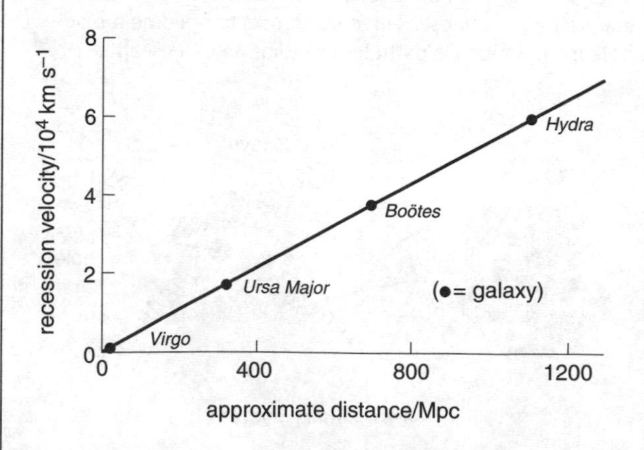

According to ***Hubble's law***, the distance d of a galaxy and its recession velocity v are linked by this equation:

$$v = H_0 d \qquad (1)$$

H_0 is called the ***Hubble constant***. Large distances are difficult to estimate accurately, so the value of H_0 has a high uncertainty. However, it is thought to lie in the range 50–100 km s^{-1} Mpc^{-1} (1.6–3.2×10^{-18} s^{-1}). Its value is important for several reasons:

- It enables the distances of the most remote galaxies to be estimated from their red shifts.
- The age of the Universe can be estimated from it (see above right). (H_0 has dimensions of 1/time.)
- The fate of the Universe depends on it (see next page).

The expanding Universe

The most generally accepted explanation of galactic red shifts is that the Universe is expanding. At zero time, all its matter and energy was together in a highly concentrated state.

Estimating the age of the Universe If a galaxy is d from our own, and has a steady recession velocity v, then separation of the galaxies must have occurred at a time d/v ago. This represents the approximate age of the Universe. From equation (1) $d/v = 1/H_0$, so

$$\text{age of the Universe} \approx 1/H_0$$

This gives an age in the range 1–2×10^{10} years (10–20 billion years).

Note:

- The above calculation assumes constant v. However, recent observations suggest that the Universe's rate of expansion may actually be increasing, although the reason for this is not yet clear.

Olbers' paradox In the 17th century, it was pointed out that, if the stars continued out to infinity, the night sky should be white, not dark – because light must be coming from every possible direction in the sky. This became known as Olbers' paradox.

Two reasons for the dark night sky have been suggested:

- In an expanding Universe, red-shifted wavelengths mean reduced photon energies (see G4), so the intensity of the light from distant stars is reduced.
- There is a limit to our observable Universe. If, say, the Universe is 15 billion years old, then we have yet to receive light from stars more than 15 billion light-years away. So everything beyond that distance looks dark.

The cosmological principle This says that, apart from small-scale irregularities, the Universe should appear the same from all points within it (i.e. the distribution of galaxies and their recession velocities should appear the same from all points).

The hot big bang theory

According to this theory, sometimes called the ***standard model***, the Universe (and time) began about 10–20 billion years ago when a single, hot 'superatom' erupted in a burst of energy called the ***big bang***. As expansion and cooling took place, particles and antiparticles formed. Further cooling meant that combinations were possible, so nuclei and then atoms formed – and eventually galaxies (see next page).

Fundamental forces In the instant after the big bang, the fundamental forces (see G8) existed as one superforce. But within 10^{-11} s, they had separated from each other.

Cosmic background There is a steady background radiation which comes from every direction in space. It peaks in the microwave region, and corresponds to the radiation from a black body (see F7) at 2.7 K. It is thought to be the red-shifted remnant of radiation from the big bang. Its presence is predicted by the big bang theory.

Inflation The standard model cannot satisfactorily explain why, on a large scale, the Universe and its microwave background radiation are so uniform. Mathematically, it is possible to overcome this difficulty by assuming that the early Universe went through a brief period of very rapid inflation, when its volume increased by a factor 10^{50}.

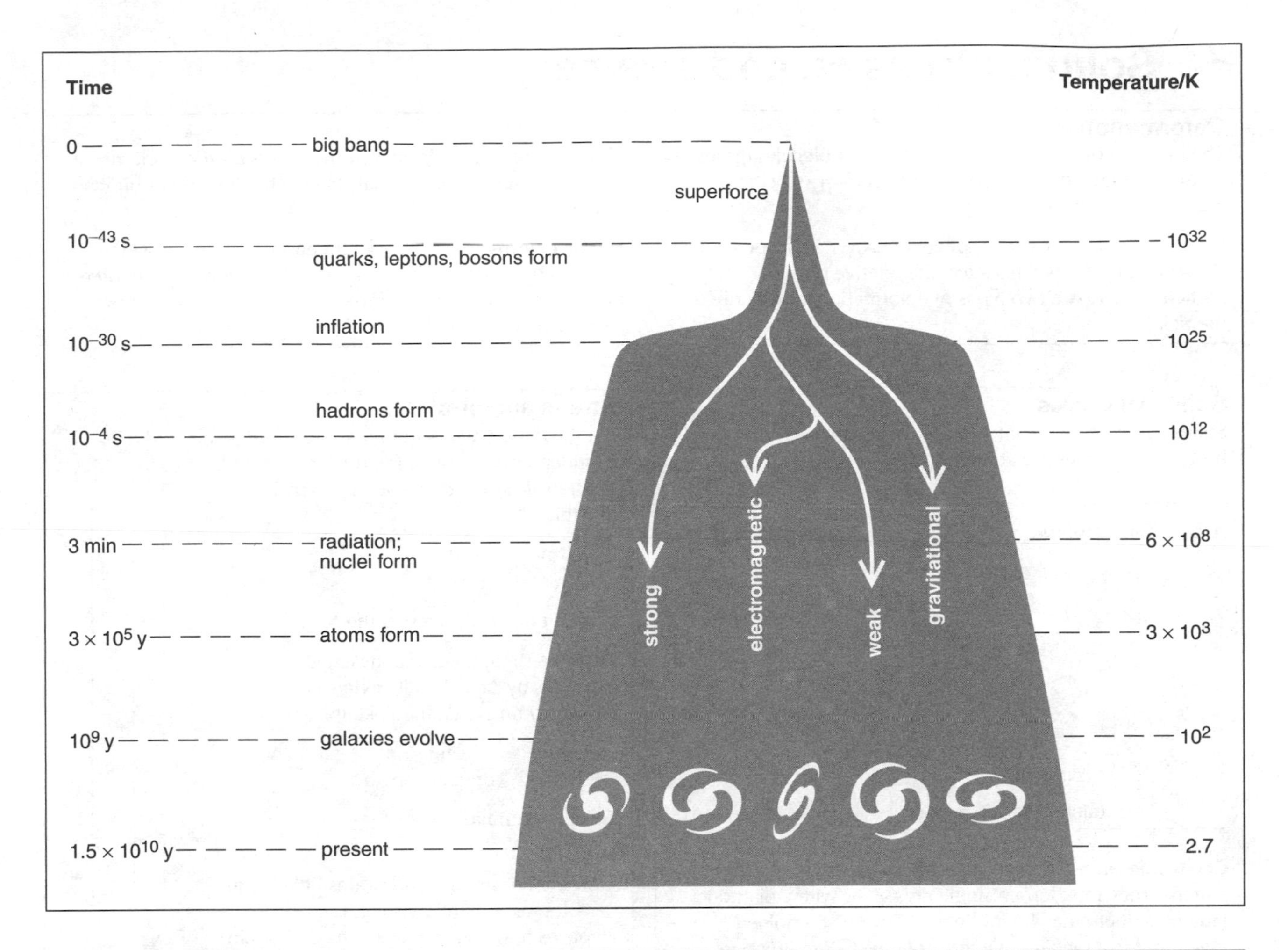

The fate of the Universe

Gravity is affecting the expansion of the Universe. The fate of the Universe depends on how its average density ρ compares with a certain **critical density** ρ_0:

- If $\rho < \rho_0$ the expansion continues indefinitely.
- If $\rho = \rho_0$ the expansion continues, but the rate falls to zero after infinite time.
- If $\rho > \rho_0$ the expansion reaches a maximum, and is followed by contraction.

The average density of the Universe is thought to be close to the critical density.

Linking ρ_0 and H_0 The critical density depends on the value of the Hubble constant. A higher H_0 means a higher recession velocity per unit separation. So a higher density is needed to stop the expansion. It can be shown that

$$\rho_0 = \frac{3H_0^2}{8\pi G}$$ where G is the gravitational constant (see E3)

This gives ρ_0 in the range $5\text{–}20 \times 10^{-27}$ kg m^{-3}.

Models of the Universe

The big bang was not an explosion into existing space. Space itself started to expand. The galaxies are separating because the space between them is increasing.

Space has three dimensions of distance (represented by x, y, and z co-ordinates) and one of time. According to Einstein's theory of general relativity, gravity causes a curvature of space-time. If gravity is sufficiently strong, it may produce a 'closed' Universe, as shown below.

To visualize the expansion of the Universe, it is simpler to use models with only two of the distance dimensions. Imagine that the Universe is on an expanding, elastic surface. Three possible models are shown below. In each case, the galaxies move apart as the surface stretches. From any position on the surface, each galaxy recedes at a velocity that is proportional to its distance away.

Note:

- It is the value of the critical density, and therefore of the Hubble constant, which decides whether we live in an open, flat, or closed Universe.

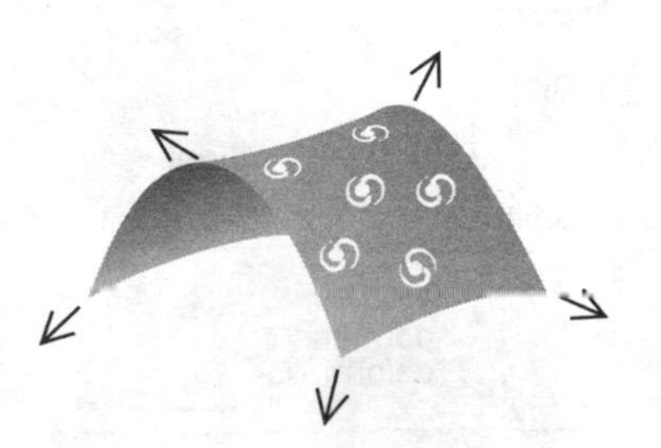

Open Universe ($\rho < \rho_0$) The surface is infinite and unbounded.

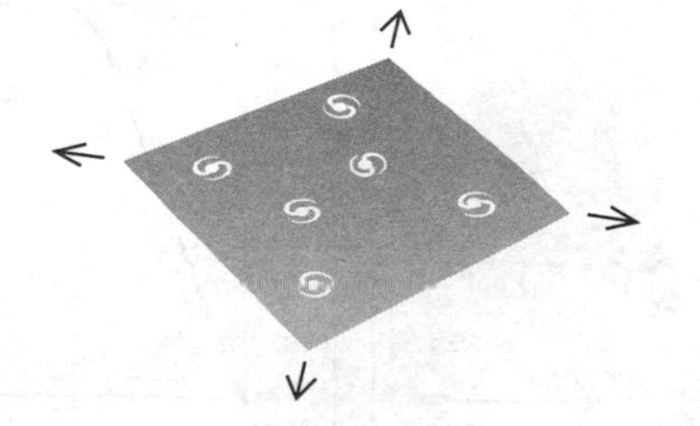

Flat Universe ($\rho = \rho_0$) The surface is infinite and unbounded.

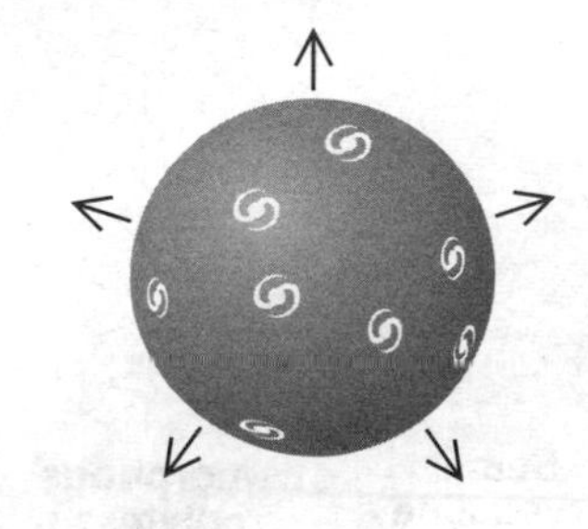

Closed Universe ($\rho > \rho_0$) The surface is finite and bounded.

H4 Solids, stresses, and strains

Deformation

The particles of a solid may be atoms, or molecules (groups of atoms), or ions (see F5). They are held closely together by electric forces of attraction.

When external forces are applied to a solid, its shape changes: ***deformation*** occurs. This alters the relative positions of its particles. There are two types of deformation, as described on the right.

Elastic deformation If the deformation is ***elastic***, then the material returns to its original shape when the forces on it are removed.

Plastic deformation If the deformation is ***plastic***, then the material does not return to its original shape when the forces on it are removed. For example, Plasticine takes on a new shape when stretched.

Solid structures

Solids can be classified into three main types, according to how their particles are arranged.

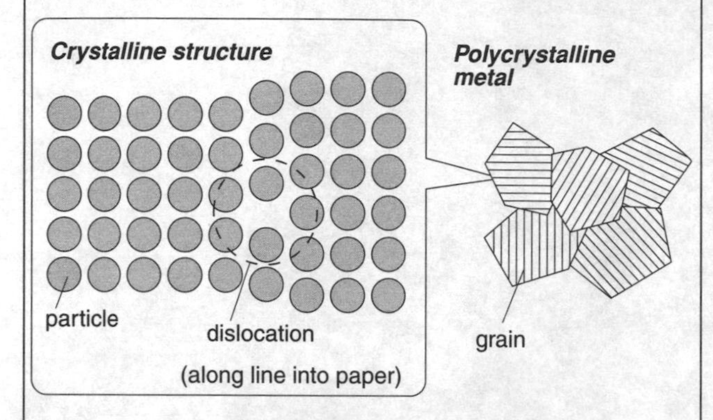

Crystalline solids The particles are in a regular, repeating pattern. They may form a single crystal, as with a diamond. However, there may be millions of tiny crystals joined together. Most metals have this ***polycrystalline*** structure. Their crystals, called ***grains***, can be as small as 10^{-2} mm.

Crystal structures normally have imperfections in them called ***dislocations***. These allow particles to change their relative positions, so the solid is more easily deformed.

Amorphous (glassy) solids The particles have no regular pattern (except over very short distances). Glass and wax have structures like this.

Polymers These materials have long-chain molecules, each of which may contain many thousands of atoms. The molecules are formed from the linking of short units called ***monomers***. In a polymer, the chains may be coiled up and tangled like spaghetti. Depending on the amount of tangling, a polymer may be described as ***semi-crystalline*** or ***amorphous***.

Rubber and wool are natural polymers. Plastics, such as nylon and artificial rubber, are synthetic polymers.

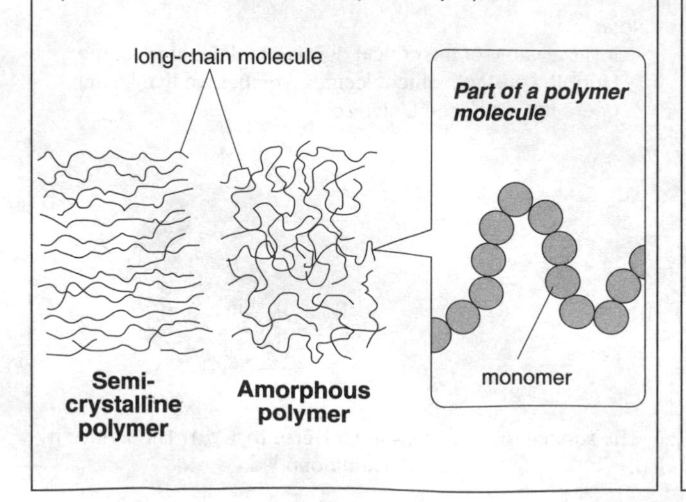

Stress and strain

On the right, a wire of cross-sectional area A is under tension from a force F (at each end). The ***tensile stress*** σ on the wire is defined like this:

$$\text{tensile stress} = \frac{\text{force}}{\text{area}} \qquad \sigma = \frac{F}{A}$$

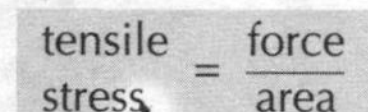

The unit of tensile stress is the N m^{-2}.

The wire stretches so that its length l_0 increases by Δl, called its ***extension***. The ***tensile strain*** ε is defined like this:

$$\text{tensile strain} = \frac{\text{extension}}{\text{original length}} \qquad \varepsilon = \frac{\Delta l}{l_0}$$

Tensile strain has no units.

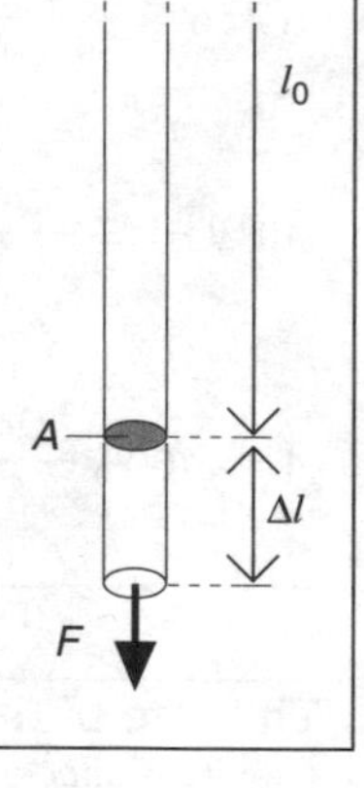

Note:

- There are stresses and strains linked with compression and twisting. On these pages however, the word stress or strain by itself will imply the tensile type.

Hooke's law

The graph below shows how stress varies with strain when a metal wire (steel) is stretched until it breaks.

Note:

- By convention, strain is plotted along the horizontal axis.
- The sequence O to E is described in detail on the next page.

If a material obeys ***Hooke's law*** then, for an *elastic* deformation, the strain is proportional to the stress.

The wire obeys Hooke's law up to point A.

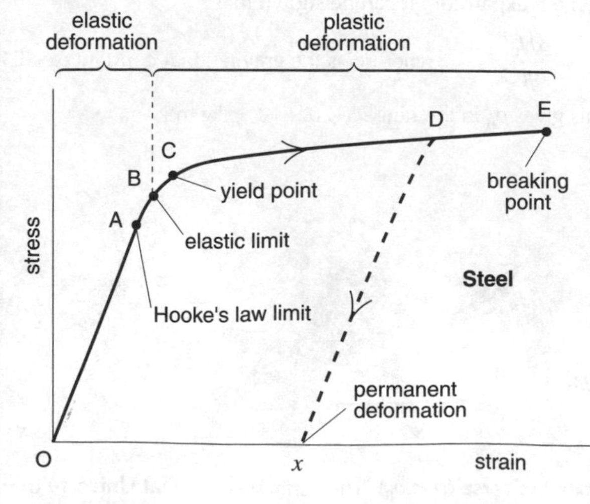

Young's modulus

For a material which obeys Hooke's law, stress/strain is a constant. This constant is called ***Young's modulus***, E:

$$\text{Young's modulus} = \frac{\text{tensile stress}}{\text{tensile strain}} \qquad E = \frac{\sigma}{\varepsilon}$$

Using the equations for σ and ε,

$$E = \frac{Fl_0}{A\Delta l}$$

Note:

- l_0 is constant. If E and A are also constant, $\Delta l \propto F$. So the extension is proportional to the stretching force.
- The above equations can also be used when the material is being compressed.

Typical values for Young's modulus, in Nm^{-2}
steel 21×10^{10} aluminium 7×10^{10}

So, steel is proportionately three times as difficult to stretch as aluminium.

Strain energy

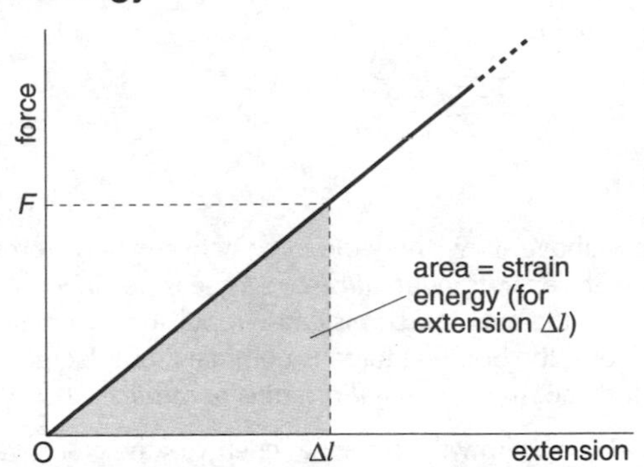

The graph above shows how the extension varies with the stretching force for a material which obeys Hooke's law. The work done for an extension Δl is given by the shaded area (see B8). The area of a triangle = $\frac{1}{2}$ × base × height. So the work done = $\frac{1}{2} F\Delta l$.

As work is done *on* the material, energy is stored *by* the material. This is its ***strain energy***. So

$$\text{strain energy} = \tfrac{1}{2} F\Delta l$$

Note:

- If Hooke's law is *not* obeyed, the work done is still equal to the area under a force–extension graph. However, the above equation does not apply.

Stretching glass

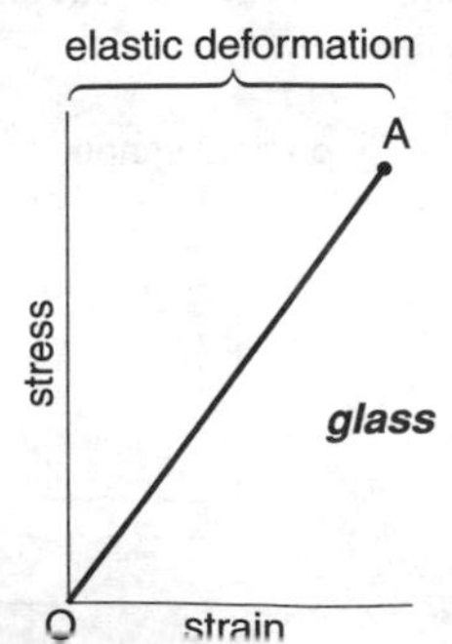

The graph above shows what happens if increasing tensile stress is applied to a glass thread. Elastic deformation occurs until, at point A, a crack suddenly grows, and the glass breaks. A material which behaves like this is said to be ***brittle***. The break is called a ***brittle fracture***.

Stretching a metal

Unlike glass, most metals do not experience brittle fracture when stretched because dislocations tend to stop cracks growing and spreading. The following descriptions refer to the graph for a steel wire on the opposite page.

O to B The deformation of the wire is elastic.

B This is the ***elastic limit***. Beyond it, the deformation becomes plastic as layers of particles slide over each other. If the stress were removed at, say, point D, the wire would be left with a permanent deformation (strain x on the axis).

C this is the ***yield point***. Beyond it, little extra force is needed to produce a large extra extension. If a material can be stretched like this, it is said to be ***ductile***.

E The wire develops a thin 'neck', then a ***ductile fracture*** occurs. The highest stress just before the wire breaks is called the ***ultimate tensile stress***.

Fatigue If a metal is taken through many cycles of *changing* stress, a fatigue fracture may occur before the ultimate tensile stress is reached. Fatigue fractures are caused by the slow spread of small cracks.

Creep This is the deformation which goes on happening in some materials if stress is maintained. For example, unsupported lead slowly sags under its own weight.

Stretching rubber

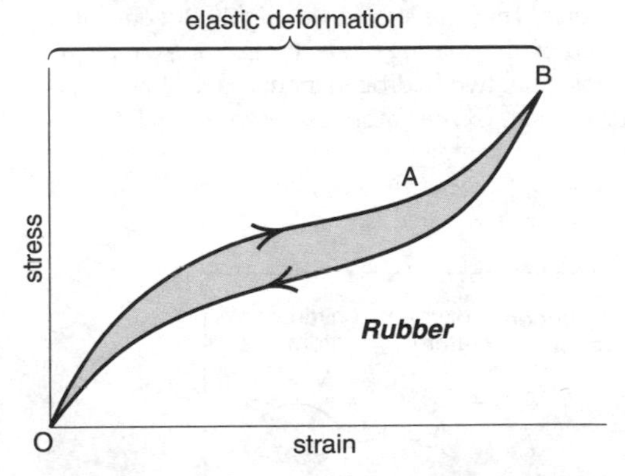

The graph shows what happens if increasing stress is applied to a rubber cord, and then released before the breaking point. Rubber does not obey Hooke's law. Also, much higher strains are possible than in steel or glass. For example, if the extension is twice the original length, the strain is 2.

O to A The molecular chains in the rubber are being uncoiled and straightened.

A to B The chains are almost straight, so the rubber is becoming proportionately more difficult to stretch.

B to A The rubber contracts when the stress is removed.

During this cycle of extension and contraction, energy is lost as heat. The effect is called ***elastic hysteresis***. The shaded area represents the energy lost per unit volume.

Bending

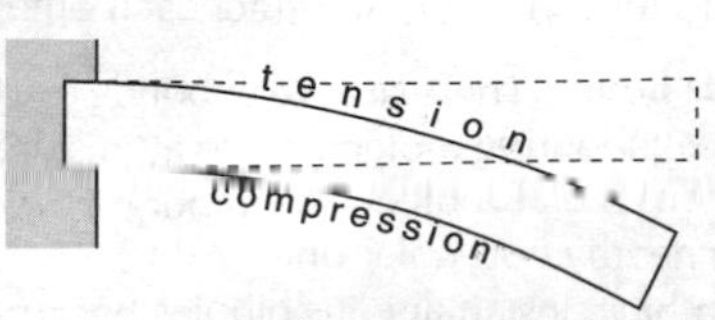

Bending is a combination of tensile and compressive strain as shown here.

H5 Materials – 1

Read H4 before studying this unit and the next one.

Bonds

The electric forces that make atoms stick together are called ***bonds***. Some different types of bond are described below.

Ionic bonds These are strong bonds, formed by the *transfer* of electrons between atoms.

For example, a crystal of sodium chloride consists of a lattice of negative (–) ions and positive (+) ions. The ions are formed by the transfer of electrons from sodium to chlorine. The attractions between opposite ions are the bonds.

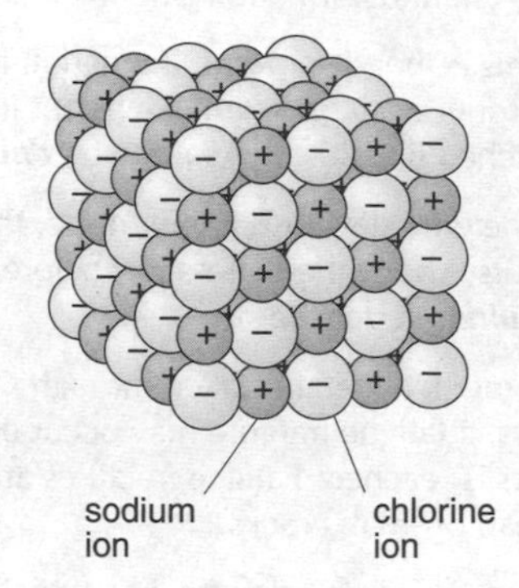

Metallic bonds In metals, some electrons are loosely held, and not tied to particular atoms. So a metal is effectively made up of positive (+) ions in a 'sea' of free electrons (–) which bind them together strongly.

Covalent bonds These are the bonds that hold atoms together in molecules. They are strong, but highly directional. They are formed by the *sharing* of electrons. For example, in a water molecule, two hydrogen atoms each share their electron with an oxygen atom as shown below.

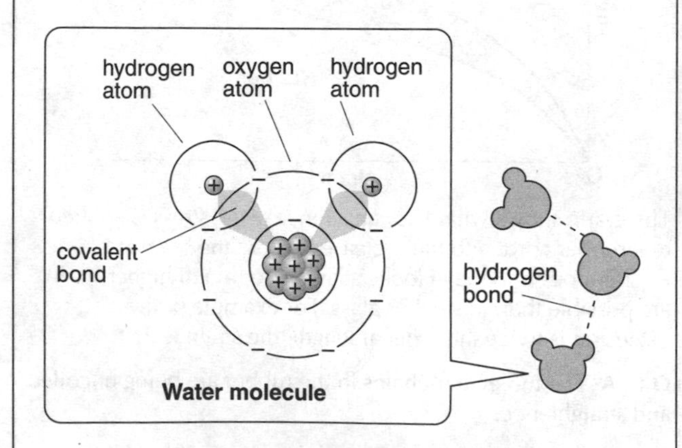

Hydrogen bonds These are the weak bonds which pull water molecules together. Although the molecules are electrically neutral, their electrons are unevenly distributed, giving them positive and negative parts which pull on other molecules.

A molecule with + and – ends is a ***polar*** molecule: it is ***polarized***. The separated charges are an ***electric dipole***. Dipoles tend to turn so that they attract each other.

Van der Waals bonds These are weak bonds existing between all neighbouring molecules (or atoms) because of dipole attractions. Polar molecules temporarily *induce* (see E1) dipoles in nearby non-polar ones. And even non-polar molecules can have instantaneous dipoles because of the random motion of their electrons.

Forces, energy, and separation

The physical properties of materials depend on how their atoms or molecules are stuck together.

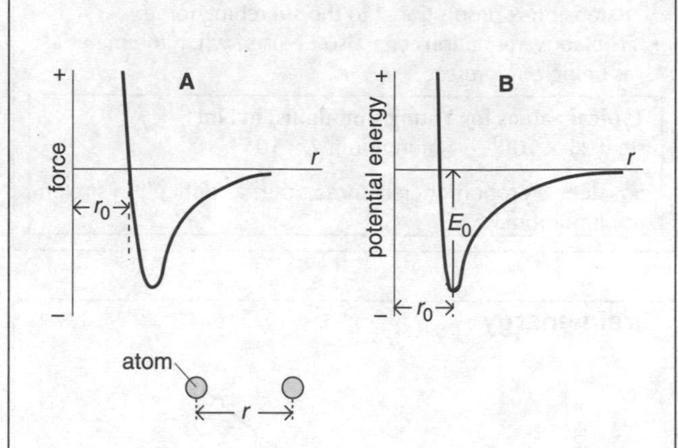

Graph A above shows how the force between two atoms varies with separation (an *attractive* force is *negative*). When very close, the outer electrons cause repulsion. At greater separations, the bonding force is dominant, but decreases with distance. n_0 is the ***equilibrium separation*** (~10^{-10} m).

Graph B shows how the potential energy of the atoms varies with separation. Minimum PE corresponds with the position for zero force in graph A. E_0 is the ***bonding energy***. It is the energy needed for complete separation. The specific latent heat of vaporization (see F3) depends on it.

Hooke's law Imagine that atoms r_0 apart are to be further separated by a force. As graph A is straight in this region, the increase in separation is proportional to the force. For a wire containing rows of billions of atoms, Hooke's law (strain ∝ stress) follows directly from this.

Energy density, and strain

A material stores energy when stretched or compressed elastically. The area under a stress–strain graph gives the ***energy density*** i.e. the energy stored *per unit volume*.

Apart from the tensile and compressive strain, other forms of strain include those below.

Bulk strain The deformation is caused by an increase in external pressure.

Shear strain The deformation is as on the right. (Twisting also causes shear strain in a material.)

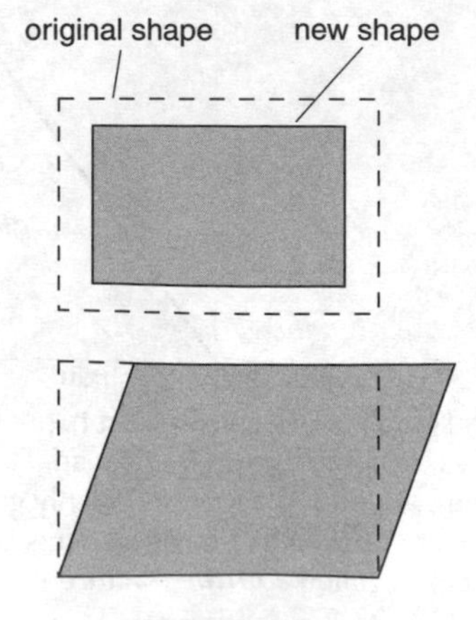

Crystal structures

Many solids are crystalline: their atoms (or other particles) are stacked in a regular pattern. Crystal structures include those on the right. The basic pattern repeats throughout the crystal.

FCC and HCP give the closest packing. The two structures are very similar. Both have layers A and B, as shown below. However, in HCP, the next layer is a repeat of A. In FCC, it is in a third site, C. (You need to study a model to visualize this.)

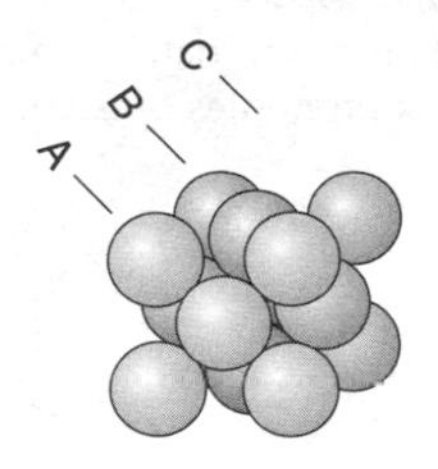

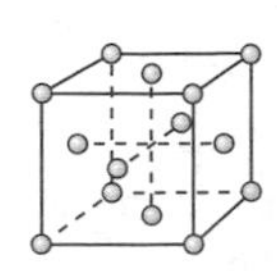

Face-centred cubic (FCC) e.g. aluminium, cobalt

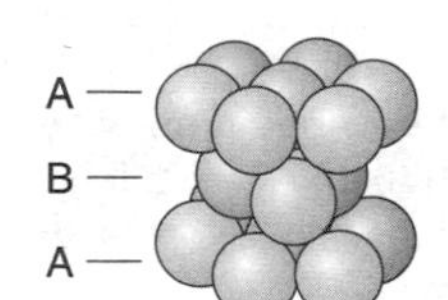

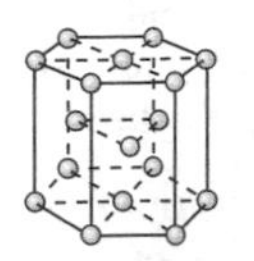

Hexagonal close-packing (HCP) e.g. magnesium, zinc

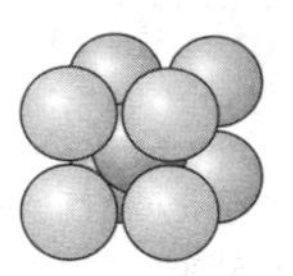

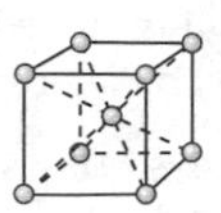

Body-centred cubic (BCC) e.g. iron, sodium

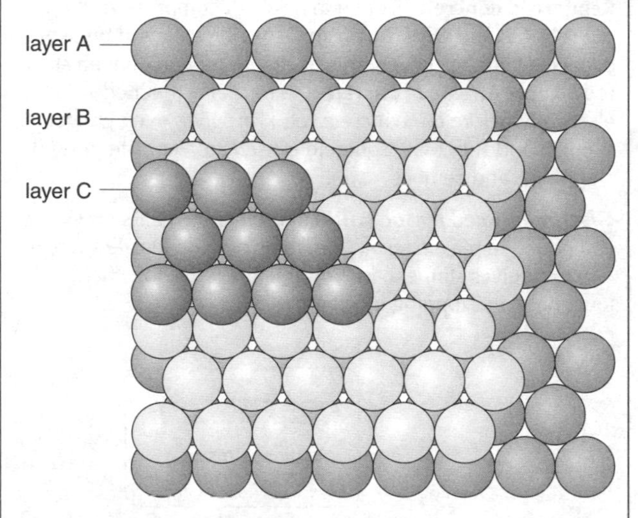

Stacking ABAB etc gives the HCP structure.
Stacking ABCABC etc gives the FCC structure.

Some substances can have more than one structure. For example, carbon can exist as *diamond* or *graphite*, as below. Diamond is very hard because of its strong bonds and rigid ***tetrahedral*** structure. Graphite is soft, slippery, and easy to break. It has layers that can slip over each other because the bonds between them are weak.

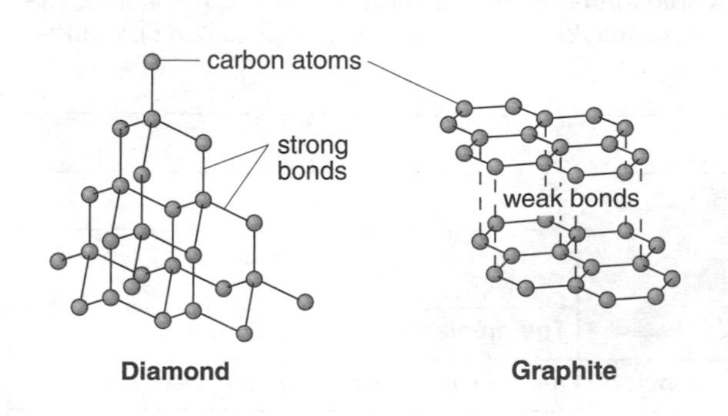

Defects

Defects occur in virtually all crystals (e.g. grains).

Line defects These are the ***dislocations*** described in H4. They are associated with ***stacking faults***: for example, layers of atoms like those on the left slipping so that they stack ABCACABC instead of ABCABCABC.

When a wire is stretched plastically, layers of atoms slip over each other. With dislocations present, much less force is needed to cause slipping. This diagram shows why.

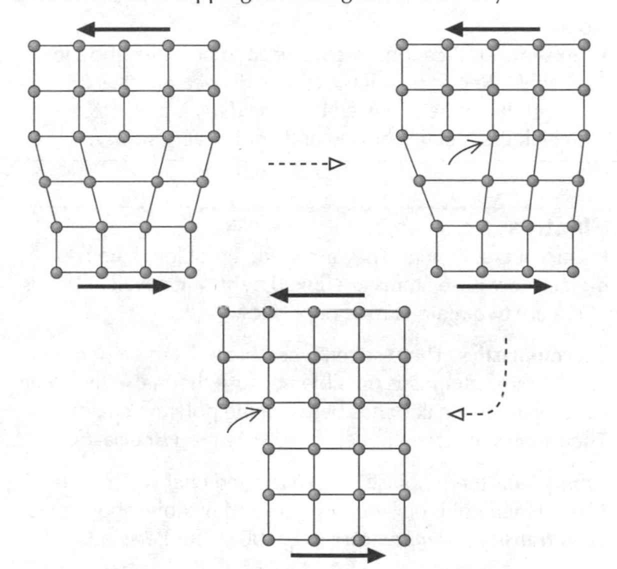

When the bottom part of the crystal is moved to the right, bonds are broken one at a time before rejoining with the next atom along. Without the dislocation, the bonds would all have to be broken at once, which takes more force.

Point defects These can be ***vacancies*** (missing atoms) or ***interstitials*** (extra atoms) in the structure. Some can move through a crystal by diffusion (see also F6).

Mechanical properties

These describe how a material behaves when forces are applied. They depend on its structure, the strength of the bonds, and the type and number of defects present.

Strength A *strong* material has a high ultimate tensile stress i.e. a high stress is needed to break it.

Ductility A *ductile* material can be drawn into wires.

Malleability A *malleable* material can be hammered into different shapes.

Stiffness A *stiff* material has a high Young's modulus, i.e. a high stress produces little strain.

Toughness A *tough* material will deform plastically before it breaks. It is not brittle, i.e. cracks do not easily spread.

H6 Materials – 2

Making metals stronger

Pure metals with large crystals are neither very stiff nor very strong. Dislocations can spread through them easily, allowing slip to take place. To improve the mechanical properties, the spread of dislocations must be reduced.

Smaller grains The smaller the grains, the more difficult it is for dislocations to spread. Smaller grains are produced if a molten metal is cooled and solidified rapidly. Grain size is also affected by later heating or mechanical treatment.

Forming alloys Alloys are mixtures of metals (or of metals and non-metals). For example, steel is an alloy of iron (about 99%) and carbon, and sometimes other elements as well. It is stronger, stiffer, and tougher than iron.

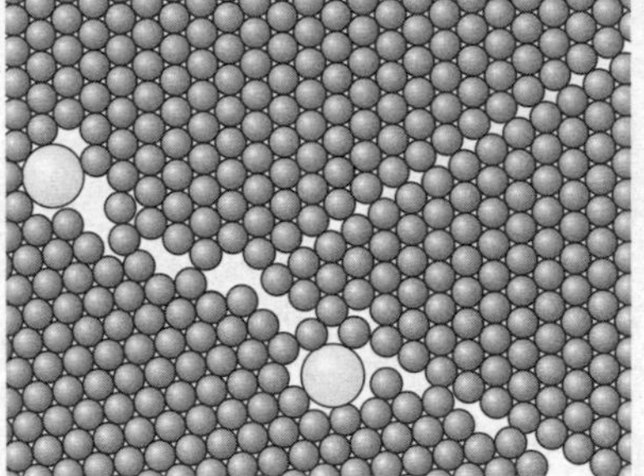

In an alloy, the 'foreign' atoms block the spread of dislocations and make slipping more difficult.

Work hardening This occurs when a metal is hammered, stretched, or bent. It produces *more* dislocations, but these become so jumbled that they block the spread of the others.

Note:

- In some applications, metals *need* to be softer and more flexible. Reducing the number of dislocations has this effect. It can be achieved by ***annealing***, a processs in which the metal is heated and cooled very slowly.

Plastics

Plastics are polymers. They are made up of long-chain molecules whose atoms are linked by strong covalent bonds. There are two main classes of plastics.

Thermoplastics These soften when heated and harden on cooling. Resoftening is possible because thermal activity can overcome the weak bonds between the polymer chains. Thermoplastics creep under stress: they are ***viscoelastic***.

Amorphous thermoplastics have tangled chains. They are *glassy* when cold, but *rubbery* (soft and flexible) above their ***glass transition temperature*** (e.g. 100 °C for Perspex).

Semicrystalline thermoplastics have regions where the chains are parallel and close, so the bonds between the chains are stronger. This produces stiffness and good tensile strength. Amorphous regions add flexibility.

Note:

- Stretching a thermoplastic makes it more crystalline, i.e. its chains uncoil and become less tangled.

Thermosets These do not soften when warmed, so they cannot be remoulded. During manufacture, they develop strong and permanent cross-links (bonds) between their chains.

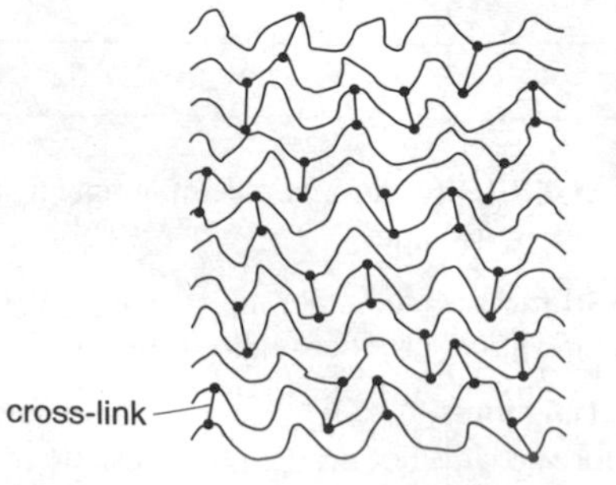

	Plastics			
	Thermoplastics		Thermosets	
Property	resoften on warming		permanent set	
Type	amorphous	semi-crystalline	elastomer	rigid
Structure	tangled chains	many chains parallel	some cross links	many cross links
Examples	Perspex	polythene nylon	artificial rubber	epoxy resins Melamine

Composites

These are combinations of materials, produced to make use of the best properties of each.

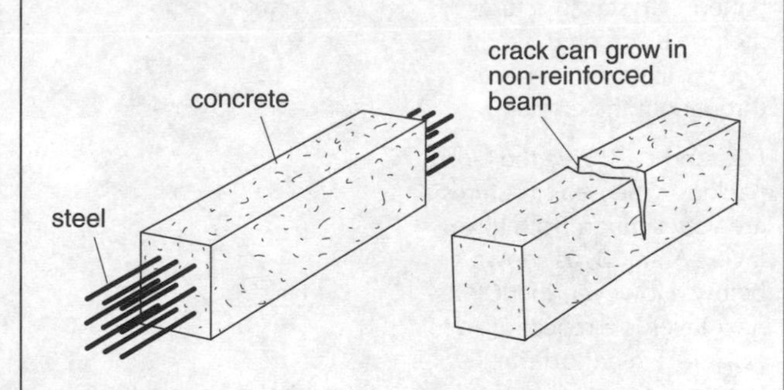

Reinforced concrete Concrete has high compressive strength but is brittle and weak in tension, so will crack and break if bent (see H4). To prevent this, it can be reinforced with steel rods. In ***pre-stressed concrete***, the rods are stretched elastically before the concrete sets. This gives even greater strength and stiffness. (Concrete is itself a composite of sand, chippings, and cement.)

Glass-reinforced plastic (GRP) ('fibre glass') Glass fibres are embedded in plastic resin. The fibres provide tensile strength. The resin gives stiffness, by bonding the fibres together, and toughness, by stopping crack growth.

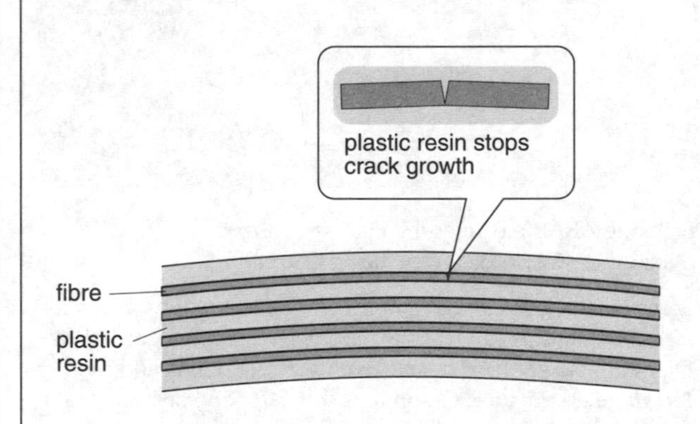

Carbon-fibre-reinforced plastic (CFRP) ('carbonfibre') This is similar to GRP, but with stronger, stiffer carbon fibres instead of glass.

Magnetic properties

Read E5 and E6 first.

If a material is placed in a field B_0, its atoms change the flux density to a new value, B. The ratio B/B_0 is the relative permeability, μ_r. Depending on the material, B may be slightly less, slightly more, or much more than B_0.

Diamagnetism ($B < B_0$) This very weak effect occurs in all materials. Orbiting electrons oppose the applied field.

Paramagnetism ($B > B_0$) This occurs in materials where the electron orbits make the atoms (or molecules) behave as tiny electromagnets. These align with the applied field and strengthen it. The diamagnetic effect is overcome.

Ferromagnetism ($B \gg B_0$) This occurs in strongly magnetic materials, such as iron, nickel, and cobalt. The atoms are grouped into tiny ***domains***, each containing magnetically aligned atoms. An applied field magnetizes the material by aligning the magnetic axes of the domains.

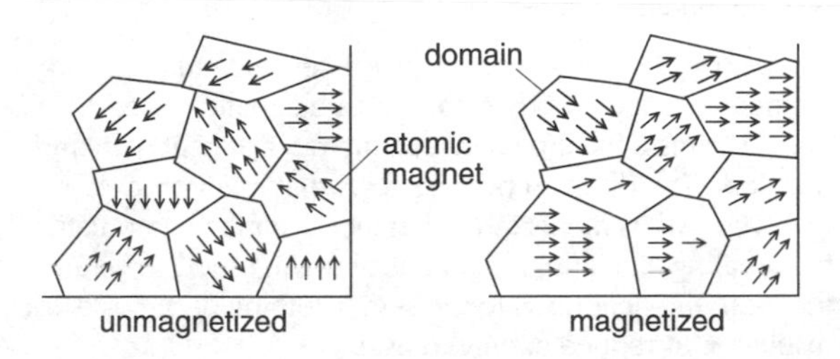

The graph below is for a ferromagnetic material. It shows how B changes when B_0 is increased from zero, reduced to zero, applied in the opposite direction, reduced, and so on. The outer loop is called a ***hysteresis*** loop.

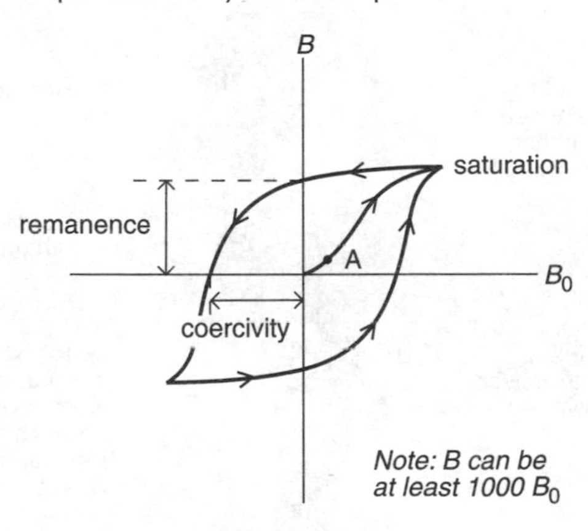

- Beyond A, the material stays magnetized when the applied field is removed.
- The ***remanence*** is the flux density remaining in the material when a strong magnetizing field is removed.
- The ***coercivity*** is the flux density needed to reduce the field from the magnetized material to zero. *Soft* magnetic materials (e.g. iron) have a low coercivity and demagnetize easily. *Hard* magnetic materials (e.g. steel) have a high coercivity and become permanent magnets.
- Above their ***Curie temperature*** (≈1000 K for iron), ferromagnetic materials become paramagnetic.

Piezoelectric effect Compressing or stretching some crystals (e.g. quartz and ferroelectrics) sets up a PD across them. The effect is used in gas lighters and in some microphones. The *reverse* effect is used to keep time in watches. A quartz crystal is made to oscillate by applying an alternating voltage across it.

Electrical conduction

Read D2 and G4 first.

When atoms are close (e.g. in a solid), their energy levels broaden into ***bands***, i.e. lots of levels close together. To take part in conduction, electrons must be able to move between atoms. This requires an energy transfer, so it can only happen if there are unoccupied levels for electrons to go to.

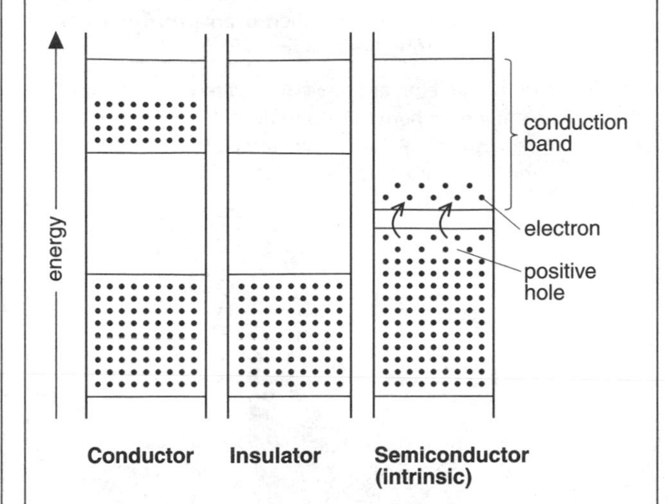

Conductors (e.g. copper) The outer band (the ***conduction band***) is only partly filled. It has unoccupied energy levels, so its electrons are free to move between atoms. (In many metals, the conduction band overlaps with the band below).

Insulators (e.g. nylon) All the electrons are in full bands, so they are unable to change energy and move between atoms.

Semiconductors (e.g. silicon) When cold, these are insulators because the conduction band is empty. However, it is so close to the band below that a temperature rise can give electrons enough energy to jump the gap. Conduction can then take place both by electron movement in the conduction band and by movement of the 'gaps' (called ***positive holes***) created in the band below.

The conductivity of an ***intrinsic*** (pure) semiconductor can be improved by ***doping*** it with small amounts of impurity. It is then an ***extrinsic*** semiconductor.

- ***n-type*** semiconductors have extra conduction electrons.
- ***p-type*** semiconductors have extra positive holes.

Polarization in insulators

Read D2 and E2 first.

Dielectrics These become polarized in an electric field, i.e. electrons and nuclei are displaced slightly in opposite directions. In a capacitor, this means that each plate has opposite charge near it. So less work is needed to put charge on the plates (for any given PD) and the capacitance is increased.

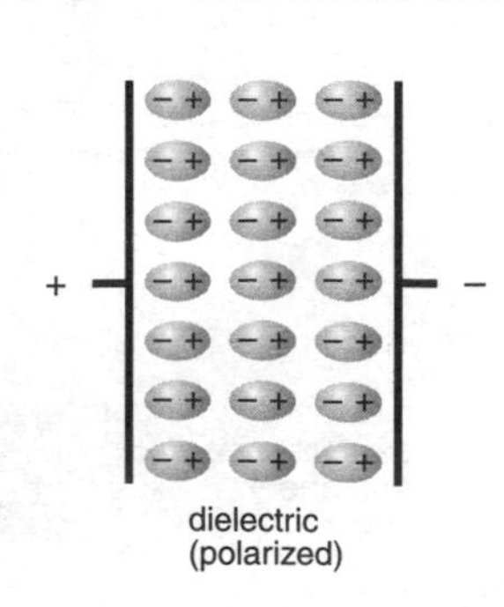

Dielectric loss is the energy transformation which occurs in a dielectric in an alternating electric field. The alternating displacement causes thermal motion: it has a heating effect.

Dielectric strength is the value of E above which a dielectric breaks down and conducts.

Ferroelectric crystals (e.g. barium titanate) These have a naturally-occurring, uneven distribution of charge. Though neutral, the whole crystal has an electric dipole.

H7 Medical physics – 1

Body mechanics

The human body is supported by a framework of bones called a ***skeleton***. The bones, which are light, stiff, and strong, are connected at ***joints***. Most joints are flexible, the bones being held together by bands of tissue called ***ligaments***. These are elastic, with similar properties to nylon.

The joints of the skeleton are moved by ***muscles***. These are the body's 'engines' where transformation of chemical energy into kinetic energy takes place. They are attached to bones by ***tendons***, which are similar to ligaments.

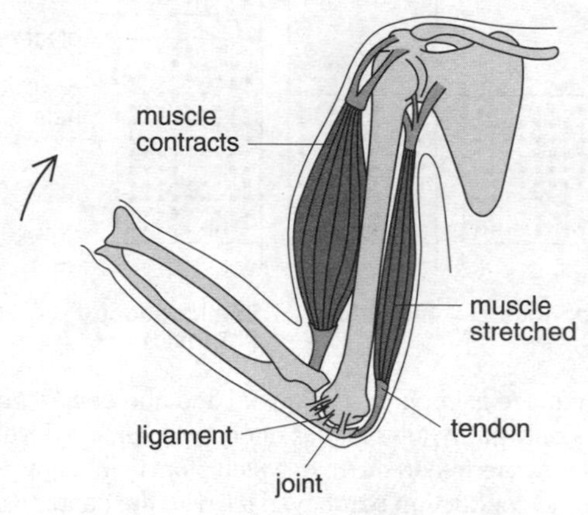

Jointed bones act as levers, with muscles providing the forces. Muscles can only contract, so they are arranged in pairs, with one muscle pulling the joint one way and its partner pulling it back again. The attachment points are close to the pivot (the joint), so high muscle forces are needed (see B5).

Spinal stresses The bones of the vertebral column (spine) are separated and cushioned by cartilage ***discs***. The lowest discs are subject to high compressive stress and also shear stress (see H5). In time, damage may occur. The situation is made worse by poor posture (body position), or by lifting heavy loads while leaning forward.

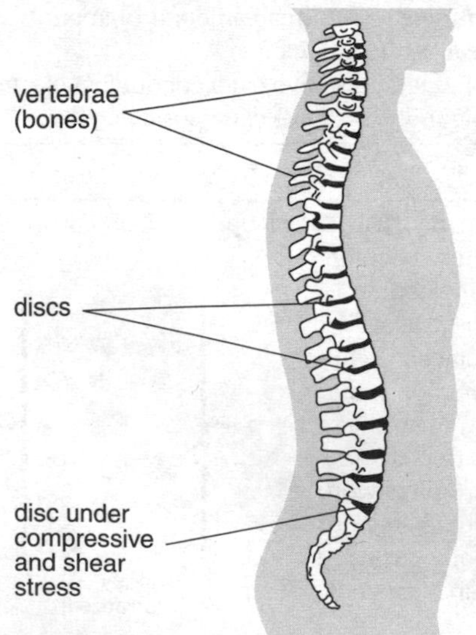

Walking This is only possible because of frictional forces between the feet (or shoes) and the ground.

Walking action tends to make the body's centre of gravity move up and down. This wastes energy, so the knees and ankles flex to reduce the vertical displacement. Also, when each foot strikes the ground, tendons and ligaments in the ankle and foot absorb strain energy, then return some of it as the leg is pushed back. This too reduces energy 'wastage'.

Body energy

Energy release The body gets most of its energy from chemical reactions in which food is combined with oxygen. The process is called ***respiration***. (Confusingly, the word 'respiration' is also commonly used for 'breathing'.)

Even when resting, the body still needs a supply of energy. The ***basal metabolic rate (BMR)*** is the minimum rate of energy release required to maintain basic life processes. As this energy is ultimately transferred from the body as heat, the BMR is normally expressed as the heat output per hour per square metre of body surface. Typical values are:

young adult female 150 kJ h^{-1} m^{-2}

young adult male 165 kJ h^{-1} m^{-2}

When the body is doing work (by moving muscles), the rate of energy transformation must rise. However, the second law of thermodynamics (see F3) limits the body's efficiency as an 'engine' to about 10%. So most of the energy results in heating the environment.

Temperature control For life processes to be maintained, the temperature of the body's ***core*** must be kept close to 37 °C. This requires a balance between the rates of heat gain and heat loss. The diagram below shows that main processes involved. Adjustments to the balance are made automatically by sweating, shivering, and control of the blood flow (and therefore the heat flow) to the skin. Clothing provides extra insulation to reduce energy transfer.

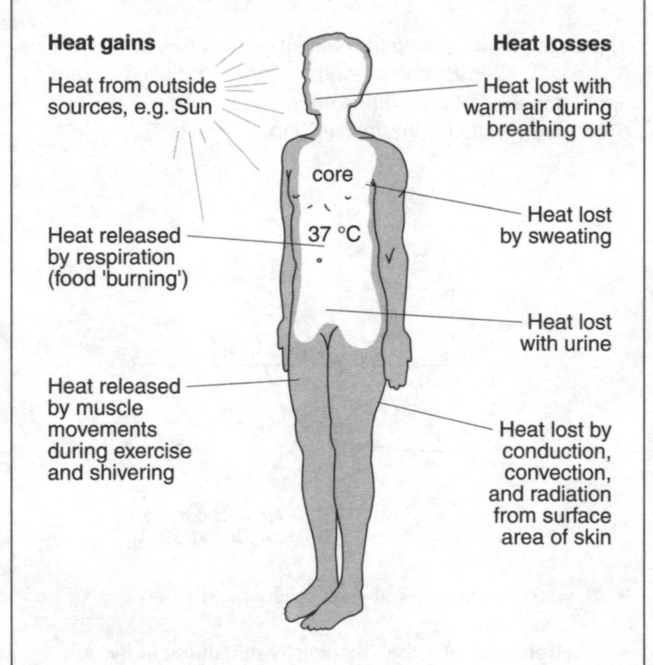

Note:

- At room temperature, an inactive body mainly transfers heat by conduction, convection, and radiation from the skin.
- In hot conditions, or when active, the body transfers most of its excess energy by sweating. The evaporation of sweat from the skin has a cooling effect (see F3 and F5).

Hypothermia This means a dangerously low body temperature. It occurs in cold conditions, when the body transfers more energy to the environment than it can replace. Babies are more at risk than adults. Being smaller, they have higher surface area per kg of body mass, so cool more quickly. Outdoors, ***windchill*** can cause hypothermia: the wind increases the energy transfer by convection – and by evaporation if the clothes are damp.

The human eye

Read C4 first.

In the eye, the cornea and lens form a real image on the ***retina***. The focus is adjusted by changing the shape of the lens – a process called ***accommodation***. This enables a normal eye to form a clear image of any object between its ***near point*** (about 25 cm away) and infinity. The amount of light reaching the retina is controlled by the ***iris***.

The retina contains millions of light-sensitive cells which send signals to the brain. ***Rods*** are the most sensitive, but do not respond to colour. Different ***cones*** respond to the *red, green,* and *blue* regions of the spectrum.

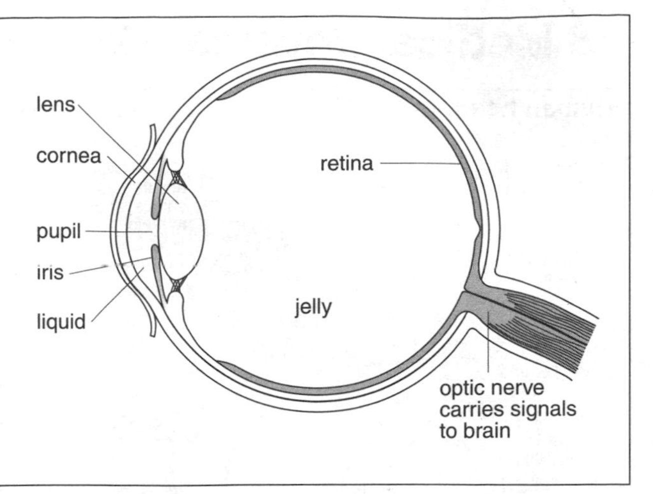

Myopia (short sight) A short-sighted eye cannot accommodate for distant objects. The rays are brought to a focus before they reach the retina. The defect is corrected by a *concave* spectacle lens.

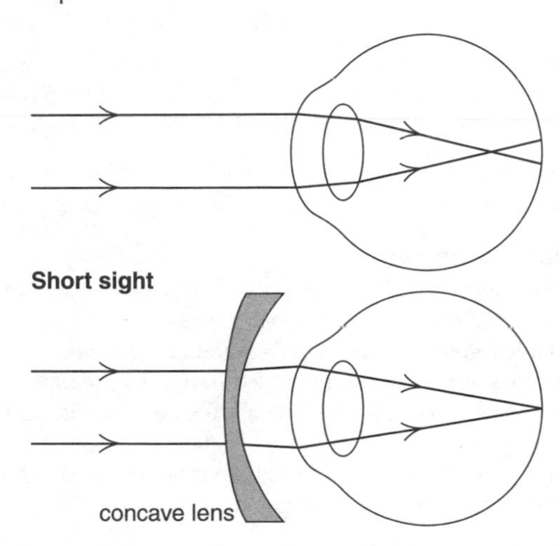

Hypermetropia (long sight) A long-sighted eye cannot accommodate for close objects. The rays are still not focused by the time they reach the retina. The defect is corrected by a *convex* spectacle lens.

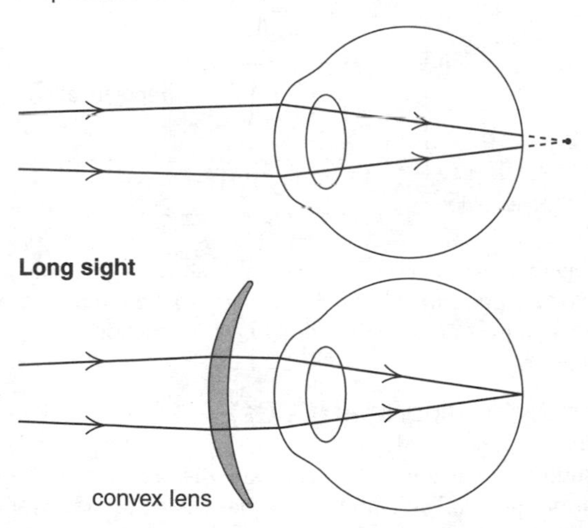

Lens power This is defined as $1/f$. With f in metres, the unit is the ***dioptre*** (D). (Note: *diverging* lenses have *negative* powers). The power of the eye's lens system varies. It is lowest when distant objects are being viewed.

The powers of close lenses can be added algebraically. For example, if a normal eye must reduce its power to 50 D to see distant objects, but a short-sighted eye cannot get beneath 54 D, then the defect is corrected by a lens of power –4 D, i.e. a concave lens with $f = -\frac{1}{4} = -0.25$ m.

Astigmatism This effect occurs when the curve of the cornea is not perfectly spherical. So, for example, vertical lines might be seen in clearer focus than horizontal ones.

Scotopic vision

This is the term that describes vision in low light or night-time conditions. Rods are not sensitive to colour but have their peak response in the green part (wavelength ≈ 500 nm) of the visible spectrum. They are very sensitive to light and are the principal receptors in dim light.

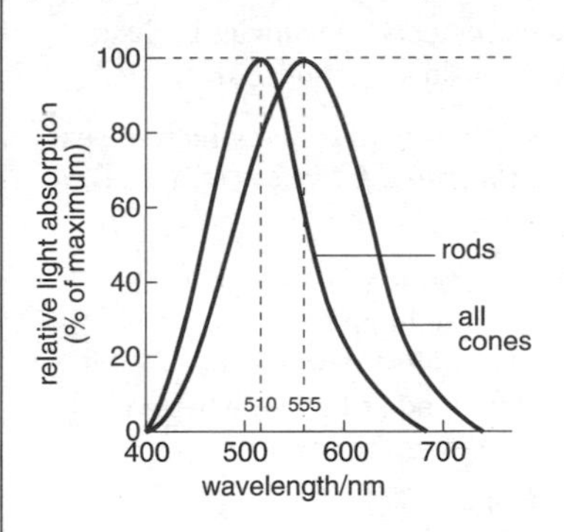

The frequency responses of the rods are shown in the diagram above.

Photoptic vision

Although they are not as sensitive as rods, cones are able to differentiate between different colours. There are three types of cone that are sensitive to different colours. The sensitivity of the cones is illustrated in the graph below.

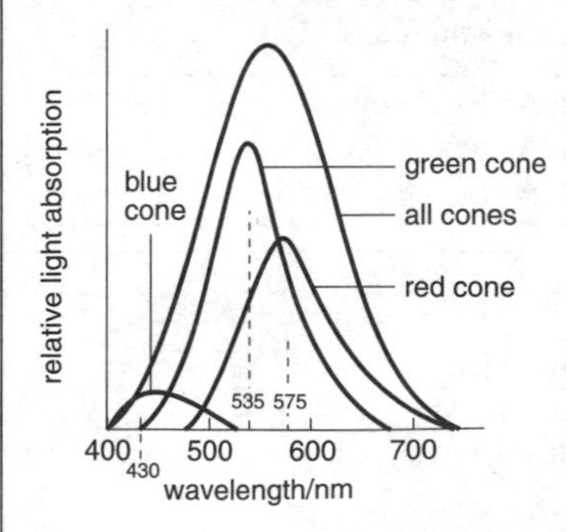

They are used in bright light or daytime conditions.

Persistence of vision

When a rod or cone has been stimulated by light it takes about 0.2 s for the image to disappear. Successive flashes merge or 'fuse' to produce a continuous image if they arrive frequently enough. In dim light 5 Hz is adequate to produce a continuous image.

In high-intensity conditions images need to arrive more frequently. In cinemas 24 frames per second are projected and in TV 60 images per second are transmitted.

H8 Medical physics – 2

Human hearing

Read C2 first.

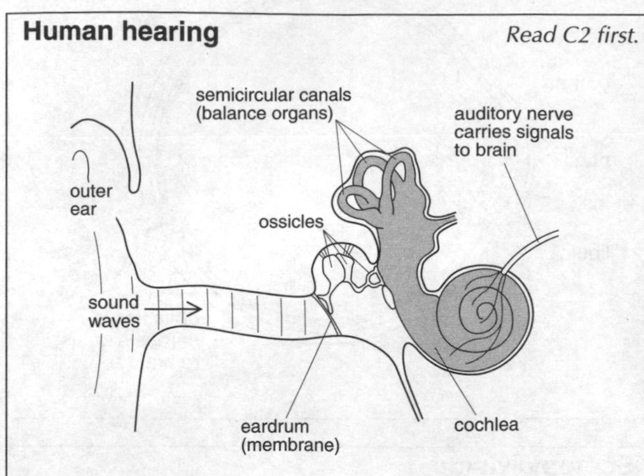

Sound waves entering the ear set up vibrations in the ***ear drum***. These are transmitted to the ***cochlea*** by ***ossicles*** (small bones) which act as levers and magnify the pressure changes. In the cochlea, sensory cells respond to different frequencies and send signals to the brain.

Frequency response The ear detects frequencies in the range 10 Hz to 20 kHz. It is most sensitive around 2 kHz.

Intensity levels The lowest intensity of sound which the ear can detect is known as the ***threshold intensity***, I_0. It is taken as 10^{-12} W m^{-2}.

The ear responds to sounds according to the *ratio* of their intensities. For example, each doubling of intensity gives the same sensation of increased loudness. For this reason, the ***intensity level*** of a sound is defined as below, where dB stands for ***decibel***, and I is the intensity of the sound:

$$\text{intensity level in dB} = 10 \log_{10}(I/I_0)$$

For two sounds of intensities I_1 and I_2,

$$\text{difference in intensity level in dB} = 10 \log_{10}(I_2/I_1)$$

Note:

- Any two sounds with the same intensity ratio I_2/I_1 have the same difference in intensity level.

dBA scale This is a dB scale, adjusted to allow for the ear's different sensitivity to different frequencies. It is used to measure noise levels. Typical values are:

hearing threshold	0 dBA	legal noise limit	90 dBA
conversation	50 dBA	near disco speakers	120 dBA
busy street	70 dBA	pain threshold	140 dBA

Hearing defects

Defects in hearing are caused by

- sound signals failing to reach the cochlea
- damage to the nerves that send signals from the cochlea to the brain.

Causes of defects

- natural ageing that gradually reduces the ability to hear high frequencies
- exposure to excessive noise such as that caused by an explosion or machinery
- an illness
- an accident
- prolonged exposure to excessively loud sounds such as that due to machines or excessively loud music. This can produce a continuous ringing sensation in the ear known as 'tinnitus'. As a health and safety measure those who work in noisy environments are required to wear ear defenders.

The heart

The diagram shows schematically the action of a heart.

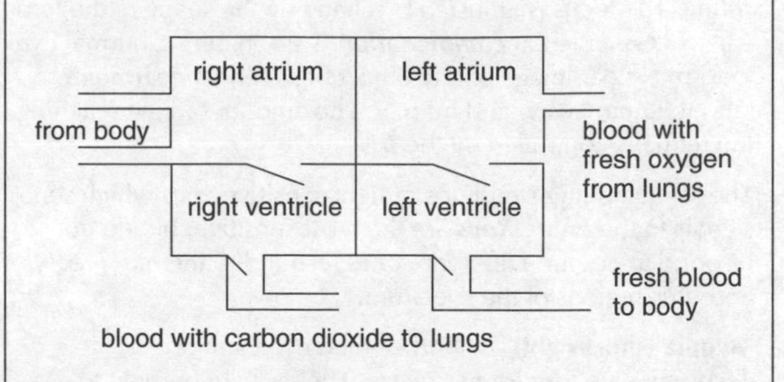

The heart is essentially a pump. There are two separate paths through which blood flows. The valves are one-way valves so that blood can only flow one way.

In path 1, blood is pumped to the body via arteries so that oxygen reaches the body tissue. At the same time the blood picks up carbon dioxide returning to the heart through veins.

In path 2 the blood is pumped to the lungs where it releases the carbon dioxide and picks up a fresh supply of oxygen.

Action potentials

Signals in the form of voltage pulses are passed round the body by the nervous system. The signals are sent by living cells called ***neurons***. The signals travel along ***axons***. The signals may stimulate routine tasks such as the heartbeat. Some nerves work as a result of an outside stimulus, such as when you touch something very hot. Some neurons send signals to the brain, others directly to muscles causing them to contract in response to an emergency.

When not transmitting the resting potential of the neuron is about –70 mV. This is the difference between the voltages inside and outside the neuron.

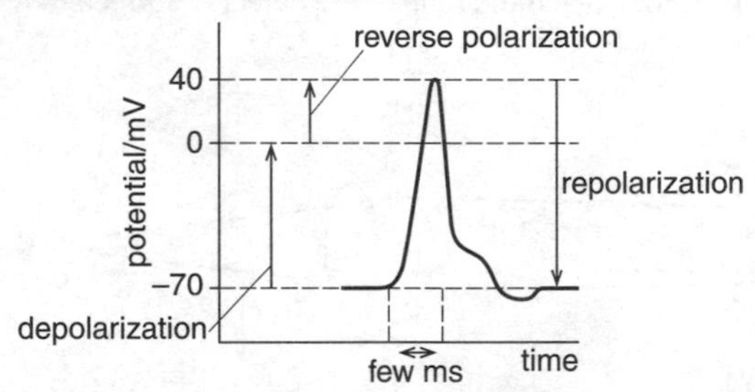

When the neuron receives a stimulus above a threshold level a voltage pulse travels along the axion to the brain or directly to the muscle causing the body to act in reaction to the stimulus.

Electrocardiograms (ECGs)

The action potentials produced during a heart beat travel round the body and, although small, they can be detected at various points. The graph shows the voltage variation for a single heartbeat.

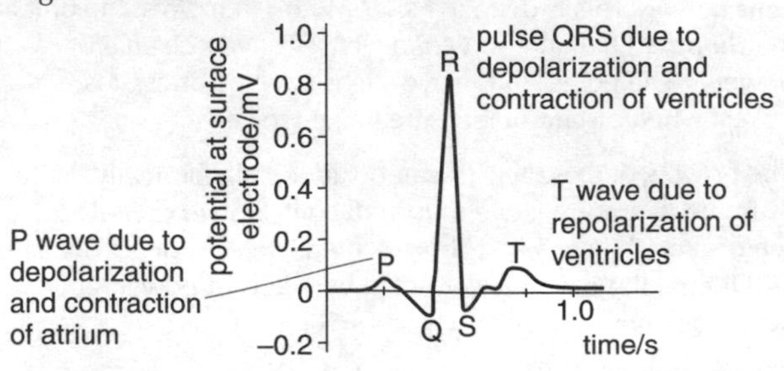

Signals detected at different parts of the body are indicative of different aspects of the body's function, so they can be used to diagnose health problems, particularly those directly related to the heart itself.

Ultrasound

Ultrasonic sound, or ***ultrasound***, has frequencies above the upper limit of human hearing, i.e. above 20 kHz. If a pulse of ultrasound is sent into the body, it is partially reflected at the boundaries between different layers of tissue, so their positions can be worked out from the time delays of the echoes received. The frequency used (in the 2–10 MHz range) depends on the depth of tissue. Increasing the frequency gives better resolution but poorer penetration.

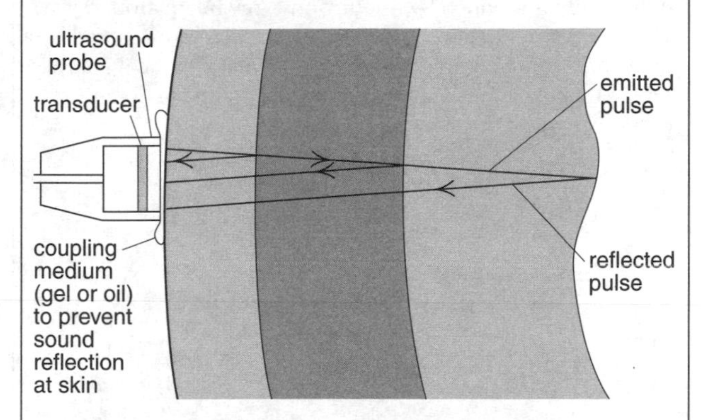

Probe This contains a ***transducer*** (see H16) which sends and receives the ultrasound pulses. The ultrasound is produced using the ***piezoelectric effect*** (see I 16): a high frequency alternating voltage is applied across a slice of crystalline ceramic so that it vibrates at its resonant frequency and emits sound waves. When the reflected waves return, they cause vibrations in the slice, which generates a small alternating voltage. Signals from the probe can be displayed on a CRO or processed by a computer.

A-scan The reflected pulses are displayed as peaks on a CRO, i.e. as **A**mplitude changes. Positions along the time axis are a measure of distances into the body.

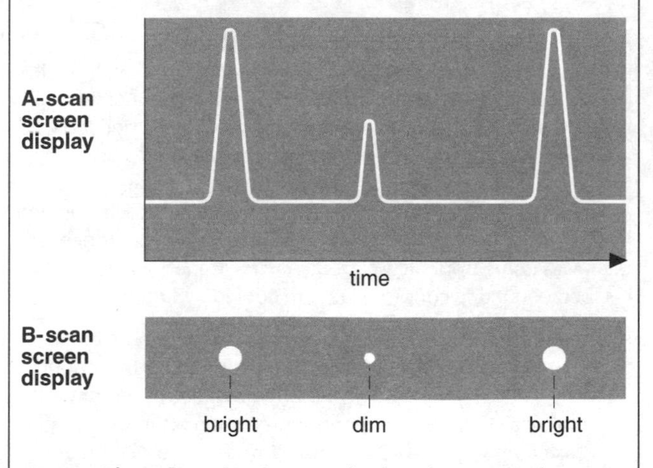

B-scan The reflected pulses are displayed as spots whose **B**rightness is a measure of their amplitude. In a ***two-dimensional B-scan***, the probe is moved around the body in order to build up a complete cross-sectional image.

- Though its resolution is poorer, ultrasound imaging is safer than X-ray imaging and cheaper than MRI.

Destructive ultrasound Focused ultrasound, at an intensity above 10^7 W m^{-2}, can be used to break up kidney stones and gallstones, so surgery is not required.

Acoustic impedance

A transmitting medium opposes the transmission of a sound pressure wave. The acoustic impedance is a measure of the resistance of a medium to the passage of sound waves.

The ***specific impedance*** of the medium $Z = \rho c$
where ρ is the density of the medium
and c is the speed of sound in the medium.

Intensity reflection coefficient α

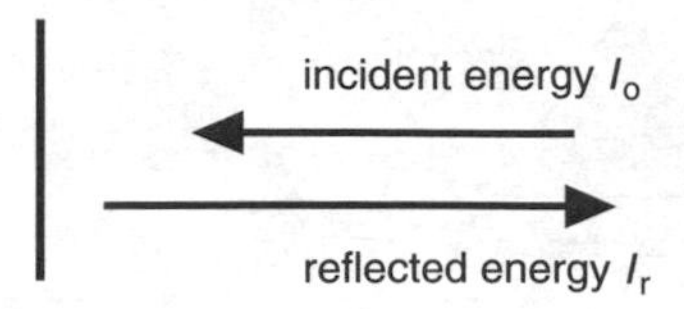

When ultrasound reaches an interface between two different media the fraction of the incident energy that is reflected is called the ***intensity reflection coefficient*** α_r and is given by

$$\alpha_r = \frac{I_r}{I_0} = \left(\frac{Z_2 - Z_1}{Z_2 + Z_1}\right)^2$$

Using ultrasound

Typical specific impedances are:

air	430 kg m^{-2} s^{-1}
fat	1.38×10^6 kg m^{-2} s^{-1}
blood	1.59×10^6 kg m^{-2} s^{-1}
bone	about 6×10^6 kg m^{-2} s^{-1}

At a muscle fat interface the reflection coefficient is 0.10 so 10% of the incident energy is reflected. Between fat and bone the figure rises to 60%. Even when only 1% of the energy is transmitted the echoes can be detected and can produce useful information.

Coupling medium

When sound is incident between air and the skin the reflection coefficient is almost 1 so that when ultrasound reaches the skin nearly all the incident sound is reflected. The coupling medium of gel or oil between the ultrasound probe and the skin reduces the amount of energy that is reflected.

Using the Doppler effect Any motion within the body causes a Doppler shift (see H1) in the reflected ultrasound. This can be used to check the heart beat or blood flow in an unborn baby. Continuous, rather than pulsed, ultrasound is used. Any difference between the outgoing and returning frequencies is heard as a tone or displayed on a screen.

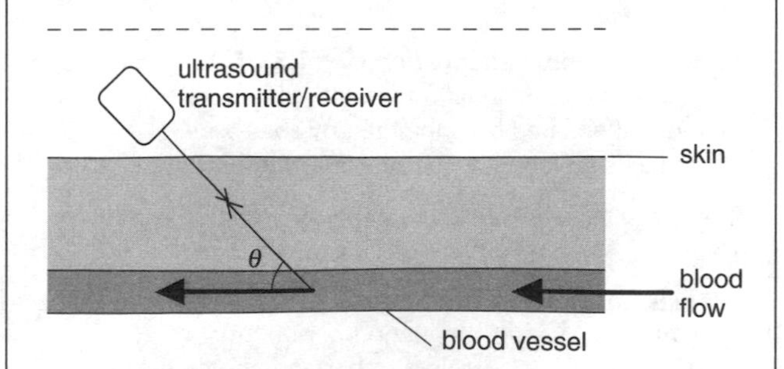

The original frequency of the ultrasound is f.
The change in frequency due to the Doppler effect is Δf.
The wave speed in the medium is c.
The speed of the blood v is given by

$$v = \frac{c \times \Delta f}{2f\cos\theta}$$

The volume flow rate of blood through the blood vessel is Av where A is the area of cross-section of the blood vessel.

H9 Medical physics – 3

X-rays

Read E8 and G5 first.

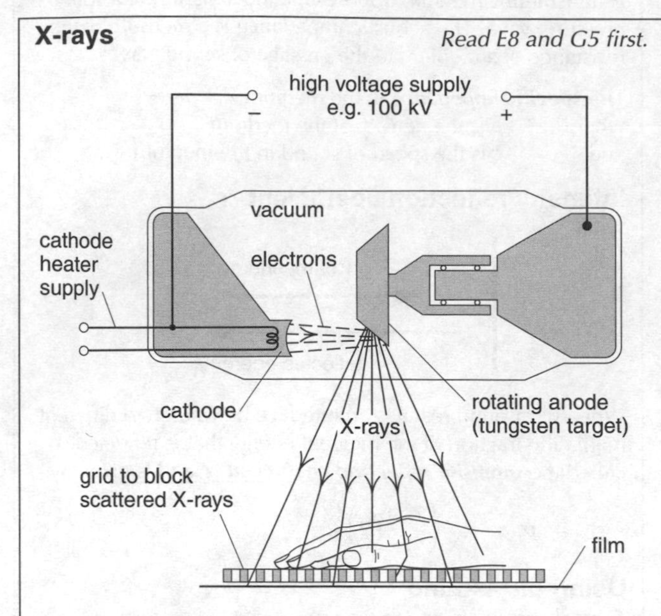

In the X-ray tube above, electrons gain high KE before striking a metal target. About 1% of the KE is converted into X-ray photons; the rest is released as heat. The anode is rotated rapidly to prevent 'hot spots'.

Quality The X-ray beam contains a range of wavelengths. The spectral spread is known as the ***quality***. The longer, least penetrating wavelengths are called *soft* X-rays; the shorter, more penetrating ones are *hard*. The beam can be hardened by using filters to absorb the longer wavelengths.

Tube current and voltage Increasing the current increases the intensity of the beam. Increasing the voltage increases the intensity *and* reduces the peak wavelength, i.e. it produces more photons, which are more penetrating.

Attenuation As X-rays pass through a material, their energy is gradually absorbed and their intensity reduced. This is called ***attenuation***. It is in addition to any reduction in intensity due to beam divergence.

If I_0 is the intensity of the incident beam (do not confuse with I_0 as used in H8), and I is the intensity at a distance x into the material, then, for a non-diverging beam,

$$I = I_0 e^{-\mu x} \quad (1)$$

μ is the ***total linear attenuation coefficient***. Its value depends on the absorbing material and on the photon energy. X-rays are more attenuated by bone than by soft tissue. The difference is greatest for photons of around 30 keV energy.

The half-thickness $x_{\frac{1}{2}}$ of an absorber is the value of x for which $I/I_0 = \frac{1}{2}$. From equation (1): $x_{\frac{1}{2}} = \log_e 2/\mu$

The ***mass attenuation coefficient*** μ_m ($= \mu/\rho$) is often used in place of μ. This is the attenuation per unit mass and is constant for a given atomic number and photon energy.

Contrast between soft tissue is poor so to yield useful information about the digestive tract patients are given a ***barium meal***. The relatively dense barium absorbs X-rays. The meal, in the form of a drink, is taken just before the X-ray is taken and this improves the contrast between the digestive system and the rest of the body.

X-ray photographs X-rays affect photographic film, but cannot be focused, so the photos are 'shadow' pictures of the absorbing areas. For a sharp image, the X-rays need to be emitted as if from a point source. To reduce the risk of cell damage, exposure times are normally less than 0.2 s.

CAT scanning (computed axial tomography) A cross-section of the body is scanned by rotating an X-ray beam around it. The intensity reduction caused by each 'slice' is measured by a detector. Then a computer carries out a mathematical analysis on the total data and uses it to construct an image of the cross-section on the screen.

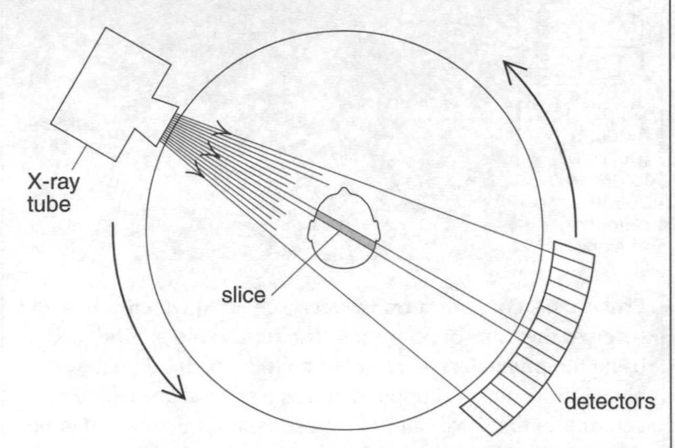

Radiotherapy Very hard X-rays (photon energies around 10 MeV) can be used to destroy cancer cells deep in the body. However, to reduce damage to the surrounding tissue, the beam is rotated around the patient so that only the target area stays in the beam all the time. (Note: gamma rays from a cobalt-60 source can be used instead of X-rays.)

MRI (magnetic resonance imaging)

This produces images of the body by scanning, but without the risks of X-rays. It uses the fact that different tissues contain different concentrations of hydrogen atoms. In a magnetic field, the spinning motions of the hydrogen nuclei are disturbed by pulses of radio waves of suitable frequency. As a result, the nuclei emit radio-frequency signals which can be detected and located electronically. The very strong magnetic fields needed for MRI come from high-current electromagnets whose superconducting coils (see D2) are cooled by liquid helium.

Using magnetic resonance techniques it is also possible to detect and measure chemicals that are in the body without having to take samples. This is called **magnetic resonance spectroscopy (MRS)**. By monitoring chemicals at the site of a disease doctors are able to conduct research to determine whether a particular treatment is effective.

Using tracers

Read G2 first.

Tracers used in medical diagnosis include the γ-emitting radioisotopes iodine-123 (half-life 13 h) and technetium-99^m (half-life 6 h). Small amounts can be carried in the bloodstream to various sites in the body.

Gamma camera γ photons from the tracer strike the sodium iodide disc, causing flashes of light whose intensity is amplified by the photomultiplier tubes. Signals from these are processed electronically, and an image built up on a CRO screen. The collimator improves image quality by only letting through γ rays travelling at right angles to the disc.

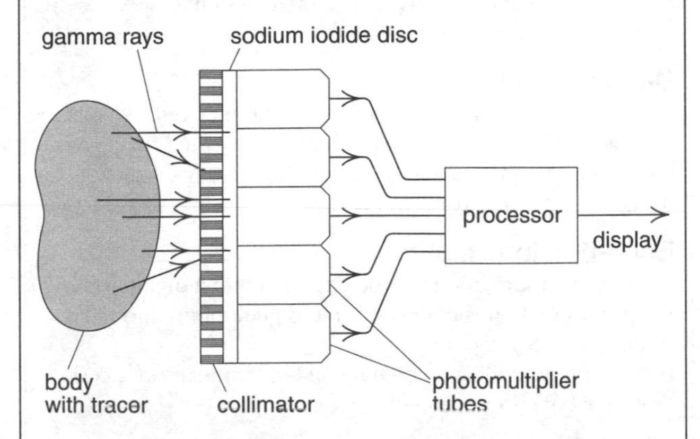

Checking blood flow in the lungs With technetium-99^m tracer present in the bloodstream, a gamma camera will reveal which parts of the lungs do not contain the tracer and, therefore, have blocked blood vessels.

Checking thyroid function The thyroid has a natural uptake of iodine. With iodine-123 tracer present in the bloodstream, the uptake can be checked by measuring the activity in the thyroid.

Radiation risks

Read G2 first.

Ionizing radiations such as X and γ rays can cause cell damage which may lead to cancer or genetic changes in sex cells. The amount of damage depends on the amount of energy absorbed. This rises with increased exposure time.

Absorbed dose This is the energy absorbed per unit mass of tissue. The SI unit is the J kg^{-1}, called the ***gray*** (Gy).

Dose equivalent For the same amount of energy delivered, α particles cause more biological damage than X or γ rays. To allow for this, the ***dose equivalent*** is defined as the absorbed dose × Q, where Q is a ***quality factor*** (e.g. 1 for X and γ rays, 20 for α particles). The SI unit of dose equivalent is the J kg^{-1}, known in this case as the ***sievert*** (Sv).

UK average dose equivalent per person per year	2.5 mSv	**Sources:**	
		natural (e.g. rocks)	87%
		medical (e.g. X-rays)	12%
		other	1%

Note:

- At low levels, radiation is ***stochastic*** (random) in its effects. There is no minimum safe level. The risk of damage increases with the dose equivalent absorbed.

Half-lives When a radioisotope is in the body, its *effective* half life T_E is less than its *physical* ('real') half-life T_P because biological processes gradually remove it from the body. If T_B is the *biological* half-life, i.e. the time for half the original radioactive material to be removed:

$$1/T_E = 1/T_P + 1/T_B \qquad (T_P = T_{\frac{1}{2}} \text{ in G2})$$

Physiological effects of radiation

Damage to cells may affect the individual by producing

- death due to cancer
- radiation sickness (nausea, loss of hair, tiredness)
- damage to genes that are transmitted to offspring and may produce mutations or other harmful effects.

The ions produced by radiation react inside the cell and interfere with the operation of the cell. For example, the ionization could cause a molecule to break apart if a bonding electron is lost. When a large number are destroyed by high doses the body may not be able to replace them quickly enough, causing health problems. The effect depends on the size of the dose.

Radiation may damage the DNA. This is serious because a cell may only have one copy of this so it is irreplaceable. The cell may die or become defective. The defective cell may then divide and produce more defective cells. This rapid production of defective cells is cancer.

Risk associated with radiation diagnosis or treatment

When deciding whether to treat a patient using radiation it is necessary to consider whether the risk associated with the use of radiation outweighs the risk of not using radiation for the treatment or diagnosis of a disease.

Maximum permitted level (MPL)

For those who work with radiation there is a maximum permitted dose of 50 mSv per year for the whole body or reproductive organs and 750 mSv per year for the hands, feet, etc. A maximum of 5 mSv per year is permitted for those who do not work with radiation.

Using light

Read C2 first.

Endoscope This is a flexible tube used for looking inside the body. Light from objects at the far end is carried to the viewing end by a *coherent* bundle of optical fibres, i.e. fibres whose positions alongside each other are the same throughout their length, so an accurate image is formed.

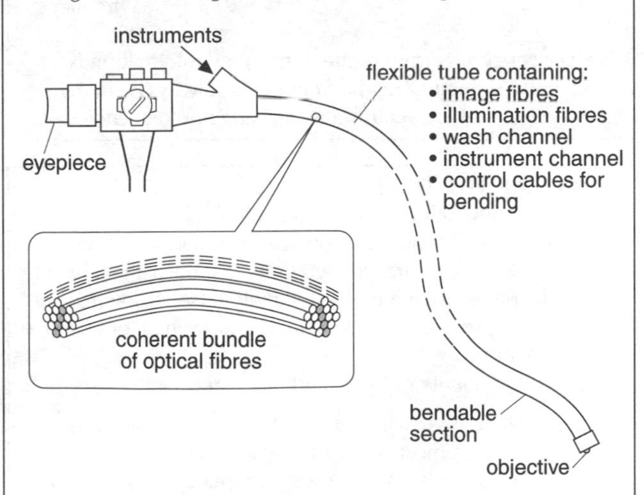

Laser This gives a fine beam of high-intensity light. It can be used like a scalpel to cut tissue, but its heating effect also seals small blood vessels at the same time. The type of laser chosen depends on the tissue – the aim being to select a wavelength which gives maximum energy absorption. In some cases, the light can be carried to the required point by an optical fibre, i.e. through an endoscope.

H10 Telecommunications – 1

Telecommunications systems

Telecommunications systems (e.g. radio, TV, and telephone) send information from one place to another using electric currents or electromagnetic radiation. The information being sent may be sounds, pictures, or computer data.

- The ***channel of communication*** may be radio waves, microwave beam, metal cable, or optical fibre.
- Signals for sounds (e.g. speech) are called ***audio*** signals.

Simple sound information: frequency spectrum

Sound information may be very simple. For example, the sinusoidal variation of the displacement of air against time, shown in the figure below, represents a pure a single note

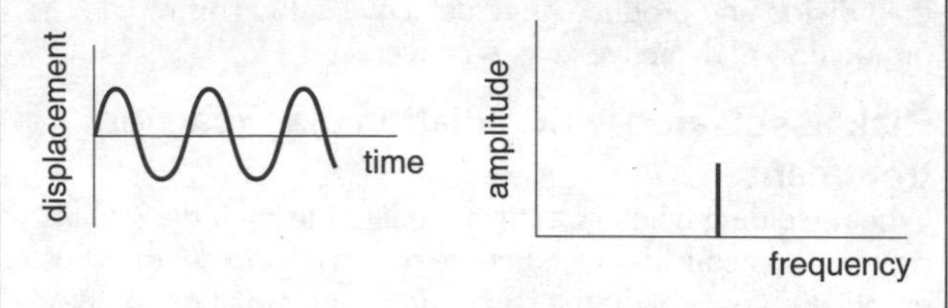

The sound can be represented by a ***frequency spectrum***. This shows the frequencies that are present, plotted on the *x*-axis, and their relative amplitudes, shown by the height of the vertical line.

More complex sounds: speech transmission

Speech consists of a succession of complex waveforms such as this:

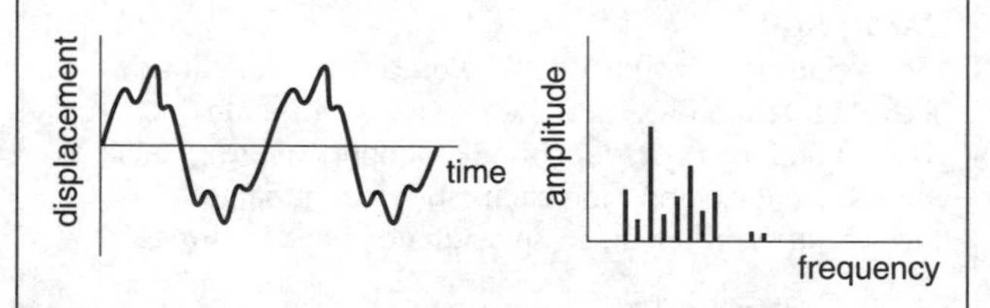

The frequency spectrum of the transmitted information is continually changing but the frequencies always lie in the same range. Music contains a wider range of frequencies.

Base bandwidth

The base bandwidth of a communication channel

- is the range of frequencies present in the original information that is transmitted (note that the actual channel bandwidth is twice the base bandwidth (see below))
- is the range of frequencies that are transmitted for effective transmission
- affects the quality of the sound that is received.

Although all frequencies to which the ear is sensitive are present in speech most of the energy lies in the frequency range below about 3.4 kHz. Adequate speech is transmitted for a base bandwidth of 300 Hz to 3000 Hz. This range is used in telephone systems.

For high-quality transmission of music, audible frequencies from 15 Hz to 15 kHz have to be transmitted.

- For reasonable-quality music and good-quality speech the base bandwidth range has to be 50 Hz to 4500 Hz.
- This range is used when transmitting signals by medium- and long-waveband radio. AM (amplitude modulated) transmissions are used in this case. This is used in long-wave and medium-waveband radio.

Video information

To transmit black and white video (TV) information with sound a much wider base bandwidth, of about 8 MHz, is needed. For colour TV a channel is needed for each of the three colours, red, blue, and green.

Analogue information

The displacement of a sound wave varies continually with time. When transmitting analogue information a voltage is transmitted that is proportional to the displacement. The transmitted voltage variation is a replica of the displacement time variation. This voltage variation is converted back to the original signal at the receiver. The radio system below is transmitting analogue signals.

Digital signals

These are pulses, produced because a circuit's output voltage is either HIGH or LOW. They can be represented by the by the logical numbers 1 and 0 (see H17). The advantages of using them are described in H11.

Encoding information

All types of information can be encoded into a digital form. Video information is converted into digital information in a similar way to audio.

Letters and numbers can be transmitted by means of a code to represent each individual character

Image production

Each image on a video display unit is made up of pixels that are arranged in an orderly way on the tube. In colour display units there are three types of pixel. One type glows red when electrons are incident on it, another blue, and another green. The brightness of each pixel depends on how many electrons hit the pixel each second. The full range of colours is obtained by suitably illuminating pixels that are close together. The eye interprets the mixture of red, blue, and green as a different colour or hue.

Resolution

If you sit close to a TV screen you will be able to see the individual pixels. In this case the eye is able to resolve (see C5) the pixels that are close together so that the effect of a full range of colours is not seen. The minimum angle for resolution is about 1×10^{-4} rad (about $6 \times 10^{-3\circ}$).

To view a screen at 0.5 m (which is typical when working at a computer) the pixels need to be closer together than 50 µm if they are not to be seen as individual pixels.

Bandwidth Mathematically, a carrier of frequency f_c, amplitude modulated at a frequency f_m, is equivalent to a pure sine wave of frequency f_c, with two ***side frequencies*** of $f_c + f_m$ and $f_c - f_m$ (i.e. at the carrier frequency $\pm f_m$).

Speech and music contains a variety of frequencies, mainly between 50 Hz and 4.5 kHz, so a modulated carrier normally has two bands of side frequencies, called ***sidebands***. The frequency range occupied by the signal is the ***bandwidth***.

- A radio station needs a bandwidth of about *twice* its AF range. In the UK, the bandwidth used for AM is 9 kHz.

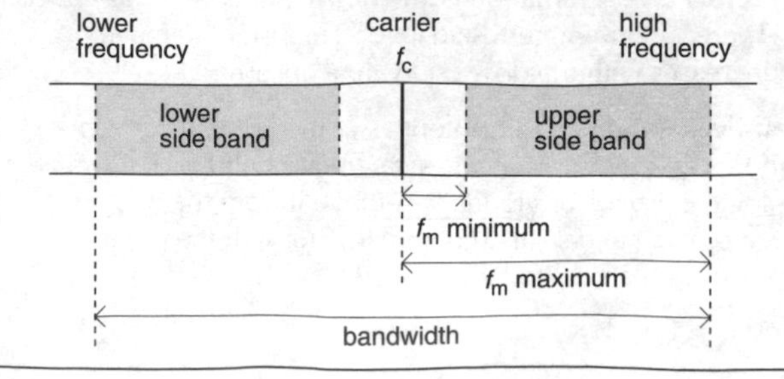

Radio communication

The principle of simple radio system is shown below:

aerial (antenna)
RF carrier
modulated RF carrier
RF oscillator
modulator
RF amplifier
AF amplifier
microphone
AF signal

Transmitter

aerial (antenna)
modulated RF carrier
tuning circuit
RF amplifier
demodulator (detector)
AF amplifier
AF signal

Receiver

Carrier Radio frequencies (e.g. 1 MHz) are much higher than audio frequencies (e.g. 1 kHz). The principle of radio transmission is to produce a steady radio signal called the carrier, and then ***modulate*** (vary) it with the audio signal.

Amplitude modulation The instantaneous amplitude of the carrier frequency depends on the displacement of the signal. This is used for medium- and long-wavelength transmissions.

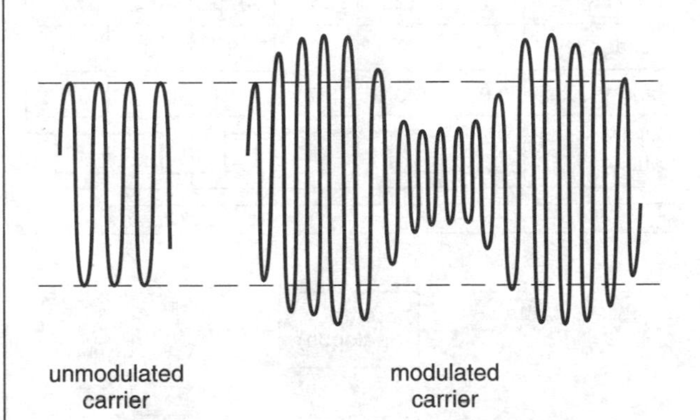

Frequency modulation The instantaneous frequency of the carrier frequency depends on the displacement of the signal. The maximum frequency change of the carrier depends on the amplitude of the signal.

The diagrams below right show how simple signals – in this case a positive signal followed by a negative – can be carried by modulating the carrier frequency. The positive signal increases the frequency, the negative signal decreases it.

Note: In practice the frequency is very high.

Frequency modulation is less affected by noise than amplitude modulation but needs a much greater bandwidth. It is used for transmission at high frequencies (UHF and VHF).

Tuning circuit Each radio station broadcasts at its own carrier frequency. The tuning circuit uses a resonant circuit (see B13) to select the incoming frequency required.

Demodulation The demodulator removes the ***RF*** (radio frequency) part of the signal, and passes on only the ***AF*** (audio frequency) part. The key component in demodulation is the diode. This rectifies (see D4) the RF signal so that only its 'forward' parts are left, as pulses of varying amplitude. The AF signal is recreated from these.

Frequencies used for radio communication

Frequency band		Examples of uses
30 kHz – 300 kHz	low (LF)	long-wave radio
300 kHz – 3 MHz	medium (MF)	medium-wave radio
3 MHz – 30 MHz	high (HF)	short-wave radio
30 MHz – 300 MHz	very high (VHF)	FM radio
300 MHz – 3 GHz	ultra-high (UHF)	TV
1 GHz – 30 GHz	microwave	telephone and TV links, satellite links, radar

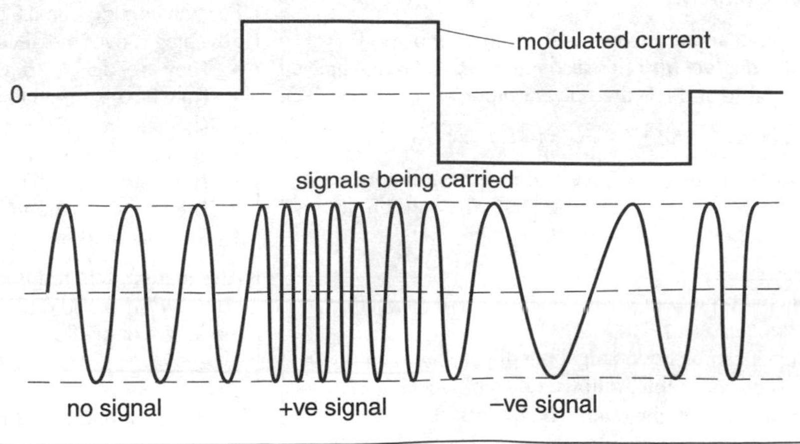

H11 Telecommunications – 2

How radio waves travel

Despite the curvature of the Earth's surface, radio waves can travel between places which are a long way apart.

Ground (surface) waves These follow the Earth's surface. Low frequencies travel furthest (up to 1000 km).

Sky waves These are reflected by the ionosphere (see H15) and the Earth. They are mostly high frequency waves.

Space waves These are waves with frequencies above 30 MHz. They are not reflected by the ionosphere. They can only be used for 'line of sight' (i.e. straight) communication, but their range can be extended by a satellite link.

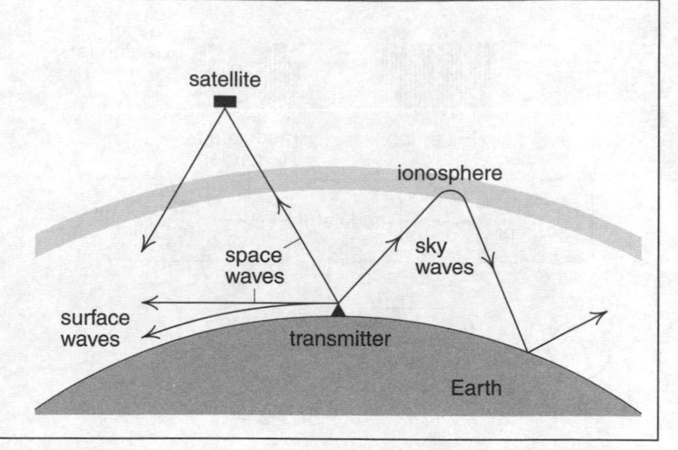

Digital transmission

Most telephone systems use digital transmission between exchanges. Audio signals from one telephone are converted from analogue to digital form, transmitted, and then changed back into analogue signals for the other telephone.

To convert the analogue signal on the right into a digital signal, it is ***sampled*** by measuring the voltage level at regular intervals of time. The levels are then changed into ***binary*** codes, consisting only of 0s and 1s, and these are transmitted as a sequence of pulses. The process is called ***pulse code modulation (PCM)***. Reversing the process produces the analogue signal again.

- Telephone systems normally use 256 voltage levels. Each requires an 8-bit binary code, e.g. 10011001.
- The ***sampling rate*** needs to be at least *twice* the highest frequency in the signal being sampled. Telephone systems normally use a sampling rate of 8000 Hz, i.e. the signal is being sampled every 1/8000 s.

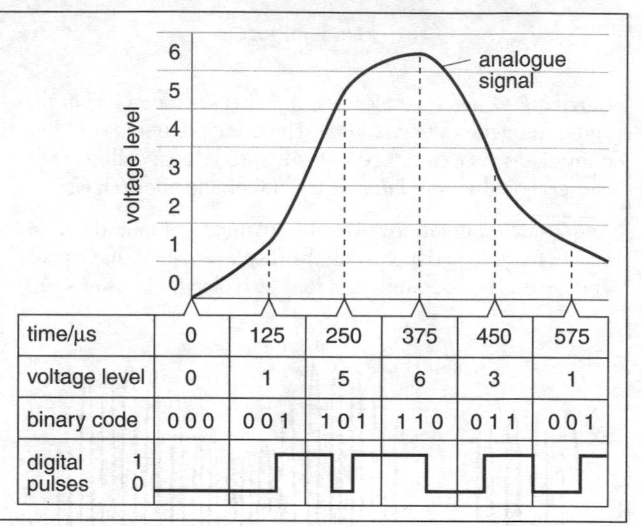

time/μs	0	125	250	375	450	575
voltage level	0	1	5	6	3	1
binary code	0 0 0	0 0 1	1 0 1	1 1 0	0 1 1	0 0 1

Advantages of digital transmission

Regeneration Signals lose power as they travel along a cable. They are also affected by ***noise*** (additions caused by electrical interference and thermal activity). To restore their power and quality, digital signals can be amplified and 'cleaned up' at intervals by ***regenerators***. (Analogue signals can be amplified by ***repeaters***, but unfortunately, the noise is amplified as well.)

Data handling Computers operate digitally, so digital transmission is ideal for long-distance computer links.

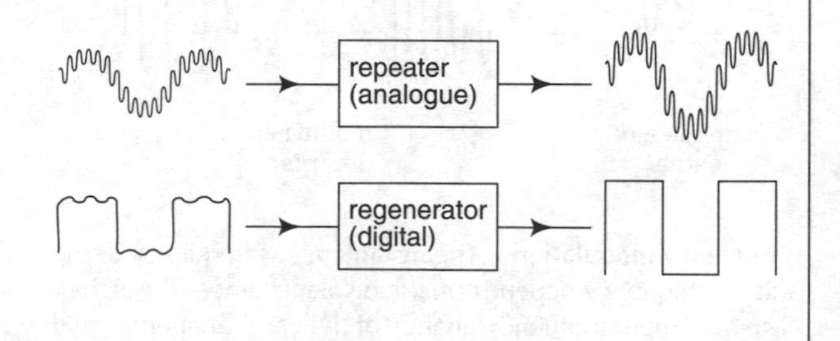

Power losses and gains

Signals lose power as they travel along a cable. The effect is called ***attenuation***. Repeaters or regenerators compensate by increasing the power.

In telecommunication, the unit used for measuring power change is the ***decibel (dB)*** (see also H8). If P_0 is the power input into a cable and P is the power output,

$$\text{power increase in dB} = 10 \log_{10} (P/P_0)$$

So, if the power input is 200 mW and the power output is 2 mW, the power increase = $10 \log_{10}(2/200) = -20$ dB. In this case, there is a power *loss* of 20 dB.

Note:

- The attenuation caused by a cable is often expressed in dB per km.
- dB changes can be added algebraically. If there is a power loss of 20 dB in a cable, followed by a power gain of 15 dB in a repeater, the overall power loss is 5 dB.

Optical fibres

Read C2 first.

Optical fibres carry infrared pulses from a laser or LED. At the far end, these are detected by a sensor (a photodiode).

For transmitting signals, optical fibre cables have many advantages over metal cables and radio waves:

- They are ideal for digital transmission,
- They have a much higher signal-carrying capacity,
- They are free of noise and crosstalk (signals crossing over from one fibre or wire to the next),
- They offer better security, e.g. they cannot be 'tapped',
- They are thinner and lighter than metal cables,
- Attenuation is low.

Attenuation Infrared is absorbed as it passes along an optical fibre. If P_0 is the power input, P is the power output, and x is the length of the fibre:

$$P = P_0 e^{-\alpha x}$$

where α is the ***linear attenuation coefficient***.

Multiplexing

This is a process that enables many signals to be transmitted at the same time, in both directions, down a single wire.

Frequency division multiplexing

This is the system commonly used in radio transmission. Each channel of communication is allotted a range of frequencies corresponding to the required bandwidth for the transmission.

No other channel can use that range of frequencies in the same reception area. This is illustrated in the diagram below where the frequency range available has been divided up into 12 separate 'chunks' each of width 8 kHz.

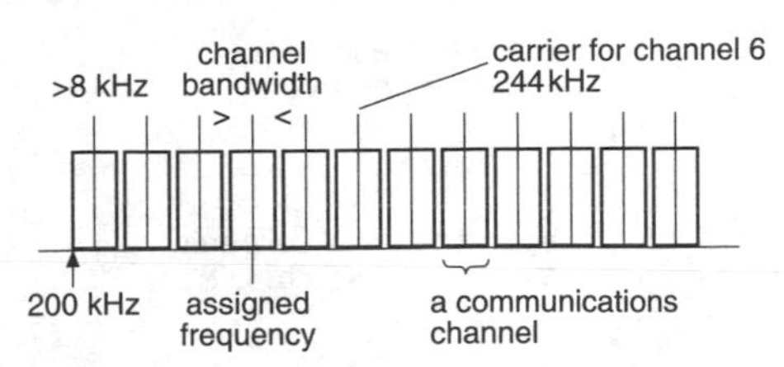

Note that because of the wider bandwidth requirement a given range of frequencies can transmit fewer video channels that audio channels.

Time division multiplexing

This is the system that enables many digitized transmissions, such as telephone conversations, to use the same communications channel. The signal of each user is sampled as explained in the section on digital transmission opposite.

To transmit a single conversation the voltage level of each signal is sampled 8000 times a second and for each sample 8 bits (0s or 1s) have to be transmitted. This is one ***byte*** of information.

When transmitting alpha-numeric data (numbers or letters) each byte will represent a number or letter.

Transmitting more channels

Using higher frequency ranges to carry the information allows more channels to be transmitted. The range available for UHF (ultra high frequency) transmission as is used for analogue TV transmission is 300 MHz to 3000 MHz, a total of 2700 MHz. This range could carry 2700/24, which is approximately 110 colour TV channels.

$$\text{number of channels} = \frac{\text{total bandwidth available}}{\text{bandwidth required for each channel}}$$

Using optical frequency ranges and transmitting information down optical fibres increases the available frequency range. The range of visible light is about 5×10^{14} Hz so theoretically $(5 \times 10^{14})/(24 \times 10^{6}) = 2.1 \times 10^{7}$ analogue TV channels could be transmitted down one optical fibre. Fewer channels are available in practice.

Bit rates

Bits can be sent at much higher rates than that required to transmit a single conversation. A coaxial cable can carry 140 Mbit s^{-1}. This is called the ***bit rate***.

This means that a single coaxial cable can carry $140 \times 10^{6}/64 \times 10^{3} = 2200$ telephone conversations.

A single cable can be shared by 2200 users with each user's signal being transmitted in turn. Each signal only uses the cable for 1/2200 of the time hence the term 'time division'.

A single transmission requires $(8 \times 8000) = 64\,000$ bits to be transmitted each second.

The diagram shows a simplified version in which four signals from users **A**, **B**, **C**, and **D** are transmitted along a single channel

The byte D_1 from user **D** was sent first, then bytes from users **A**, **B**, and **C**. Then user **D**'s signal was sampled again to transmit byte D_2, and so on.

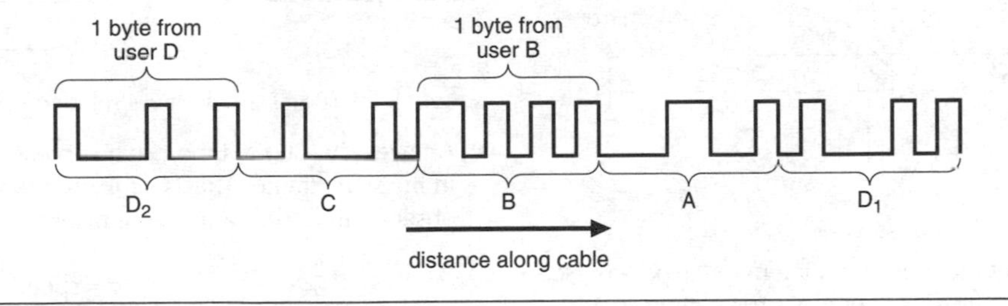

H12 Turning points in physics

Waves and particles

Read C2 and G4 first.

Corpuscles or waves? Newton suggested that light might be made up of high-speed 'corpuscles' (particles). Huygens (in 1680) proposed an alternative wave theory which satisfactorily accounted for reflection and refraction.

The wave theory could not explain how 'empty' space could transmit waves. So it was argued that space must be filled with an invisible medium, called the ***ether***.

The corpuscular theory predicted that light should speed up when passing from air to water. The wave theory predicted the opposite. In 1862, the speed of light in water was measured and found to be less than in air.

Electromagnetic waves Maxwell's electromagnetic theory (1864) predicted that oscillating charges should emit waves which would travel through an electromagnetic field at a speed of $1/\sqrt{\varepsilon_0\mu_0}$ (= speed of light). Later (in 1888), Hertz demonstrated the existence of radio waves.

Note:

- Maxwell's theory is an example of a ***classical*** theory – one that does not deal with quantum effects.

Black body radiation The graph shows the energy distribution in a black body's spectrum (see F7). The dotted line is the distribution predicted by classical theory. In classical theory, oscillating particles are assumed to emit waves continuously, over a continuous range of energies. But at higher frequencies, there is a huge discrepancy between the predicted and experimental results. This is known as the ***ultraviolet catastrophe***.

Quantum theory This was proposed by Planck (in 1900). It solved the black body spectrum problem by restricting energy changes to multiples of hf. Later, it led to the concepts of the ***photon*** and ***wave–particle duality***.

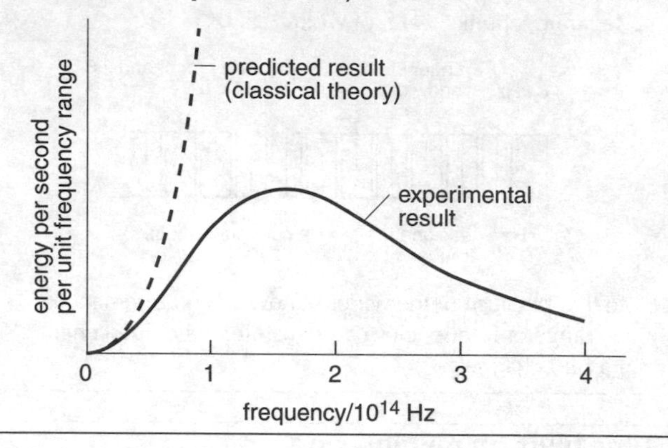

The special theory of relativity

Looking for the ether As the Earth moves through space, the ether should flow past it. In 1887, Michelson and Morley tried to detect this flow with an ***interferometer***.

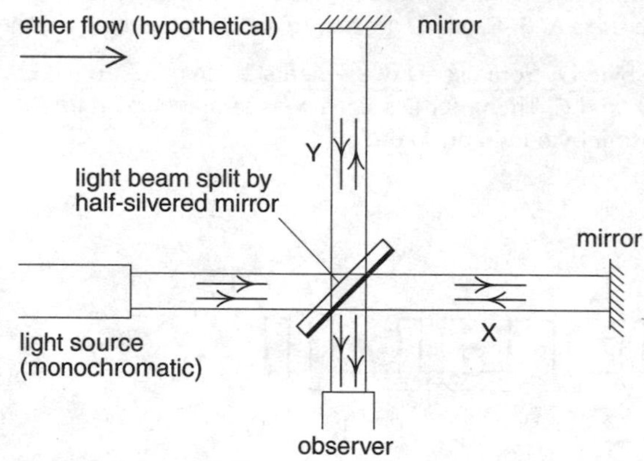

Beams X and Y recombine to form an interference pattern (see C3). If there is an ether flow as above, X's travel time should be slightly longer than normal. But when the apparatus is rotated 90°, Y's travel time should be slightly longer than normal, so the interference pattern should shift. No shift was detected. All experiments indicated that the measured velocity of light in a vacuum is *invariant* (always the same).

Postulates Einstein rejected the idea of absolute motion through space: motion could only be relative to the observer's ***frame of reference***. He developed his ***special theory of relativity*** (1905) from the following two postulates (assumptions):

1. Physical laws (e.g. the laws of motion) are the same in all *inertial* (unaccelerated) frames of reference.
2. The speed of light in a vacuum has the same measured value in all inertial frames of reference.

Deductions From these postulates, Einstein deduced that length and time measurements could not be absolute. They must depend on the relative motion of the observer. Some results of his mathematical analysis are given on the right.

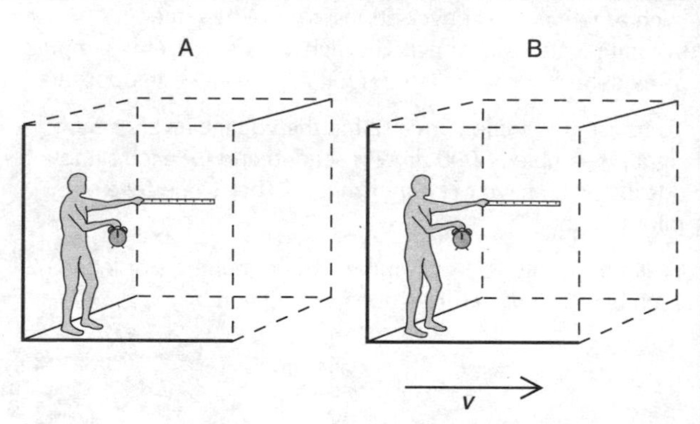

Above, frame B has a velocity v relative to frame A.

- An event in B takes time t_0 when measured by an observer in the same frame. This is its ***proper time***. But to an observer in A, the same event takes time t, where:

$$t = \frac{t_0}{\sqrt{1 - v^2/c^2}}$$

(c = speed of light in a vacuum)

Note: $t > t_0$. To the observer in A, a clock in B runs slow: there is ***time dilation*** (enlarging).

- If an object in frame B has a ***proper length*** l_0 (in the v direction), its length as observed from A is given by

$$l = l_0\sqrt{1 - v^2/c^2}$$

Note: $l < l_0$. To the observer in A, the object in B is shortened: there is ***length contraction***.

- The object's mass as observed from A also depends on v:

$$m = \frac{m_0}{\sqrt{1 - v^2/c^2}}$$

Note: $m > m_0$. As v increases, m increases. When $v = c$, m is infinite, and so is the KE. This effectively makes the speed of light a universal speed limit.

- Observed increases in mass and energy are linked by the equation $\Delta E = \Delta mc^2$ (see G3).

Discovering the electron

Read E8 first.

Cathode rays 19th century experiments with electric discharges through gases at very low pressure (as below) indicated that invisible rays were travelling from the cathode to the anode in the discharge tube. These were called ***cathode rays***. Perrin (in 1895) demonstrated that they were negatively (–) charged.

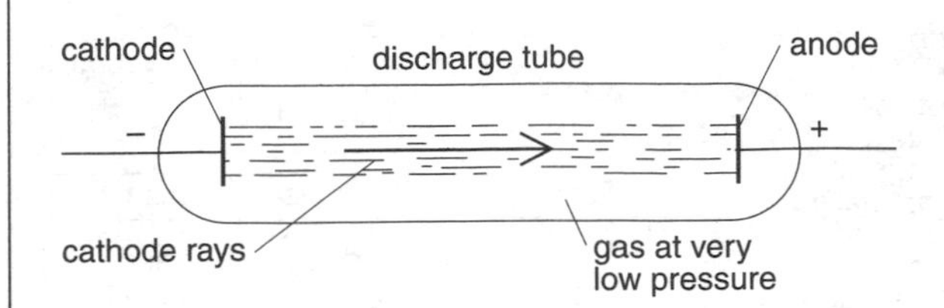

Specific charge (charge/mass) J. J. Thomson suggested that cathode rays were particles, which he called ***electrons***. In 1897, he measured their specific charge (e/m_e). A modern version of his experiment is shown below:

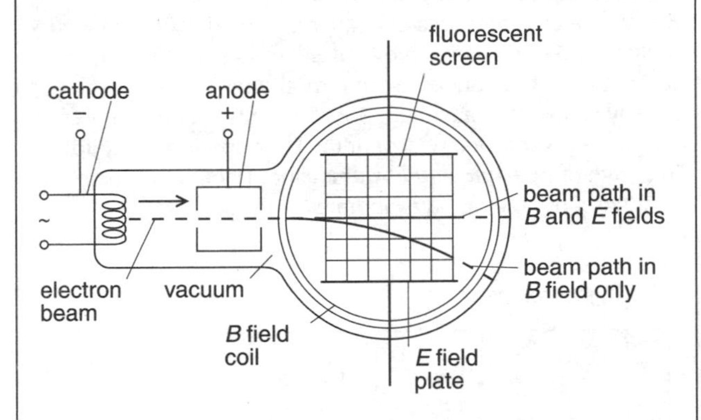

The electron beam is deflected by a magnetic field (B), then restored to a straight path using an electric field (E).

When the magnetic field alone is acting

$$Bev = m_e v^2/r$$

So $e/m_e = v/rB$ (quantities defined as in E8)

With r and B measured, e/m_e can be calculated if v is known. When the electric and magnetic forces on the electron beam are equal, $Bev = Ee$. So $v = E/B$.

Electronic charge *e* Millikan (in 1909) observed oil droplets as they fell at terminal velocity through air, then measured the *change* in terminal velocity (Δv) which occured when a droplet gained charge (q) and a vertical electric field (E) was applied. In this case, electric force = Eq.

So $Eq = 6\pi\eta a\Delta v$ (see B9)

giving $q = 6\pi\eta a\Delta v / E$

(a was deduced from terminal velocity measurements when E was zero). Millikan took hundreds of measurements and found that q was always a multiple of a basic charge e.

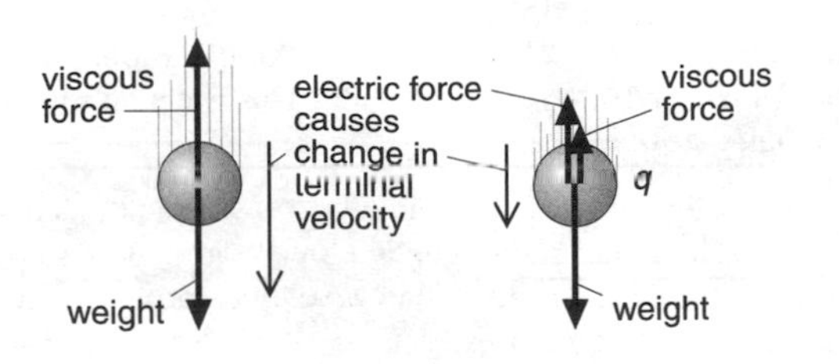

Lowering temperatures

Read F2–F4 and F6 first.

Predicting absolute zero The expansion of gases was investigated by Charles (in 1787) and Gay-Lussac (in 1802). Their results suggested that all gases should have zero volume at –273.15 °C, now called ***absolute zero***.

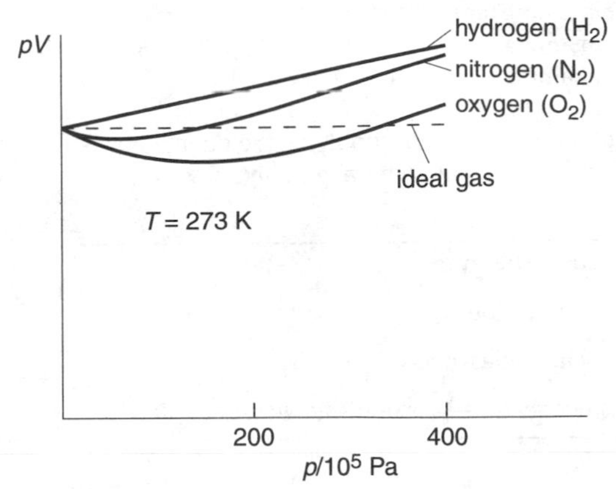

Real and ideal gases Real gases deviate from Boyle's law (1) because of intermolecular forces and (2) because the molecules themselves occupy a finite volume. The graph shows how pV varies with p for three gases. A gas's ***Boyle temperature*** is the temperature at which pV is most nearly constant.

In an ideal gas, the molecules occupy zero volume and there are no intermolecular forces. Cooling an ideal gas would never liquefy it. It would remain a gas at absolute zero.

Gases and vapours A real gas can be liquefied by compressing it – provided it is beneath its ***critical temperature***. The pressure needed to liquefy it at this temperature is the ***critical pressure***. A gas below its critical temperature is called a ***vapour*** (see also H13).

Critical temperatures		
helium 5 K	hydrogen (H_2) 33 K	oxygen 155 K

The Joule–Kelvin effect Below, a gas expands as it is passed through a porous plug. For an ideal gas, there would be no change in temperature. However, with most real gases, there is a slight *decrease* in temperature, because work is done against intermolecular attractions at the expense of the molecules' KE. The effect (discovered in 1852) only produces cooling below a gas's ***inversion temperature.*** Above this, intermolecular repulsions cause slight heating.

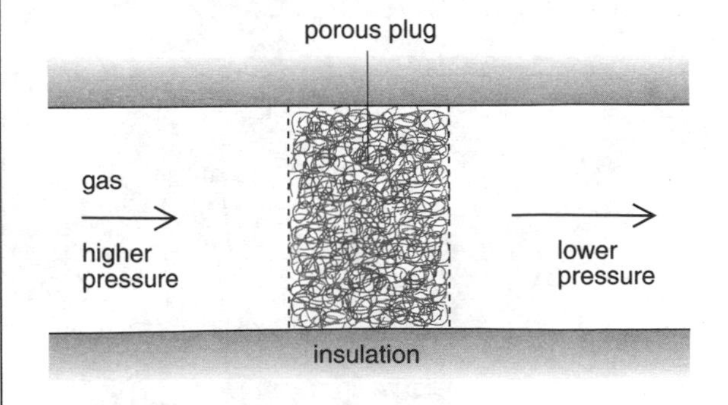

Liquefying gases Techniques for this can make use of:
- the cooling effect of evaporation,
- the cooling effect of adiabatic expansion,
- the cooling effect of a Joule–Kelvin expansion.

Superconductivity at low temperatures: see D2.

Superfluids Near absolute zero, liquid helium loses all viscosity. The effect, called ***superfluidity***, occurs below the ***lambda point*** (2.172 K for ^{4}He; 0.000 93 K for ^{3}He).

H13 Energy and the environment – 1

Energy for the Earth

The Earth's prime source of energy is ***solar radiation*** (radiation from the Sun). Its effects include:

- maintaining temperatures on Earth
- maintaining winds, ocean currents, and weather systems
- maintaining plant and animal life
- maintaining atmospheric composition
- millions of years ago, supplying the energy now stored in fossil fuels (e.g. oil, natural gas, and coal).

Non-solar energy

Heat from inside Earth

Energy in nuclear fuels

Tidal energy (due to Moon's gravitational pull)

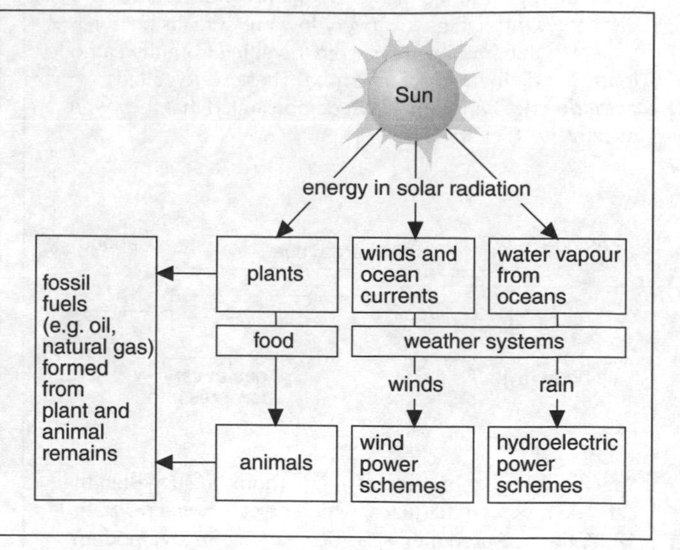

Maintaining atmospheric composition

Animals take in oxygen. They use it to 'burn up' their food and obtain energy. This process is called ***respiration*** and it produces carbon dioxide and water. (Burning fossil fuels also produces carbon dioxide and water.)

Plants take in carbon dioxide and water. With these, they make food using energy from sunlight. This process is called ***photosynthesis*** and it gives out oxygen.

Note:

- Plants also use oxygen for respiration. But overall, they make more oxygen than they consume.
- For photosynthesis, plants absorb light at the red and blue ends of the visible spectrum. They transmit or reflect green light, which is why they look green.

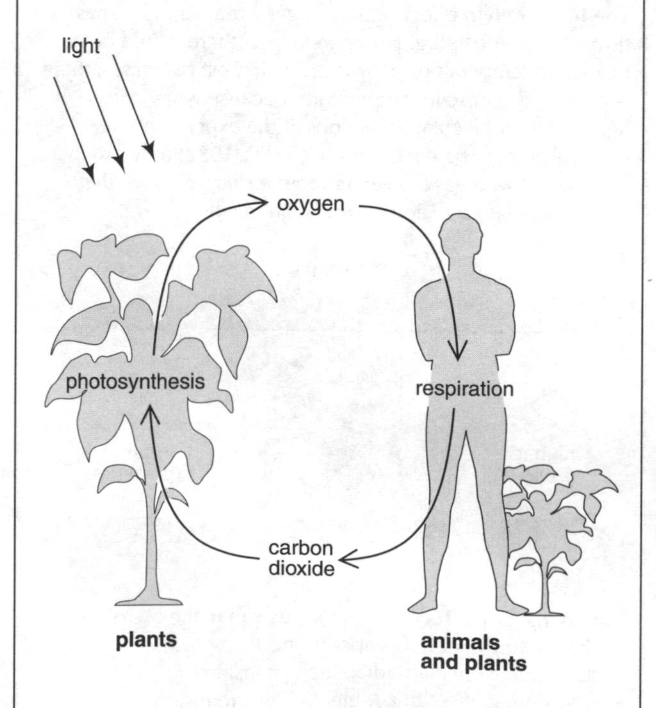

Together, plants and animals maintain the amounts of oxygen and carbon dioxide in the atmosphere. By various chemical processes, some nitrogen is also incorporated in the bodies of living organisms, but this is recycled.

Water vapour and rising air

Below, water vapour in a container is cooled. With reduced molecular motion, some molecules have stuck together, i.e. some vapour has condensed to form liquid. The vapour density is now the maximum possible for that temperature: the space is ***saturated*** with vapour. The vapour and liquid are in a state of ***dynamic equilibrium***: vapour continues to condense, but the liquid evaporates at the same rate.

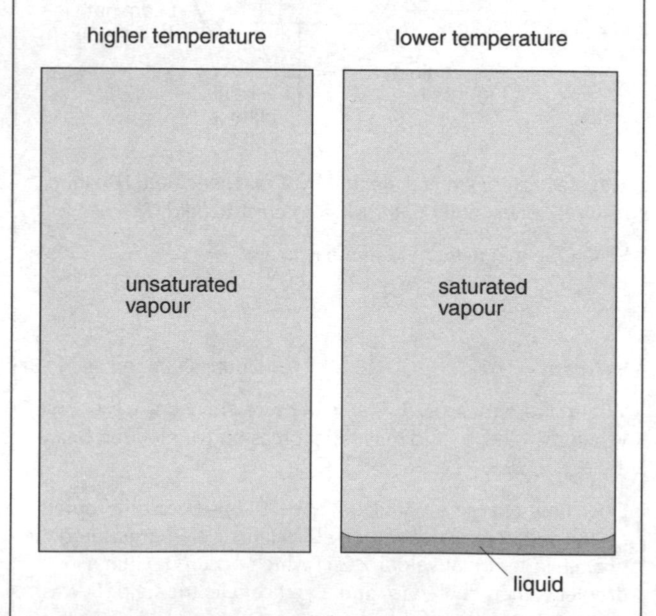

The temperature to which an unsaturated vapour must be cooled to become saturated is called its ***dew point***. If unsaturated vapour in the atmosphere is cooled beneath its dew point, the condensation is seen as millions of tiny droplets (clouds, fog, and mist), or as dew on the ground.

Rising air If warm, dry air rises above the ground, it expands adiabatically and cools (see F4). The fall in temperature is about 1 °C per 100 m of height gain. This is called the ***dry adiabatic lapse rate***.

If unsaturated air rises, it eventually cools beneath its dew point and becomes saturated. So clouds form. With saturated air, the lapse rate is reduced because the vapour releases latent heat as it condenses.

Water balance

Overall, the rate at which water evaporates from the oceans and land regions is equal to the rate at which water is returned by precipitation (rain, snow, and hail). On average, there is a *water balance*. However, at any given time, each region may have an imbalance – in which case, water is either going into storage or coming out of it.

The oceans are the main water storage system. Others include the polar ice caps, soil, plants, and porous rocks. Underground water-bearing rocks are called ***aquifers***. Water can stay locked away in them for thousands of years.

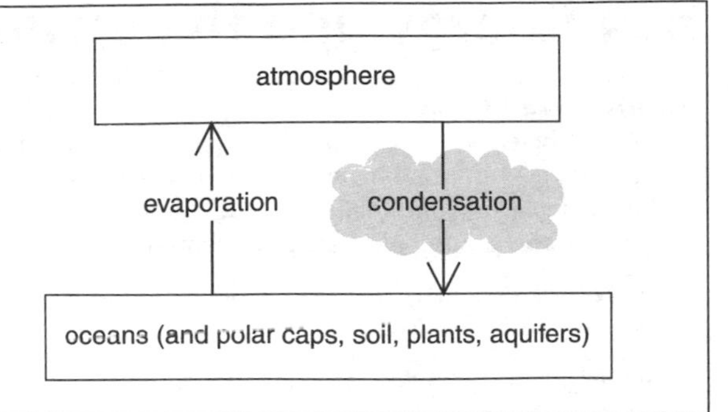

Solar constant

This is the amount of solar energy arriving at the Earth's outer atmosphere per second per square metre (at right angles to the radiation). It is equal to 1.35 kW m^{-2}.

- About 25% of the incoming solar radiation is reflected back into space by the atmosphere and clouds. It does not reach the Earth's surface.
- The amount of solar energy per second striking each square metre of the surface, depends on the time of day, season, atmospheric conditions, and latitude. For example, the radiation reaching region B on the right is spread over a larger area than that reaching A.

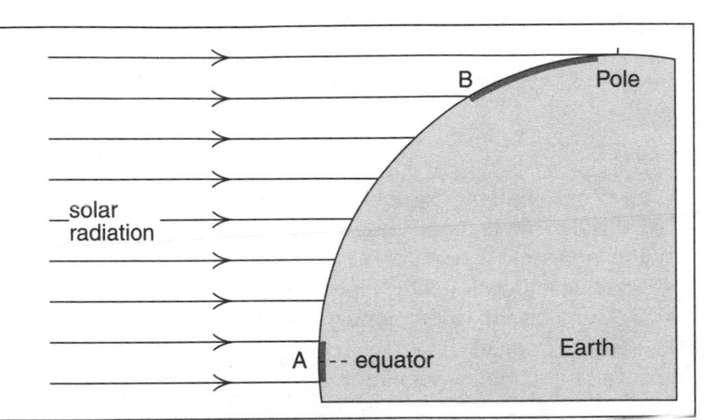

The natural greenhouse effect

Read F7 first.

The Earth gains energy from solar radiation, and loses it by emitting radiation into space. But overall, the *rates* of energy gain and loss are the same. This is another example of dynamic equilibrium.

If the Earth had no atmosphere, its average surface temperature would be about –18 °C. However, the 'heat trapping' effect of the atmosphere, called the ***greenhouse effect***, means that dynamic equilibrium occurs at about 15 °C. The raised temperature is caused like this:

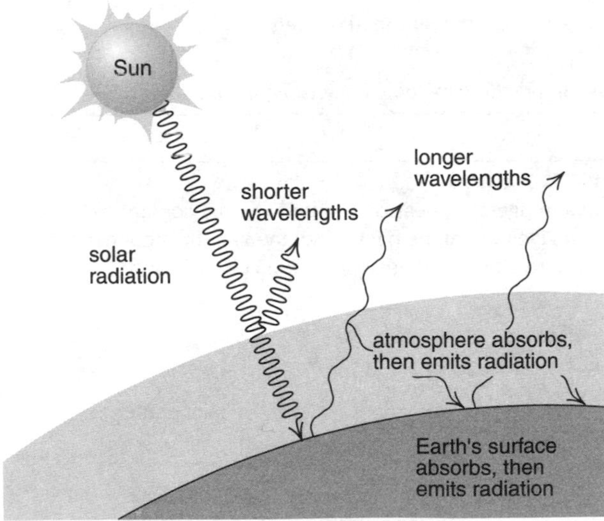

Visible and short-infrared radiation from the Sun pass easily through the atmosphere and warm the Earth's surface. However, being cooler than the Sun, the surface radiates back at longer infrared wavelengths (see F7). Some of these wavelengths are absorbed by molecules of water vapour and carbon dioxide in the atmosphere. These then emit infrared in all directions - including downwards. So the atmosphere and surface are both warmed, and dynamic equilibrium is reached at a higher temperature.

Global warming

By burning fossil fuels, industrial societies are putting carbon dioxide into the atmosphere at a faster rate than plants can absorb it. This is adding to the greenhouse effect and may be causing ***global warming***. In the past 100 years, the average global temperature has risen by almost 1°C.

- Though often called the 'greenhouse effect', global warming is an *addition* to the natural greenhouse effect.
- Extra carbon dioxide may not be the only cause of global warming. Global temperatures have always fluctuated.
- Other 'greenhouse gases', include methane (from paddy fields, animal waste, and oil and gas fields), and CFCs.

The increase in carbon dioxide in the atmosphere is less than half the extra emitted. Possible reasons are:

- With extra carbon dioxide in the air, plant growth is increased, so more of the gas is absorbed.
- Some carbon dioxide dissolves in the oceans.

Future effects of global warming These cannot be predicted with any certainty. Melting of polar ice may cause a rise in sea level. Evaporation from the oceans will increase, so some regions may get more rain. But shifts in climate may mean that some regions are drier.

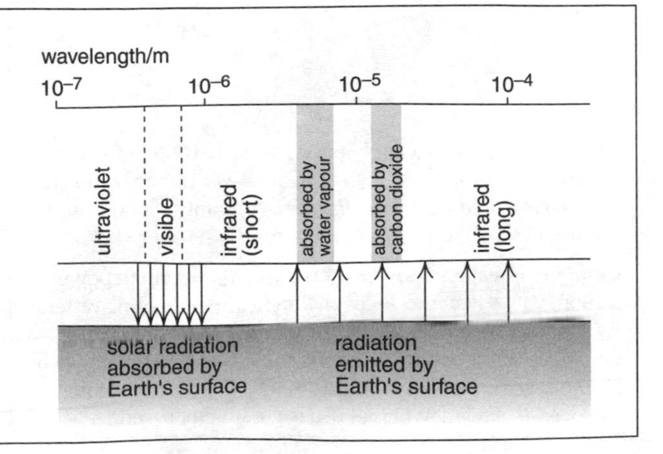

H14 Energy and the environment – 2

Using fossil fuels

Industrial societies get over 90% of their energy by burning oil, natural gas, and coal. However, these fossil fuels took millions of years to form in the ground. They are effectively ***non-renewable*** and supplies are ***finite*** (limited).

Resources of fossil fuels are the total quantities known to exist in the Earth.

Reserves are the quantities which could now be extracted commercially. They are less than resources.

World reserves (1993): energy/10^{20} J	
oil	20
natural gas	6
coal	200

Future consumption This will decide how long reserves will last. It is difficult to predict but will be affected by:
- the increase in world population
- economic growth in developed countries
- industrialization of developing countries
- efficiency of fuel use
- levels of building insulation.

Problems from fossil fuels Major problems include:
- atmospheric pollution, and risk of global warming
- risk of marine pollution from oil
- economic debt in developing countries which import oil.

Power stations: efficiency of fuel use

As with all heat engines, the efficiency of a fuel-burning power station is limited by the second law of thermodynamics (see F3). With additional energy losses (e.g. frictional), the efficiency is reduced to about 40%. So more heat is produced (60%) than electrical energy.

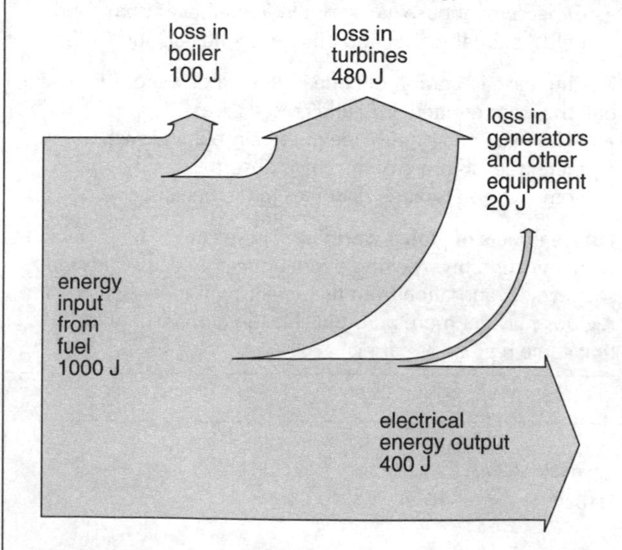

Above, you can see what happens to each 1000 J of energy released by burning fuel in a typical power station. A diagram like this is called a ***Sankey diagram***. Amounts of energy are represented by the widths of the arrows.

Combined heat and power (CHP) In a fuel-burning power station, waste heat produces slightly warmed cooling water. By running the power station at *reduced* generating efficiency, it is possible to supply the local district with water hot enough for heating systems. In a CHP scheme like this, the *overall* efficiency of fuel use is greatly improved.

Pumped storage

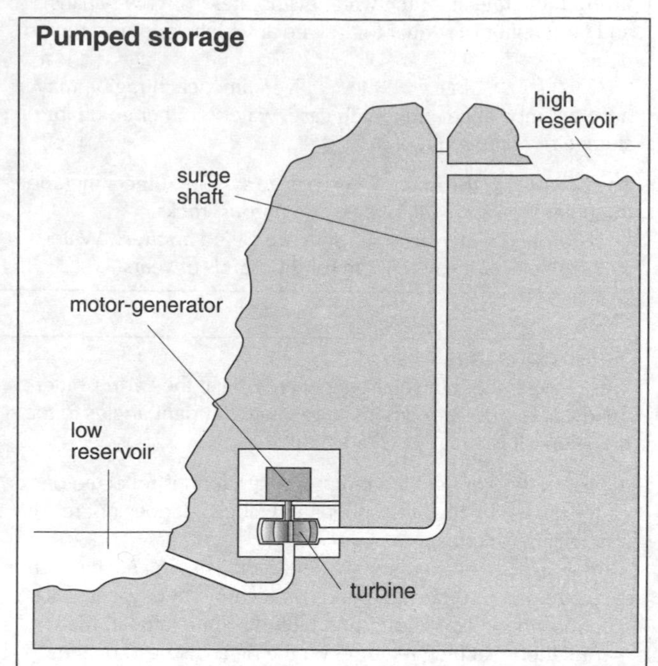

The electricity supply industry has to respond instantly to changes in demand for mains power. A ***pumped storage scheme*** aids this process by acting as an energy store. When demand is high, water flows from the top reservoir and turns the turbines which drive the generators. When the demand is low, the generators are used as motors. Power is taken from stations with spare generating capacity, and water is pumped back up to the top reservoir.

Calculating power output The following calculation assumes that gravitational PE is converted into electrical energy with an efficiency of 100%. If a mass m of water flows from the top reservoir in time t and loses a height h,

electrical energy output = loss of PE = mgh

Dividing by t gives

electrical power output = mgh/t

Note:
- In practice, the efficiency is, typically, about 75%.

Heat pumps

In a refrigerator, heat is absorbed when a coolant evaporates, and given out at the back when the vapour is condensed by compression. Work is done by the compressor.

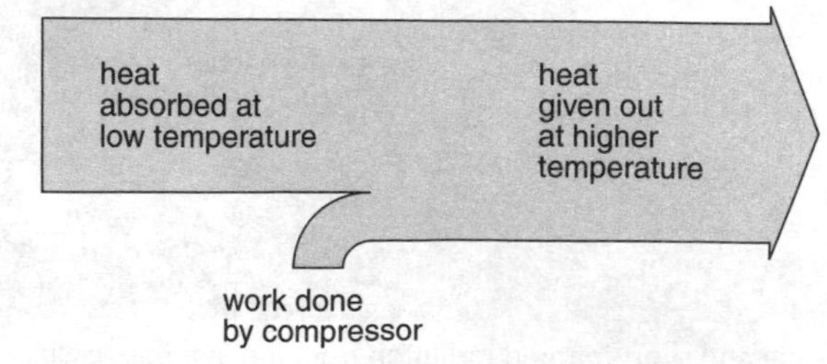

A refrigerator is a ***heat pump*** – a heat engine (see F3) in reverse. Work is done *on* it in order to transfer heat from a *low* temperature source to a *higher* temperature sink.

Some heating systems use heat pumps. The building is heated by cooling the air, ground, or nearby stream outside. Much less energy is needed than is given out as heat. But the disadvantages are that (1) the local environment is affected and (2) the system works *less* well on a *cold* day.

Energy (in kWh) and power

For practical reasons, energy is sometimes measured in units other than the joule (J). For example:

1 ***kilowatt hour*** (kWh) is the energy supplied when delivered at the rate of 1 kW (i.e. 1000 J s^{-1}) for 1 hour.

energy = power × time (see B2)

So 1 kWh = 1000 J s^{-1} × 3600 s = 3.6×10^6 J

The main alternatives to fossil fuels

Scheme	Details
hydroelectric	Rainwater fills lake behind dam. Flow of water from lake drives generators.
tidal	Lake behind dam fills and empties with tides. Water flow drives generators.
nuclear	See G3.
solar	See panel on right.
wind	See panel below.
biofuel	Wood (for burning) Methane gas from plant and animal waste Alcohol (fuel) produced from sugar cane
geothermal	Using hot water from natural geysers and springs in volcanic areas. Hot rocks underground used to produce steam for turbines in power station.

Scheme	Renewable energy	Fuel costs	Greenhouse gas emissions
hydroelectric	YES	NO	NO
tidal	YES	NO	NO
nuclear	NO	YES	NO
solar	YES	NO	NO
wind	YES	NO	NO
biofuel	YES	YES	YES*
geothermal	YES	NO	NO

* With managed crops, there is no overall addition to global warming because of the carbon dioxide absorbed.

Solar power

Solar power systems make *direct* use of solar radiation. This is done in two main ways:

Solar cells These use the ***photovoltaic effect***: a small voltage is generated when light falls on a slice of doped semiconductor. Such cells are expensive and their efficiency of energy conversion is normally less than 20%.

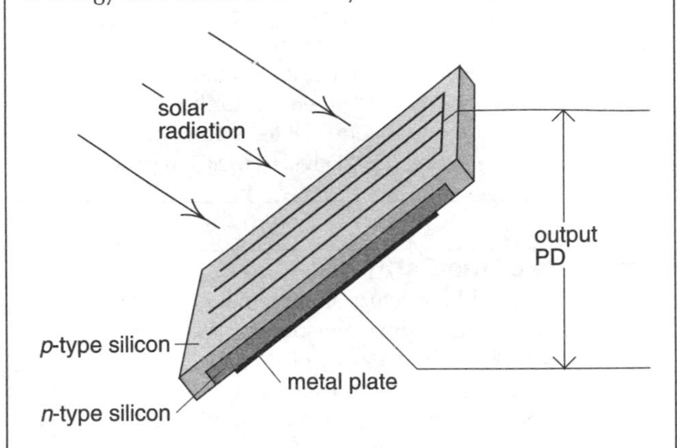

Solar panels These use the heating effect of solar radiation – for example, to pre-warm the water in a domestic hot water system.

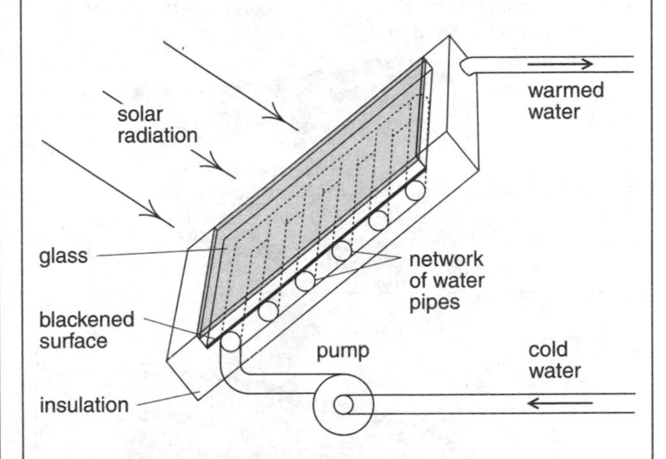

The maximum amount of available energy per second per square metre of panel is equal to the solar constant less corrections for atmospheric absorption, cloud cover, latitude, and angle of panel to the incoming radiation.

Wind power

Generators driven by wind turbines ('windmills') are called ***aerogenerators***. The largest ones have power outputs of about 7 MW – compared with over 3000 MW for a large fuel-burning station. For increased power from one site, aerogenerators are grouped together in ***wind farms***.

Calculating power output The following calculation assumes that the wind loses all its velocity v when striking the turbine blades and transfers all its KE to them.

In the diagram on the right, all the air in the shaded region will transfer its KE to the blades in time t. As this air has a volume of Avt, its mass is $A\rho vt$, so

$$\text{KE of air} = \tfrac{1}{2} \times \text{mass} \times v^2 = \tfrac{1}{2} \times A\rho vt \times v^2 = \tfrac{1}{2} A\rho v^3 t$$

If the energy conversion efficiency is 100%, this KE is also the electrical energy output. So dividing by t gives

$$\text{electric power output} = \tfrac{1}{2} A\rho v^3$$

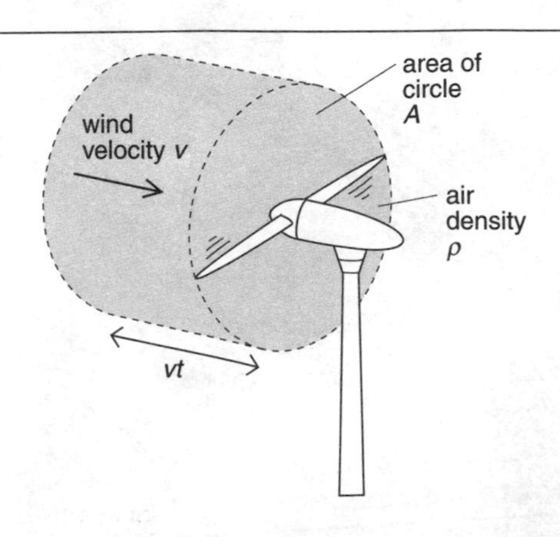

Note:

- In practice, a turbine cannot extract all the wind's KE, and there are other energy losses as well. Overall, the efficiency is reduced to about 40% or less.

H15 Earth and atmosphere

Age of the Earth

Read G2 first.

The Earth is believed to have formed, along with the other planets, about 4.6×10^9 years ago (see H2). Its oldest rocks are around 3.8×10^9 years old. Age estimates like this come from ***radiometric*** data, i.e. data from radioactive decay measurements (see G2). Methods include:

- Comparing the proportions of uranium-235 and uranium-238 now found in natural uranium with those which would have existed when the Earth was formed
- Finding the ratio of uranium-238 to trapped helium (formed from emitted α particles) in rock samples.

Variations in *g*

Read E3 first.

Across the Earth's surface, the measured value of g varies slightly, though by less than 1%. It is lowest at the equator
(1) mainly because of the effects of the Earth's rotation
(2) also because the Earth has a slightly greater radius towards the equator than towards the poles. g is also less where the crust is thick, because crust is less dense than mantle.

Variations in g can be detected using a ***gravimeter*** – a very sensitive spring balance with a mass attached. Local anomalies (unusual variations) in g can give clues about the presence of mineral deposits.

Structure of the Earth

The Earth's probable structure and internal conditions are shown in the diagram. The main sources of data are:

- analysis of ***seismic waves*** (see next page)
- gravitational measurements (see also E3)
- measurement of the heat flow out of the Earth.

Radioactive decay provides the energy which maintains the high temperatures within the Earth.

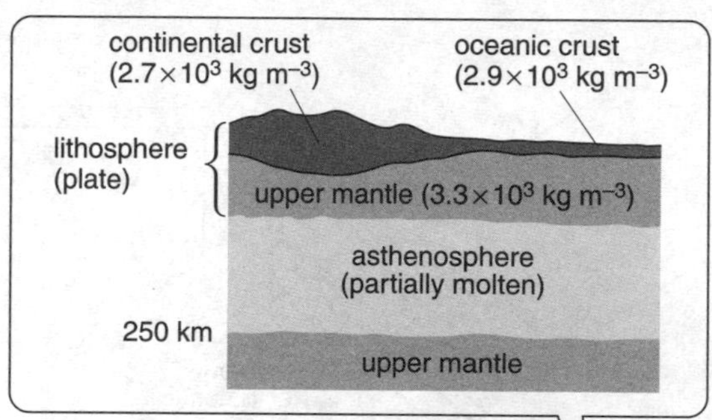

Earth's magnetism The Earth has a weak magnetic field. However, the Earth's core is above the Curie temperature (see H6), so cannot be a permanent magnet. The field is thought to be due to currents generated by the circulation of molten metal in the outer core. The Earth's magnetic axis is inclined at about 12° to its spin axis, but 'wobbles' slightly. Magnetic surveys of rocks indicate that there have been several complete pole reversals in the past.

Plates The Earth's crust and part of its upper mantle form huge 'rafts', called ***plates***, which 'float' on the material beneath them. Pushed by convection currents which cause creep in the mantle, the plates move against each other. At their boundaries, this can cause (1) mountain building,
(2) volcanoes, because of the frictional heating
(3) earthquakes.

Mountains have an increased thickness of crust beneath them. This means that the crust floats in a state of equilibrium, called ***isostasy***. As higher material is worn away by erosion, the crust underneath rises to compensate.

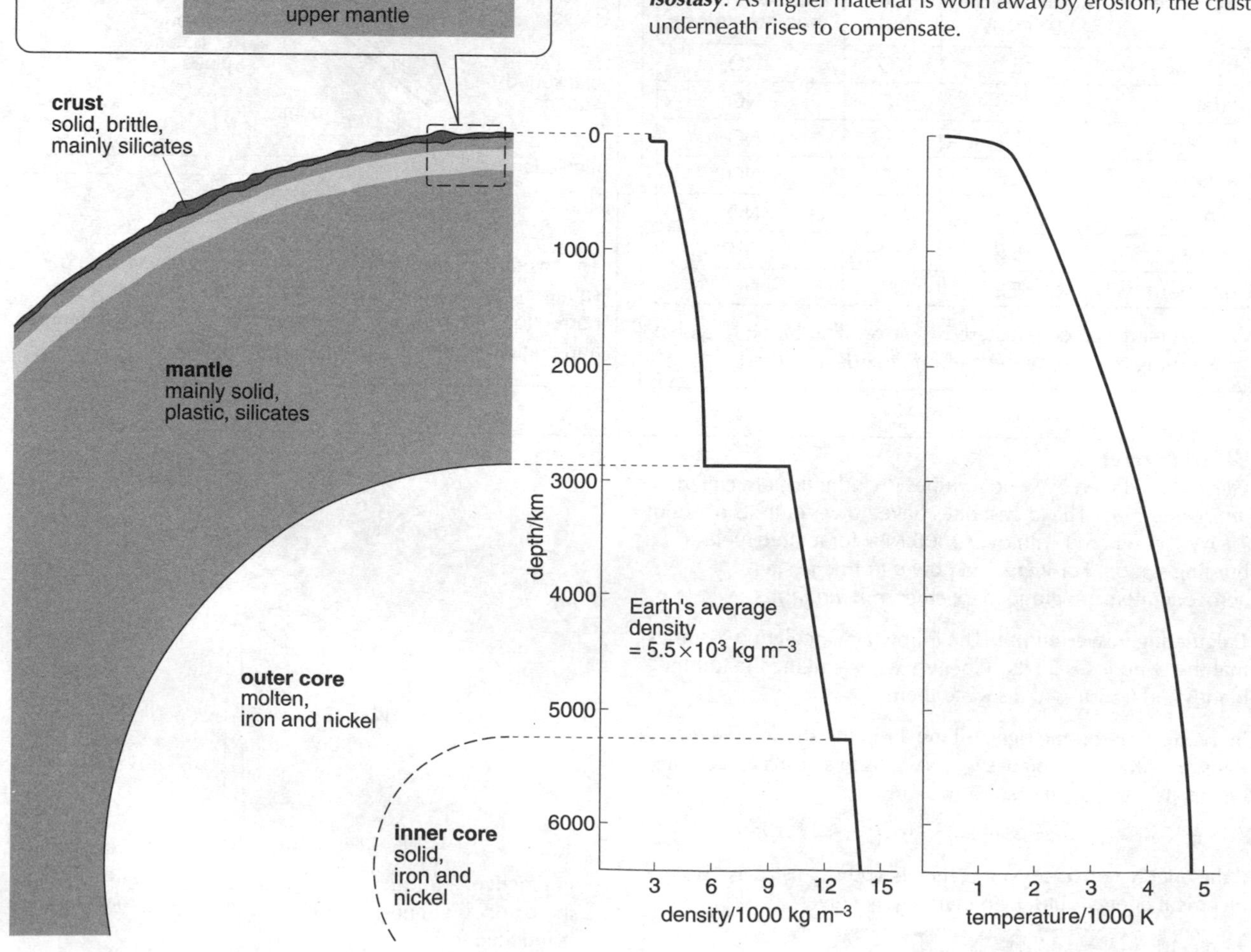

Seismic waves

Read C2 first.

When stresses build up in rocks because of plate movements, the rocks may fracture and release strain energy. The result is an earthquake, which sends ***seismic waves*** through the Earth. These include:

P (primary) waves These are longitudinal waves. They travel through solid rock and molten material.

S (secondary) waves These are transverse waves. They are slower than P waves and cannot travel through molten material.

Note:
- P and S waves travel faster through denser rock, so density changes in the Earth cause refraction.

Seismic waves are recorded using a ***seismograph*** in which sensors detect vibrations transmitted to a suspended mass. The paths of the waves can be worked out from their travel times to different detectors around the world.

Note:
- S waves do not travel through the Earth's core, which suggests that part of this must be molten.

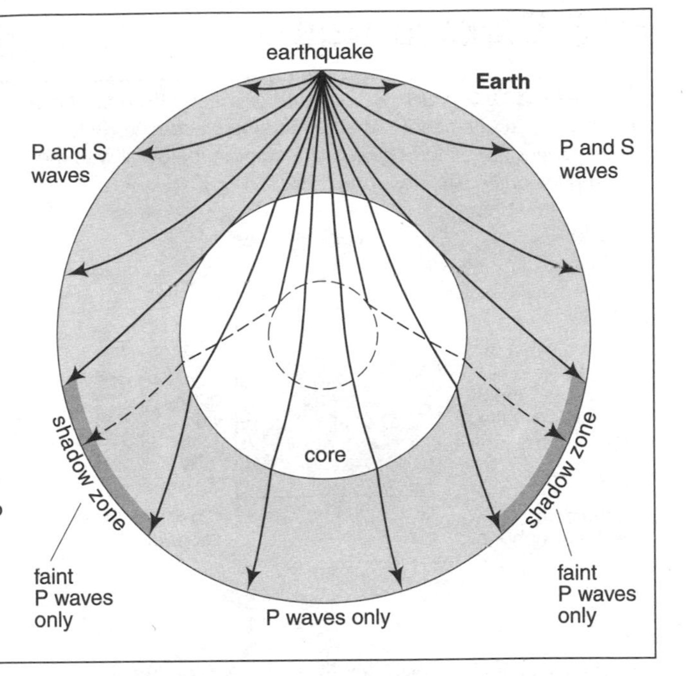

Structure of the atmosphere

Air is mainly nitrogen (78%) and oxygen (21%), with much smaller amounts of other gases, including carbon dioxide (0.03%), and variable amounts of water vapour.

Note:
- Hydrogen and helium escaped from the atmosphere early in its formation when it was still hot. Their faster molecules had speeds that exceeded the Earth's escape speed (see E3 and F6).

The atmosphere can be divided into four main layers, with temperature changes marking their boundaries. (The boundary heights vary with latitude, season, and other factors.)

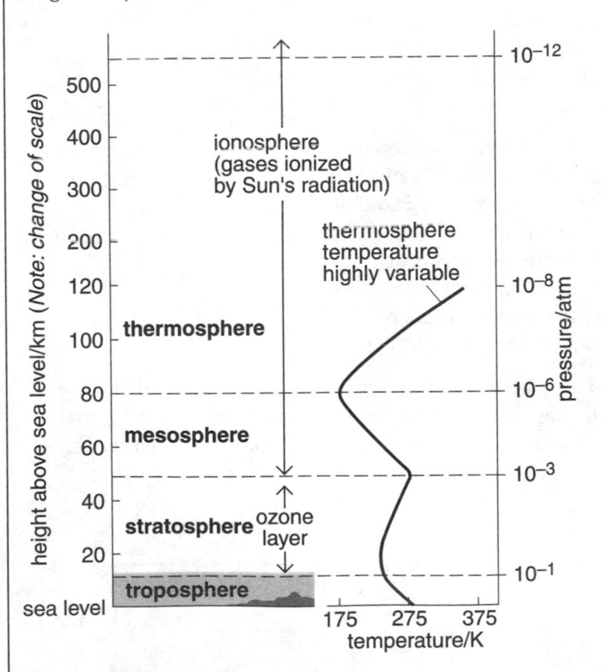

Pressure variations with height Atmospheric pressure p decreases with height h above sea level. For an 'ideal' atmosphere at constant temperature:

$$p = p_0 e^{-kh}$$

where k is a constant and p_0 is the pressure at the surface.

The dynamic atmosphere

Being 'square on' to the Sun's radiation, the equator receives more radiant power per m^2 than regions to the north and south. Hot air rising above the equator creates huge convection currents in the atmosphere, called ***cells.*** These, and the Earth's rotation, give the Earth its wind belts.

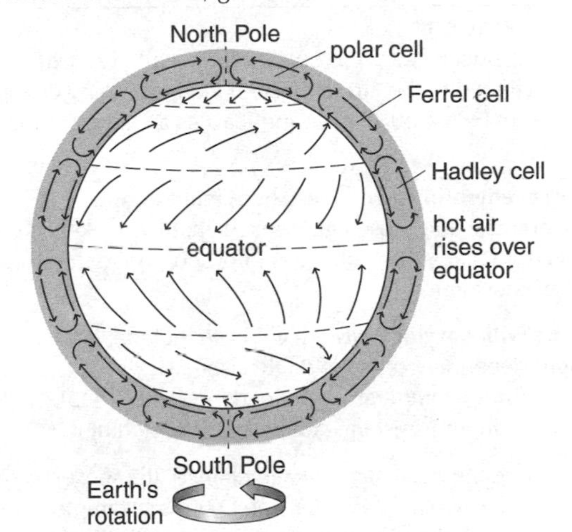

Winds blowing from the equator veer away from a north-south line. This is called the ***Coriolis effect.*** It is due to the motion of the air relative to the rotating Earth.

Note:
- The Earth also has ocean currents which are driven by convection, winds, and the Coriolis effect.

Troposphere Weather effects mostly occur in this layer which contains the bulk of the atmosphere.

Ozone layer The stratosphere contains ozone gas. Biologically, it is important because it absorbs most of the harmful ultraviolet radiation arriving from the Sun. (The absorbed energy warms the stratosphere.)

Ozone (O_3) is an unstable form of oxygen. It is constantly being made and destroyed by chemical processes in the atmosphere. However, pollutants such as CFCs upset the balance by increasing the rate of destruction.

H16 Electronics – 1

Parts and process

Read D2 and D3 first.

Electronic circuits handle small, changing electric currents. The changes are called ***signals***. The diagram below shows the main parts of a basic electronic system.

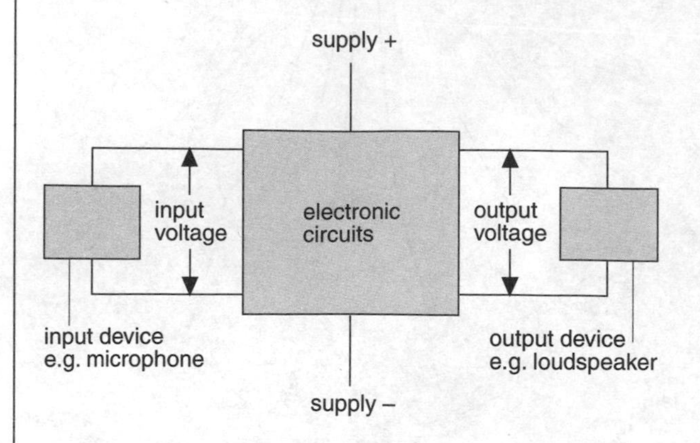

Note:
- Devices which change signals from one form to another (e.g. sound to electrical) are called ***transducers***.
- Input signals can be very weak. But by means of electronic circuits, they can control output signals which can be much stronger. The output power is provided by the supply (usually low voltage DC).

Two possible functions of electronic circuits are described below.

Switching Depending on the input signals received, the output voltage is either HIGH (close to the supply voltage) – or LOW (zero), so the output device is either ON or OFF.

Amplification The output signals are an amplified (magnified) version of the input signals. For example, very low voltage AC from a microphone causes a higher voltage AC output for a loudspeaker.

The ***voltage gain*** of an amplifier is defined like this:

$$\text{voltage gain} = \frac{\text{output voltage}}{\text{input voltage}}$$

Note:
- In electronics, the term 'voltage' is commonly used for potential, PD, and EMF.

The key component in electronics is the ***transistor***. It can amplify or act as a switch. An ***integrated circuit (IC)***, in a package as on the right, may have thousands of transistors and other components formed on a single chip of silicon.

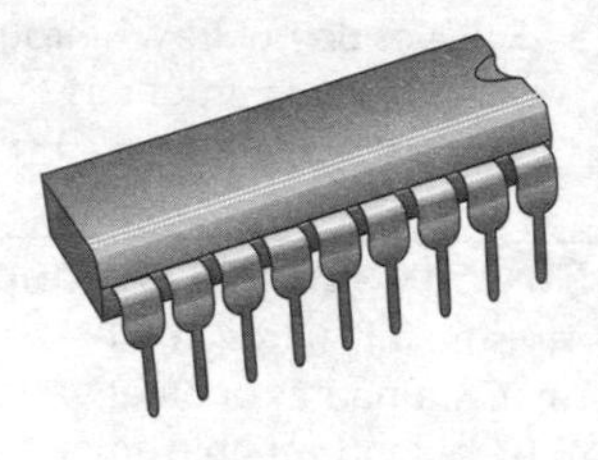

Input devices

Input transducers (e.g. microphones) are called ***sensors***. They must be linked to the circuit in such a way that any change they detect (e.g. a pulse of sound) causes a change in input voltage.

Sensors generating a voltage These include some microphones, and those light sensors that work like solar cells (see H14). Ideally, the voltage is in proportion to the external change causing it.

Sensors with varying resistance These include:
- light-dependent resistors (LDRs) (see D1),
- thermistors (temperature-dependent: see D1, D2),
- strain gauges (resistance changed by stretching).

To produce the necessary voltage change, the sensor forms one part of a ***potential divider*** (see D3). For example, when light falls on the LDR below, its resistance falls. The LDR therefore takes a smaller share of the supply voltage, so the voltage *V* falls.

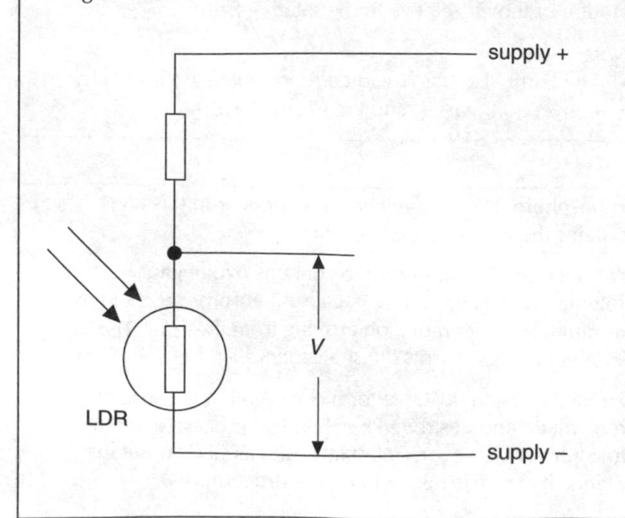

Output devices

These include loudspeakers, buzzers, LEDs, and relays.

Light-emitting diodes (LEDs) These emit light when a small current passes through. They can be used as indicator lamps to show the presence of an output voltage. Like all diodes, they only conduct in one direction (see D1 and D2).

To avoid damage, a resistor must be placed in series with an LED to limit the current through it.

The LED on the right can take a current of 0.01 A, which produces a 2 V drop across it. If a circuit's output voltage is 9 V, there needs to be a 7 V voltage drop across the resistor, so its resistance should be 7/0.01 = 700 Ω.

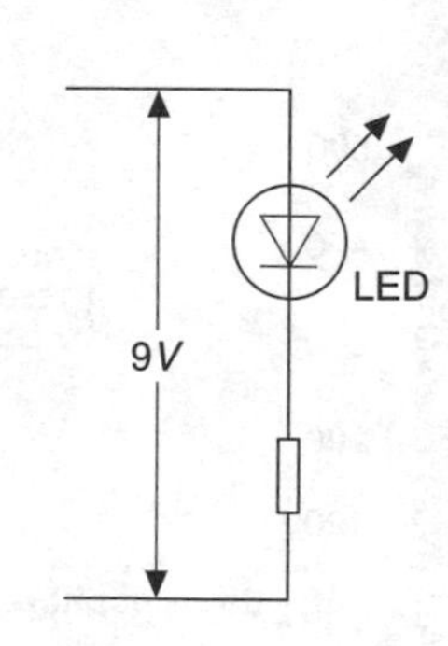

Relays These are electromagnetic switches. A small current activates an electromagnet. This closes (or opens) a switch which can control the flow of a larger current in a separate circuit, e.g. a circuit with a mains heater or motor in it.

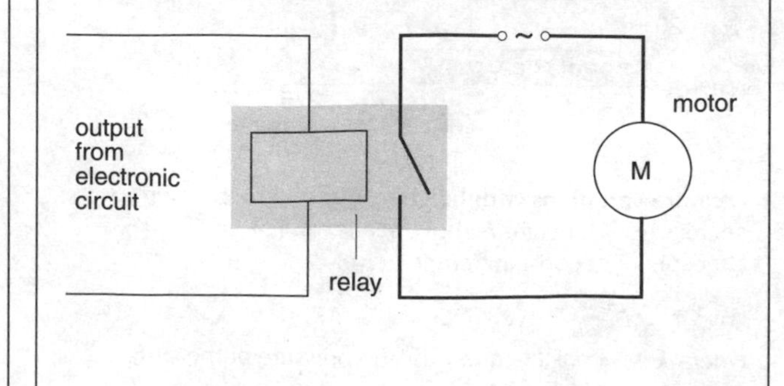

Impedances and matching

Resistance is one form of impedance (see D4). The microphone and loudspeaker below each have impedance. The amplifer has ***input impedance*** and ***output impedance*** rather as a battery has internal resistance (see D3).

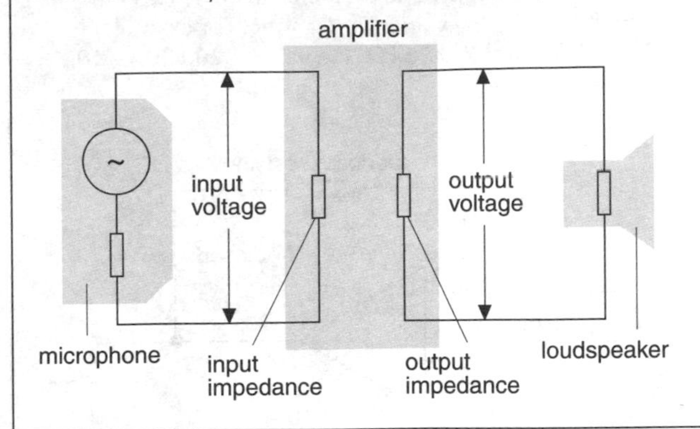

In the diagram, the various impedances are represented by resistors in the input and output circuits.

In each circuit, it is the *higher* impedance that has the *higher* voltage drop across it. So

- **For maximum input voltage**, the microphone's impedance should be as *low* as possible.
- **For maximum output voltage**, the loudspeaker's impedance should be as *high* as possible. However, this gives a very low output current and almost no sound!
- **For maximum output power**, the loudspeaker's impedance should be the *same* as the output impedance. (Output voltage × output current is then at its highest.)

A ***transformer*** can be used to match impedances and give maximum power transfer. A transformer with an impedance Z in its primary circuit is equivalent to an impedance of $Z(N_2/N_1)^2$. (See E5 for diagram and equation symbols.)

The operational amplifier (op amp)

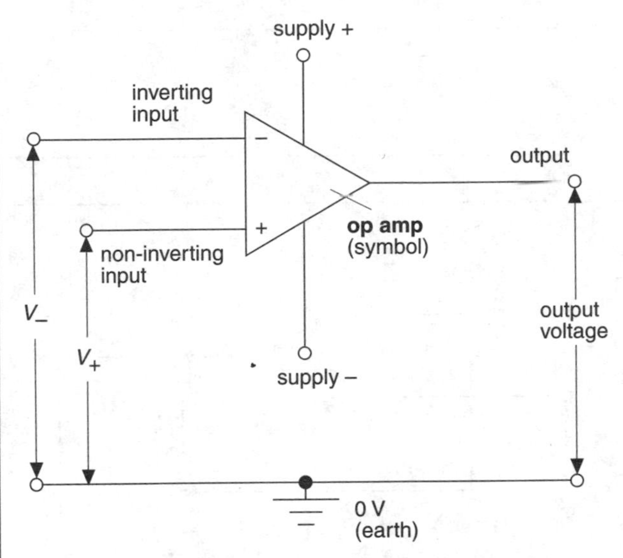

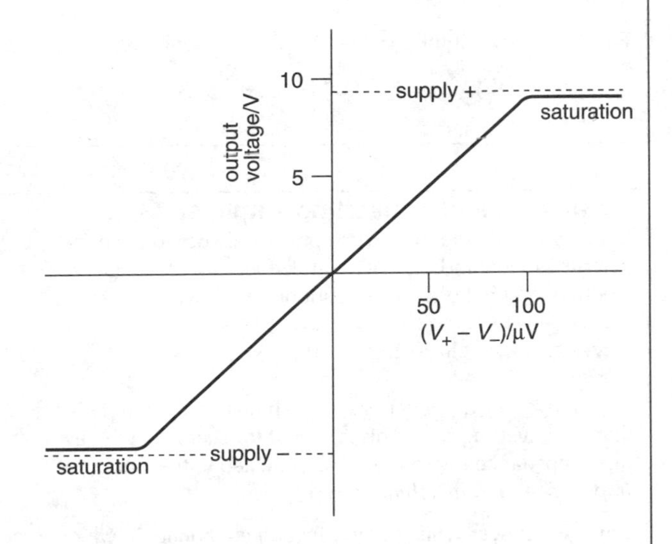

This IC has one output and *two* inputs. It amplifies the *difference* between the two input voltages. It is a ***differential amplifer***.

- If $V_+ > V_-$ the output voltage is positive (+).
- If $V_+ < V_-$ the output voltage is negative (–).
- If $V_+ = V_-$ the output voltage is zero.

If only *one* input is in use (and the other is at zero voltage),

- a *positive* (+) voltage on the ***inverting input*** causes a *negative* (–) output voltage,
- a *positive* (+) voltage on the ***non-inverting input*** causes a *positive* (+) output voltage.

Note:

- The + and – signs on the inputs show whether they invert or not. They do *not* indicate + or – voltage.

Power supply A three-terminal DC supply is required, giving 0 V (earth) and typically, ±9 V. The supply connections are often excluded from diagrams.

Op amp features These include:

- an extremely high voltage gain, typically around 10^5, though less at high frequencies,
- a very high input impedance (e.g. 10^{12} Ω),
- a very low output impedance (e.g. 10^2 Ω).

To reduce gain, op amps are used with a resistor or wire linking the output to one input (see H17). This is called a ***closed loop***. (The op amp above has an ***open loop***.)

The graph above is for an open-loop op amp.

Note:

- The output voltage cannot exceed the supply voltage.
- The amplification is *linear*, provided the output voltage is not close to the supply voltage,
- At saturation, the output and input voltages are no longer in proportion, so the output is *distorted*.

Advantages of reducing gain These include the following:

- Input signals can be stronger without causing saturation and, therefore, distortion.
- The gain is less affected by temperature changes.
- The ***bandwidth*** is greater, i.e. the gain is constant over a wider range of input frequences (see below).

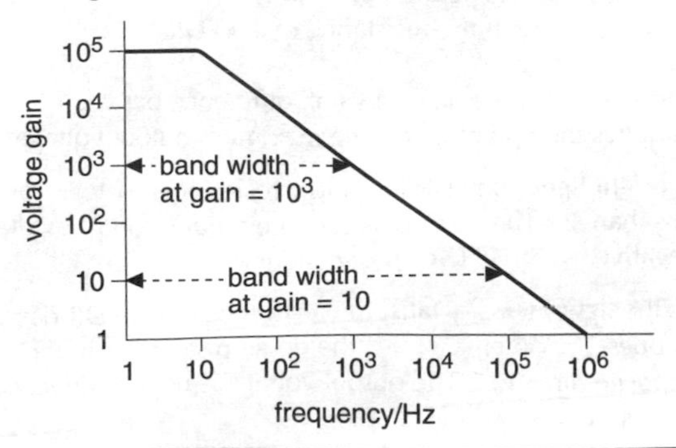

H17 Electronics – 2

Op amp as an inverting amplifier

The amplifier circuit on the right uses a closed loop to feed back a set fraction of the output voltage to the inverting input. This is called ***negative feedback*** because the signal being fed back partly cancels the input signal.

To calculate the amplifier's closed-loop voltage gain (V_{out}/V_{in}), the following assumptions are made:

- The op amp is not saturated. It has such a high open-loop gain that the voltage difference between its inverting and non-inverting inputs must be negligible. Therefore P is effectively at 0 V. (P is called a ***virtual earth***.)
- The current through P is negligible. Therefore, the current through R_f is the same as through R_i (see D3).

If I is the current through R_f and R_i,
voltage drop across $R_f = IR_f$
voltage drop across $R_i = IR_i$

But if P is at 0 V,
voltage drop across $R_f = 0 - V_{out} = -V_{out}$
voltage drop across $R_i = V_{in} - 0 = V_{in}$

From the above, it follows that $V_{out}/V_{in} = -R_f/R_i$

So $$\text{closed-loop voltage gain} = -\frac{R_f}{R_i}$$

From the closed-loop voltage gain equation below:

- An *inverting* amplifier has a *negative* gain.
- The gain does not depend on the characteristics of the op amp. It is set by the values of R_f and R_i. For example, if $R_f = 200\ \text{k}\Omega$ and $R_i = 20\ \text{k}\Omega$, the gain $= -200/20 = -10$.

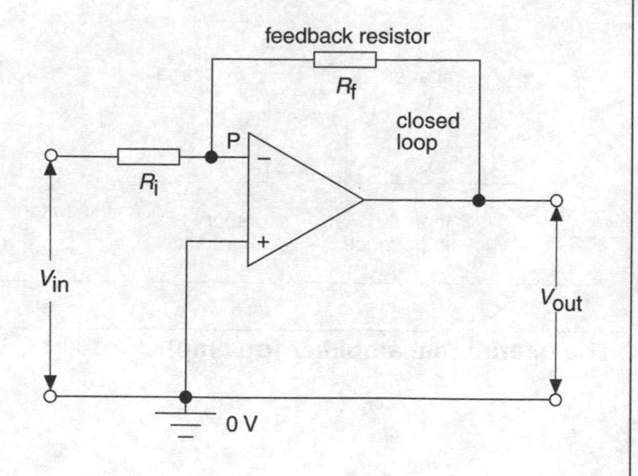

Op amp as a non-inverting amplifier

In the circuit on the right, the input signal goes to the non-inverting input, and the potential divider provides negative feedback. With the same assumptions as above,

$$\text{closed-loop voltage gain} = 1 + \frac{R_f}{R_i} \qquad (1)$$

A non-inverting amplifier has a much higher input impedance than the inverting type. This is useful, for example, where a high-impedance microphone is connected to the input (see *Impedances and matching* in H16).

Voltage follower This is a non-inverting amplifier in which R_f is zero (i.e. a direct connection) and R_i is infinite (i.e. removed). In this case, the gain is 1, so the output voltage is the same as the input voltage. However, the output impedance is much less than the input impedance, which makes the circuit useful for impedance matching.

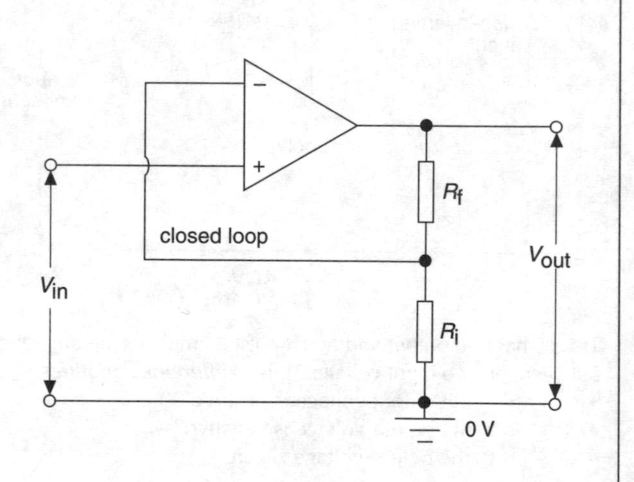

Op amp as a comparator

On the right, the LED is switched on automatically when darkness falls. The switching is done by an op amp which *compares* the voltages from two potential dividers:

V_- is fixed by the ratio of R_2 to R_1,
V_+ varies with the resistance of the LDR.

There is no feedback, so a small difference between V_- and V_+ saturates the op amp, causing maximum output voltage.

In bright light, the LDR has a low resistance, so V_+ is low and less than V_-. The op amp is saturated, but its output voltage is negative, so the LED cannot conduct.

As the light intensity falls, the resistance of the LDR rises, and so does V_+. When $V_+ > V_-$, the op amp saturates in the opposite direction. The output voltage is now positive, so the LED conducts and lights up.

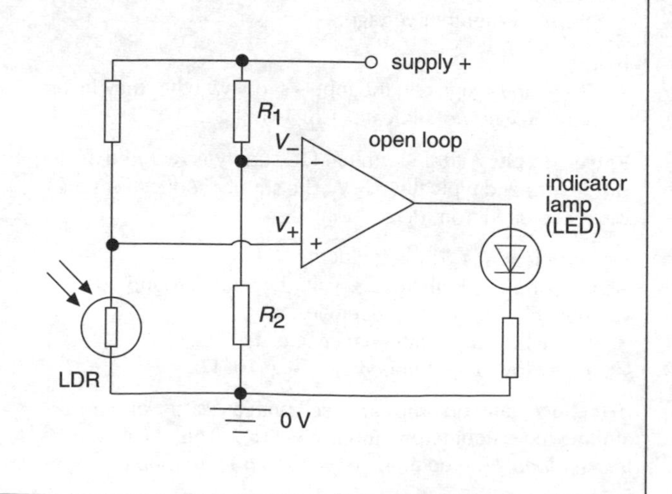

Logic gates

These are electronic switches, with one output and one or more inputs. A typical logic IC has several gates on the same chip, and needs a DC supply providing +5 V and 0 V.

Each input is made either HIGH (e.g. +5 V) or LOW (0 V). As a result, the output is either HIGH or LOW – depending on the type of gate and its input state(s).

The outputs produced by different inputs are shown in a ***truth table*** (as below). The output and input states are represented by the ***logic numbers*** 1 (HIGH) and 0 (LOW).

NOT gate (inverter)

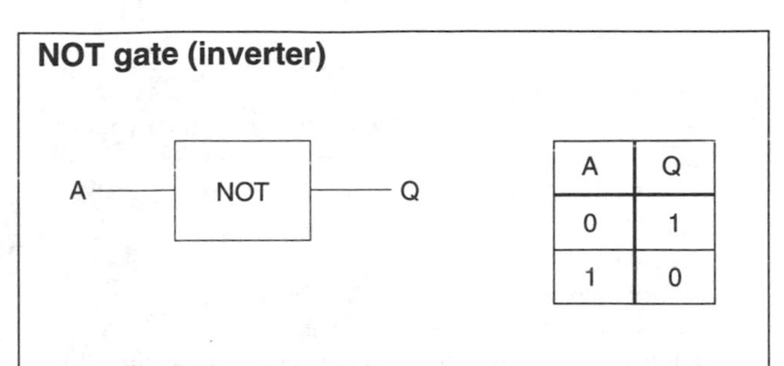

A	Q
0	1
1	0

The ouput is HIGH if the input is *NOT* HIGH, and vice versa.

AND gate

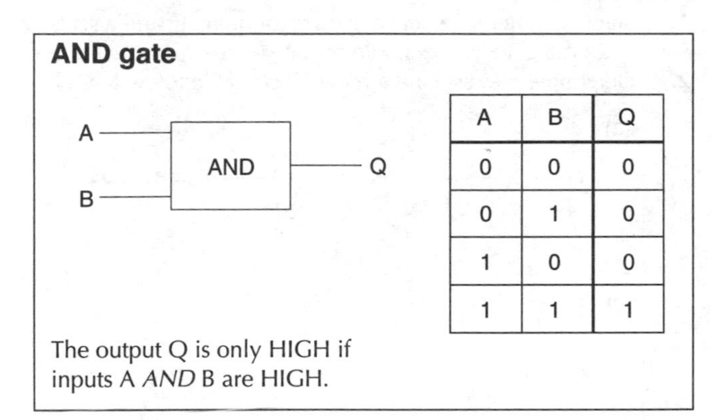

A	B	Q
0	0	0
0	1	0
1	0	0
1	1	1

The output Q is only HIGH if inputs A *AND* B are HIGH.

OR gate

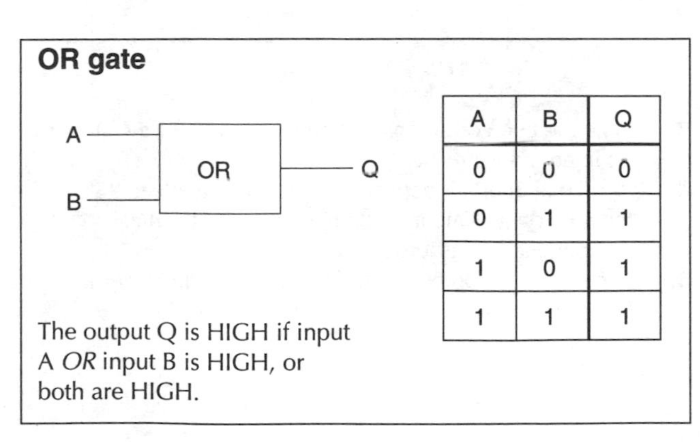

A	B	Q
0	0	0
0	1	1
1	0	1
1	1	1

The output Q is HIGH if input A *OR* input B is HIGH, or both are HIGH.

Exclusive-OR gate

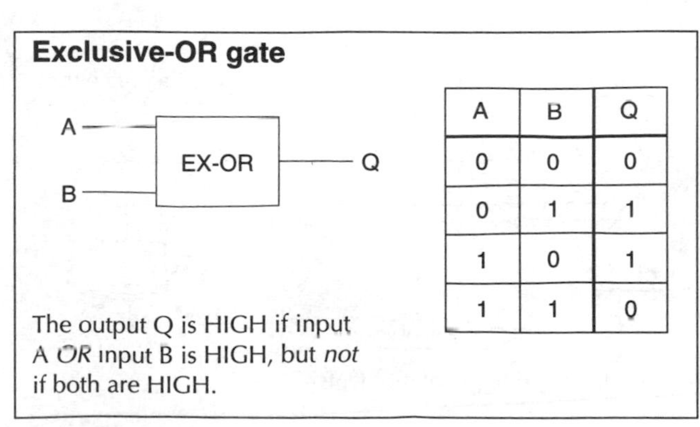

A	B	Q
0	0	0
0	1	1
1	0	1
1	1	0

The output Q is HIGH if input A *OR* input B is HIGH, but *not* if both are HIGH.

Using gates

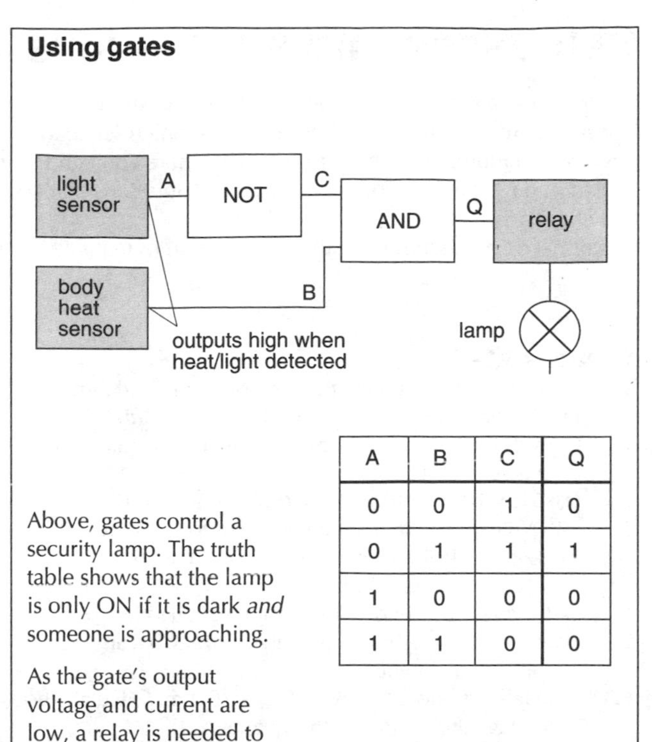

A	B	C	Q
0	0	1	0
0	1	1	1
1	0	0	0
1	1	0	0

Above, gates control a security lamp. The truth table shows that the lamp is only ON if it is dark *and* someone is approaching.

As the gate's output voltage and current are low, a relay is needed to switch the lamp on and off.

NAND and other gates

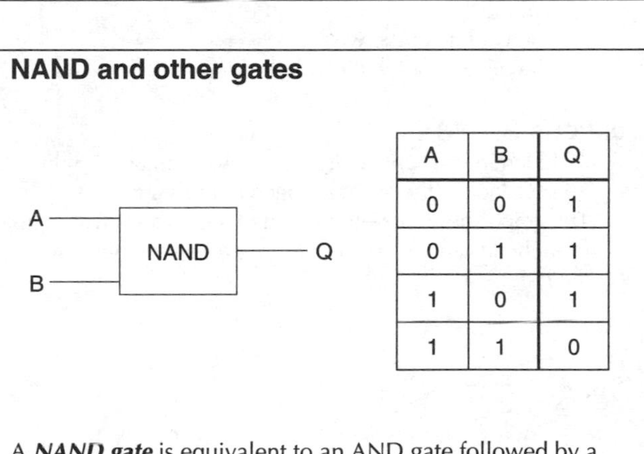

A	B	Q
0	0	1
0	1	1
1	0	1
1	1	0

A ***NAND gate*** is equivalent to an AND gate followed by a NOT gate.

A ***NOR gate*** is equivalent to an OR gate followed by a NOT gate. Its output is that of an OR gate *inverted*.

An ***exclusive-NOR gate*** is equivalent to an exclusive-OR gate followed by a NOT gate.

Made from NAND gates

All gates can be made from NAND (or NOR) gates. e.g.

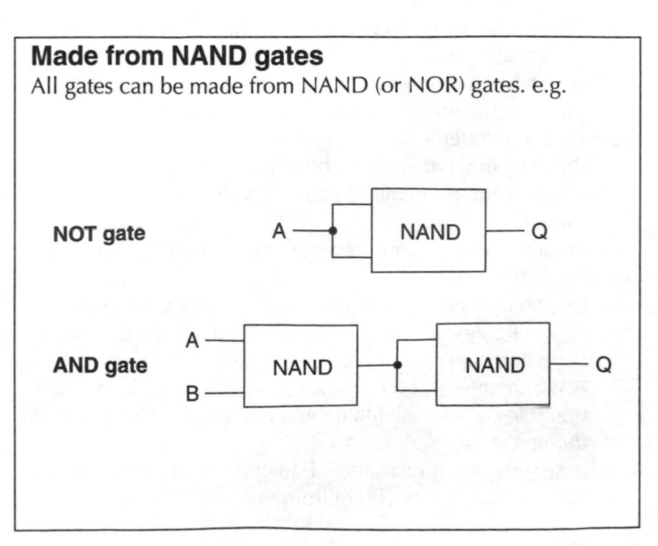

Self-assessment questions

After revising a section you should try these questions.
Questions are only given for those sections which relate to compulsory material in the specifications.
Answers, including references to sections where you can find more detail, begin on page 151.
You should review the work relating to any questions that you were unable to do or when you obtain incorrect numerical answers.
Where necessary assume that the acceleration of free fall g = 9.8 m s^{-2}.

Sections A1–A2

1. Write down the following quantities in standard form:
   (a) 3.5 MΩ **(b)** 220 pF **(c)** 15 mm s^{-1} **(d)** 25 mm^2
2. Convert the following quantities to include a number with a suitable prefix:
   (a) 5.0×10^7 m **(b)** 3.2×10^{-3} A **(c)** 39×10^{-9} s
3. **(a)** Write down the defining equation for pressure.
   (b) Use the defining equation to arrive at a unit for pressure in terms of base units.
4. State what is meant by a dimensionless quantity.
5. State the two types of uncertainty that may occur in scientific experiments.
6. A quantity is quoted as $(3.7 \pm 0.2) \times 10^{-3}$ m. Calculate the percentage uncertainty in this quantity.
7. The diameter of a wire is measured as 1.2 ± 0.1 mm. The length of the wire is 73.0 ± 0.5 cm.
   Determine
   (a) the volume of the wire
   (b) the uncertainty in the volume.

Sections B1–B3

1. A student runs round a circular track of radius 40 m in 30 s. Calculate the average speed of the student.
2. The graph shows how the speed of a car varies with time from the instant when a driver sees a dog running into the road:

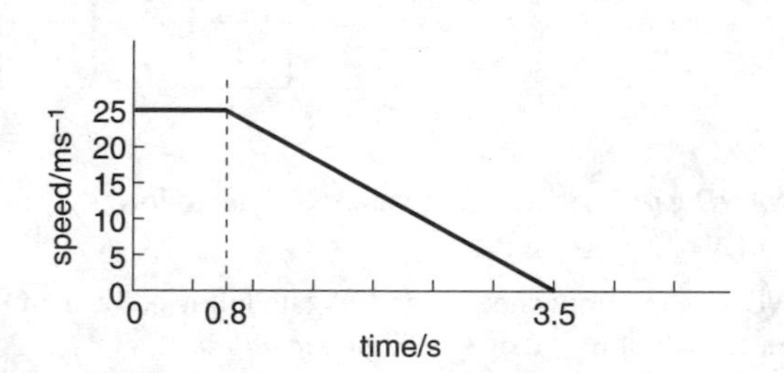

   (a) Determine the distance travelled before the driver applies the brakes.
   (b) Calculate the deceleration produced when the brakes are applied.
3. **(a)** State the difference between mass and weight.
   (b) State the unit in which each is measured.
4. **(a)** List 6 different forms of energy.
   (b) Explain what is meant by internal energy.
   (c) State the principle of conservation of energy.
5. Calculate
   (a) the kinetic energy of a car of mass 850 kg travelling at 110 km h^{-1}
   (b) the change in potential energy when a skier of mass 70 kg skis from a height of 1500 m above sea level to a height of 1210 m above sea level.
6. A weight lifter lifts a mass of 110 kg through a height of 0.5 m in 0.40 s. Calculate the useful power developed during the lift.
7. Assuming no air resistance, how long would it take a ball to fall a distance of 100 m from rest?
8. A car travelling at 10 m s^{-1} accelerates uniformly at 2.5 s^{-2} for 3.5 s.
   (a) How far does it travel while accelerating?
   (b) Calculate its final speed.
9. A dart player throws a dart that hits the dartboard at the same height as that from which it was thrown.
   The dart was thrown from a distance of 3.0 m from the board and takes 0.25 s to reach the board.
   Calculate
   (a) the maximum height reached by the dart
   (b) the horizontal speed of the dart
   (c) the vertical speed of the dart when it reaches the dartboard.

Sections B4–B5

1. State the difference between a vector quantity and a scalar quantity. Give one example of each.
2. Determine the resultant force in each of the following:

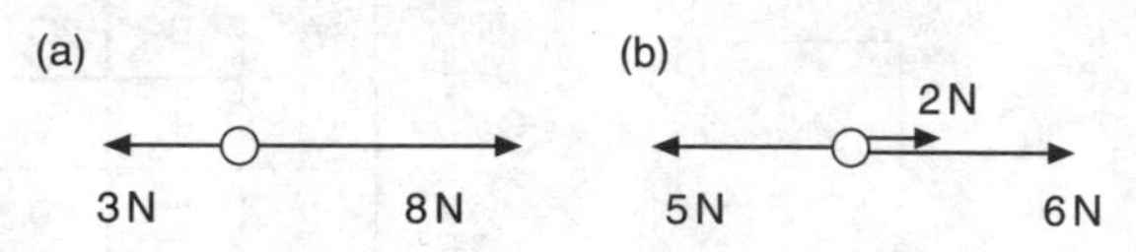

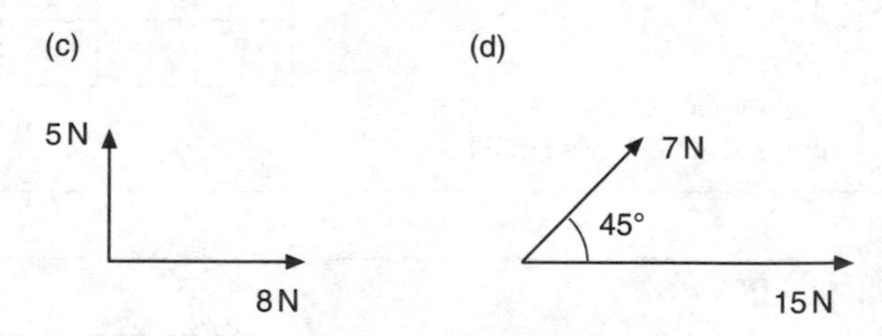

3. State the two conditions that are necessary for a system to be in equilibrium.
4. A force of 250 N is applied to a box at an angle of 35° to the horizontal. Calculate the horizontal and vertical components of this force.
5. The system of forces in the diagram is in equilibrium. Calculate the tension in each of the strings.

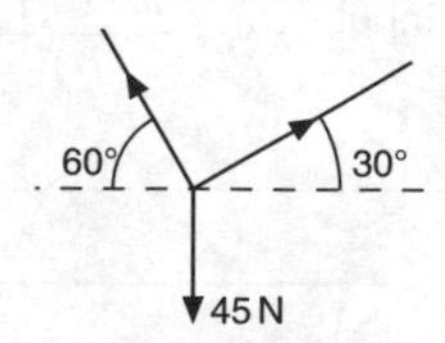

6. The diagram shows one arm of a 'mobile'. Calculate the weight of the ship.

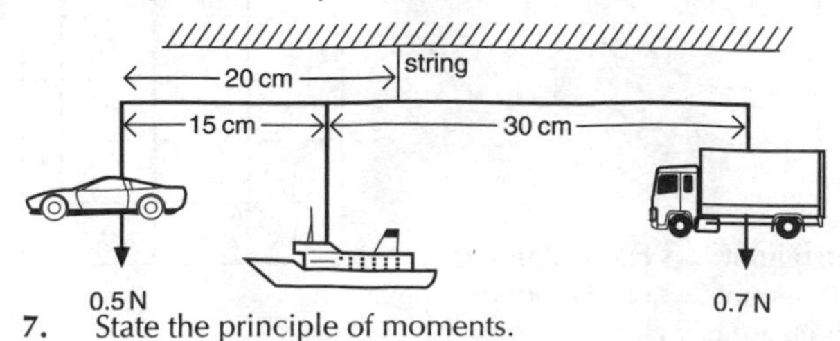

7. State the principle of moments.

8. A person of weight 550 N stands on a plank of weight 150 N in the position shown in the diagram:

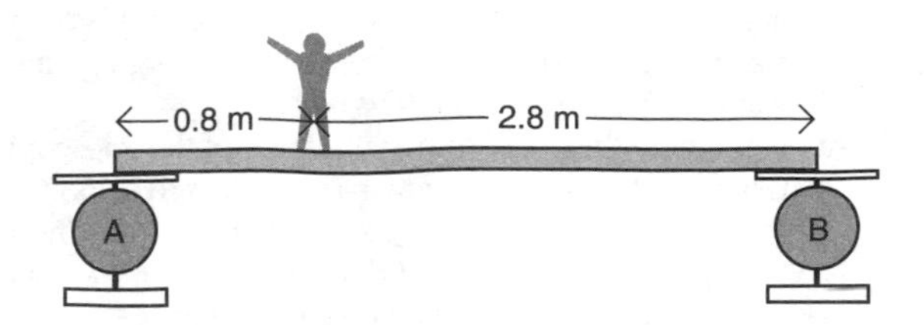

Calculate the reading in N of each of the scales A and B.

9. Explain what is meant by stable equilibrium. Explain how you would design a table lamp so that it will be in stable equilibrium.

Sections B6–B10

1. The kinetic energy of a body changes from 120 J to 40 J. It travels 3.5 m as this change occurs. Calculate the force acting on the body.
2. State the similarities and differences between an elastic and an inelastic collision.
3. A truck of mass 5000 kg travelling at 30 m s^{-1} collides with the rear of a car of mass 800 kg travelling in the same direction at 20 m s^{-1}. The two vehicles move together after the collision. Calculate the final velocity.
4. A nucleus of mass 4.00×10^{25} kg splits up into two parts. One of mass 0.07 kg moves off at a speed of 1.0×10^7 m s^{-1}. Calculate the speed of the other particle.
5. A van that develops an output power of 3.0 kW moves at a steady velocity of 30 m s^{-1}.
   (a) Calculate the total forward force developed between the tyres and the road.
   (b) Explain why the air resistance must be exactly equal to the forward force in this case.
6. State what is meant by viscosity.
7. A plumber wants to double the rate of flow of water down a pipe. The pressure difference and the length of pipe are unchanged. The original pipe has an internal diameter of 10 mm. Determine the diameter required for the new pipe.
8. A ball bearing of mass 3.1×10^{-5} kg and diameter 2.0 mm falls through glycerol. Glycerol is a liquid of viscosity 1.5 N s m^{-2}. Calculate
   (a) the weight of the ball bearing
   (b) the terminal velocity of the ball bearing.
9. The graph shows how the speed of an athlete varies with time during the run-up to a long jump.

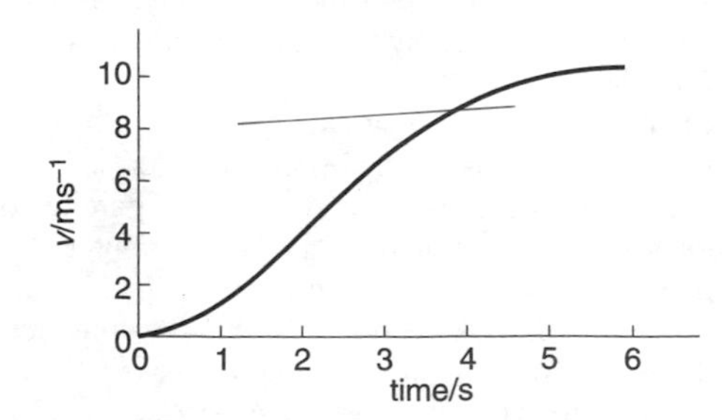

Determine
(a) the maximum acceleration of the athlete
(b) the length of the run-up.

Sections B11–B13

1. **(a)** State what is meant by angular velocity.
   (b) State the unit in which it is measured.
2. A disc of radius 0.080 m rotates at a rate of 1500 revolutions per minute. Calculate
   (a) the period of rotation of the disc
   (b) the speed of the rim of the disc.
3. A proton of mass 1.7×10^{-27} kg and speed 2.0×10^6 m s^{-1} moves in a circular path of radius 0.40 m. Determine the magnitude of the force acting on the proton. State the direction of this force at any instant.
4. Explain why a body moving at constant speed in a circular path is accelerating.
5. How is the centripetal force provided in the following situations?
   (a) A cyclist moving round a bend.
   (b) A car going round a bend on a very slippery banked track.
   (c) An orbiting satellite.
6. Draw two cycles of a displacement–time graph for SHM. Show on your diagram the amplitude and period of the oscillation.
7. A particle is released and moves with SHM of period 5.0 Hz and amplitude 0.040 m.
   Calculate
   (a) the period of the motion
   (b) the maximum acceleration of the particle
   (c) the maximum velocity of the particle.
8. Show that the period of a simple pendulum is simple harmonic.
9. A car and its suspension system behaves as a mass–spring system. A car has a mass of 1250 kg. The period of oscillation after going over a bump in the road is 1.5 Hz. Calculate
   (a) the total stiffness of the suspension
   (b) the stiffness of each of the 4 'springs'.
10. The period of a simple pendulum is 1.2 s. It swings with a maximum displacement of 0.018 m. The mass of the bob is 0.050 kg. Calculate
    (a) the total energy of a pendulum
    (b) the kinetic energy when the bob is at half its maximum displacement
    (c) the potential energy in this position.
11. **(a)** Explain what is meant by a forced oscillation.
    (b) Under what conditions does resonance occur?

Sections C1–C2

1. Draw a sketch to show how the displacement of the medium varies with distance from the source for a sinusoidal wave. Indicate on your sketch the wavelength and the amplitude of the wave.
2. Distinguish between
   (a) refraction and diffraction
   (b) a longitudinal and a transverse mechanical wave
   (c) a progressive and a stationary wave
   (d) a node and an antinode in a stationary wave.
3. Magenta (red and blue) light is incident on a parallel-sided glass block. Draw a diagram to show the passage of the light through the block.
4. Place the following parts of the electromagnetic spectrum in order of increasing frequency.
   ultraviolet, UHF radio, gamma rays, visible light, microwaves
5. Explain what is meant by total internal reflection and state the conditions under which it takes place.
6. A ray of light is incident on an air–glass interface at an angle of incidence of 25.0°. The refractive index of the glass is 1.55. Calculate the angle of refraction.
7. Calculate the critical angle for water of refractive index 1.33.
8. The core of a fibre optic cable has a refractive index of 1.52 and the cladding a refractive index of 1.51. Calculate the critical angle for a ray of light travelling in the core.

9. In an experiment, it is found that it is possible to polarize the radiation used. State and explain the conclusions that can be drawn from this observation.
10. The intensity of the radiation reaching the outer atmosphere of the Earth is 1400 W m^{-1}. Calculate the intensity on Pluto.
Distance from Earth to Sun = 1.5×10^{11} m.
Distance from Pluto to Sun = 5.9×10^{12} m.

Section C3

1. **(a)** Explain what is meant by superposition of waves.
(b) State the conditions necessary for observation of an interference pattern from two sources.
2. Explain the role of diffraction in setting up an experiment to observe interference using light.
3. Calculate the fringe spacing produced using laser light of wavelength 650 nm, at a distance of 2.5 m from two sources that are separated by 1.5 mm.
4. An observer notices that if she walks 1.8 m in the direction shown in the diagram, the sound intensity falls from a maximum to a minimum. The speakers are emitting notes of the same frequency.

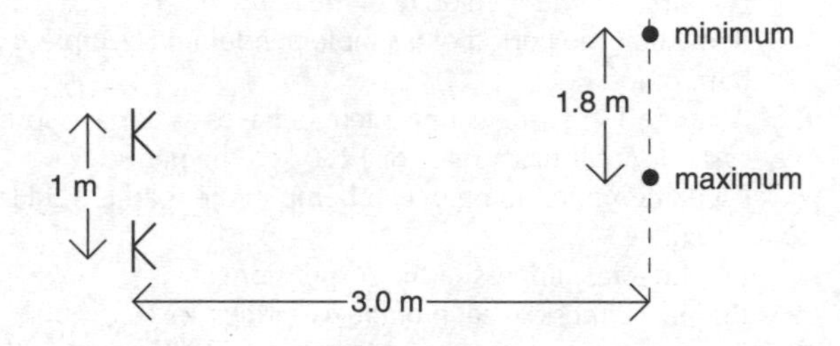

(a) Explain this observation.
(b) By means of a scale diagram determine the wavelength of the sound.
5. A diffraction grating has 300 slits per mm. Calculate the angular separation of the second-order maxima for red light (wavelength = 650 nm) and blue light (wavelength = 450 nm).
6. A string resonates at a fundamental frequency of 200 Hz. The length of the string is 0.65 m.
(a) Sketch a diagram to show this mode of vibration.
(b) Calculate the frequency and wavelength of the third harmonic vibration for this string.
(c) By how much would you need to increase the tension to double the fundamental frequency?
7. A pipe that is closed at one end has a length of 0.45 m. It resonates at a fundamental frequency of 190 Hz. Estimate the speed of sound in air.

Sections C4–C5

1. State the difference between a real and a virtual image.
2. Draw a diagram showing the passage of a parallel beam of light through a converging lens. Indicate the position of the principal focus of the lens and the focal length.
3. An object is 0.35 m from a converging lens of focal length 0.12 m. Determine the position of the image formed by the lens.
4. A lens of focal length 0.045 m is used as a magnifying glass. The virtual image is formed 0.25 m from the lens. Determine
(a) the position of the object
(b) the magnification produced.
5. State and explain two reasons why lenses may not produce a sharp image.
6. **(a)** State the position of the image formed by the objective of a refracting telescope.
(b) What is the position of the final image formed by the eyepiece when the telescope is in normal adjustment?
7. A refracting telescope is made from two converging lenses of focal length 0.90 m and 0.030 m.
(a) Determine the angular magnification of the telescope.
(b) Calculate the distance between the objective and the eyepiece when the telescope is in normal adjustment.
8. A telescope has an aperture of diameter 0.25 m. Two stars viewed through a red filter that transmits light of wavelength 650 nm are just resolved. Determine the angular separation of the stars.

Sections D1–D4

1. Calculate the number of electrons passing each point in a circuit when the current is 30 mA.
2. **(a)** Explain what is meant by a supply having an EMF of 12 V and an internal resistance of 0.5 Ω.
(b) Calculate the current that flows when a 4.0 Ω resistor is connected between the terminals of the supply.
(c) Determine the current that would flow if the supply terminals were short circuited.
3. Calculate the total resistance of each of the following combinations of resistors:

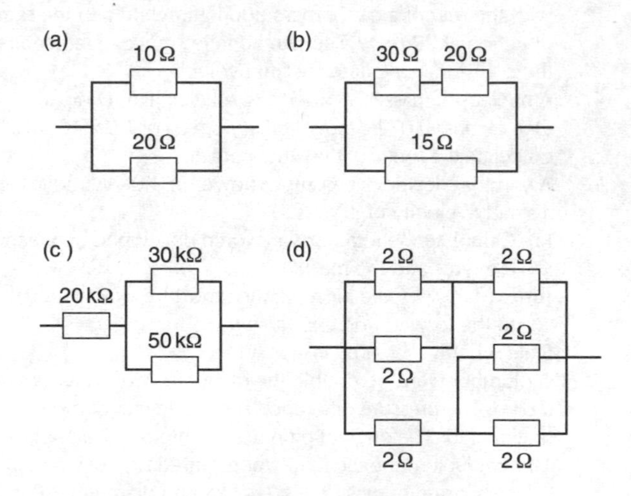

4. A car headlamp is connected to a 12.0 V supply of negligible internal resistance. The lamp then works at its rated power of 24 W. Calculate
(a) the current in the circuit
(b) the resistance of the filament when the lamp is working normally.
5. **(a)** Two components of resistance 4.0 Ω are connected in series to a 5.0 V supply of negligible internal resistance. Calculate
(i) the current in the circuit
(ii) the power dissipated by each resistor.
(b) The components are now connected in parallel to the same supply. Calculate the new current and power dissipated for each resistor.
6. Explain in microscopic terms why as temperature increases
(a) the resistance of a filament lamp increases
(b) the resistance of a thermistor decreases.
7. A copper wire of diameter 0.80 mm carries a current of 1.5 A. The copper contains 8×10^{28} electrons per m^3. Determine the drift speed of the electrons down the wire.
8. When a potential difference of 9.0 V is placed across a wire there is a current of 1.6 A in the wire. The wire has a length of 1.8 m and an area of cross-section of 0.15 mm^2. Calculate the resistivity of the material from which the wire is made.

9. (a) Determine the readings of the meters in the following circuit:

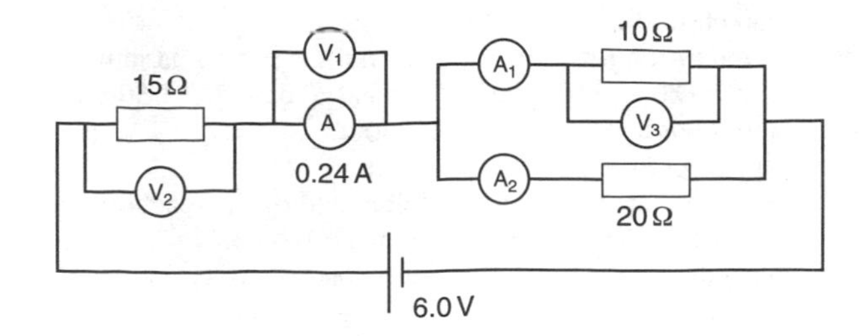

(b) State the terminal potential difference of the supply.
(c) Explain why this is lower than the EMF of the supply.

10. (a) Determine the magnitude of the output voltage in the circuit below.

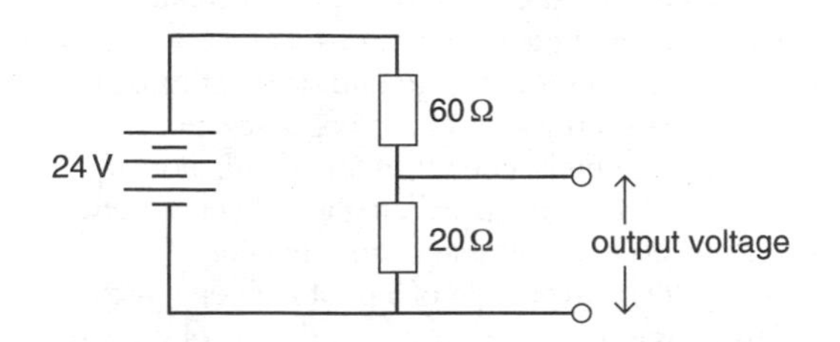

(b) A 4.0 Ω resistor is connected across the output terminals. Determine the new value of the output voltage.
(c) Determine the current drawn from the supply in **(a)** and **(b)**.

11. (a) Determine the peak value of the voltage in a supply that is producing 12 V RMS.
(b) What will be the average power dissipated in a 8.0 Ω resistor connected between the terminals of the supply?
(c) Calculate the maximum power dissipated by the resistor.

12. A 7.0 V RMS supply is produced from 230 V mains.
(a) Determine the turns ratio of the transformer needed.
(b) The current in the circuit connected to the 7.0 V supply is 30 mA. Calculate the current drawn from the mains assuming that the transformer
(i) is 100% efficient
(ii) has an efficiency of 85%.

Sections E1–E4

1. State two similarities and one difference between electric and gravitational fields.
2. Calculate the force between two point charges of magnitude 3.2×10^{-19} C and 1.6×10^{-19} C separated by a distance of 2.5×10^{-10} m.
3. How much energy is required to separate completely the two particles in question **2**?
4. An electron of charge -1.6×10^{-19} C is placed in a uniform field of strength 20 000 V m^{-1}. Calculate the force on the electron.
5. (a) State what is meant by a capacitance of 220 μF.
(b) List the design factors that affect the capacitance of a capacitor.
6. A 100 μF capacitor is charged to a potential difference of 6.0 V. It then discharges through a 2.2 kΩ resistor. How long will it take for the voltage to **(a)** halve **(b)** fall to 2.0 V?
7. A 100 μF capacitor is in series with a 200 μF capacitor. The PD across the combination is 12 V. Calculate
(a) the total capacitance of the combination
(b) the energy stored.
8. Mars has a mass of 6.4×10^{23} kg and a radius of 3.4×10^{6} m. Determine the acceleration of free fall at its surface.
9. (a) Explain why gravitational potentials have negative values.
(b) Sketch a graph to show how the gravitational potential varies with distance from the centre of the Earth over a range of R to 4R.
10. The acceleration of free fall at the surface of the Moon is 1/6 that at the surface of the Earth. Calculate the speed and period of a satellite in orbit close to the surface of the Moon. Radius of the Moon is 1.7×10^{6} m.
11. (a) State the factors that affect the moment of inertia of a body.
(b) A torque of 3.0 N m is applied to a disc of moment of inertia 0.5 kg m^2. The disc is free to rotate and starts from rest. Calculate
(i) the angular acceleration
(ii) the angular speed after 5 seconds
(iii) the number of revolutions during the first 5 s.

Section E5–E8

1. Sketch the magnetic field of a bar magnet.
2. Explain what is meant by a neutral point in a magnetic field.
3. (a) Define magnetic flux density.
(b) Calculate the force on a straight wire of length 0.20 m when it carries a current of 2.5 A in a magnetic field of flux density 50 mT.
4. Two wires of length 1.0 m are separated by a distance of 0.80 mm. They each carry a current of 1.5 A but in opposite directions. Determine the magnitude and direction of the force exerted by one wire on the other.
5. Explain the operation of a brake that depends on electromagnetic induction.
6. The flux density through a 200 turn coil of area 8.5×10^{-4} m^2 changes from 0.03 T to 0.12 T in 15 ms. Calculate the induced EMF in the coil.
7. A 50 mH inductor and a 300 Ω resistor are connected in series. They are connected to a battery of EMF 12 V and negligible internal resistance. Calculate
(i) the initial rate of rise of current
(ii) the maximum current in the circuit.
8. Calculate the magnitude of the force on an electron moving at a speed of 1.5×10^{6} m s^{-1} in a magnetic field of flux density 0.20 T.
9. Explain why the path of an electron in a magnetic field is circular while that in an electric field is parabolic.
10. Calculate the speed of an electron when it is accelerated through a potential difference of 2.5 kV.
11. In a mass spectrometer a designer wants to select ions with a velocity of 1.2×10^{7} m s^{-1}. The magnetic field available has a flux density of 0.8 mT. Determine the strength of the electric field required to select this velocity.

Sections F1–F3

1. Determine the pressure at the bottom of a tower containing a depth of 25 m of water (density 1000 kg m^{-3}). The tower is open to the atmosphere and the barometric pressure is 0.75 m of mercury. (Density of mercury = 13 600 kg m^{-3}.)
2. List the advantages and disadvantages of a thermocouple for measuring temperature.
3. The temperature of a 300 g block of metal is raised from 20 °C to 35 °C when 590 J of energy are supplied. Show that the specific heat capacity of the metal is about 130 J kg^{-1} K^{-1}.
4. The metal in **3** has a molar mass of 197 g. Calculate the molar heat capacity for the metal.

5. A cook forgets about a saucepan on a cooker. The pan initially contains 0.250 kg of boiling water. The power supplied to the saucepan from the cooker was 0.600 kW. How long will it be before the saucepan boils dry? (Specific latent heat of water = 2.3×10^6 J kg^{-1}.)

6 **(a)** Calculate the maximum efficiency of an engine that is taking energy from a source at a temperature of 500 K when the temperature of the sink is 300 K.
(b) When working at maximum efficiency the power from the source is 20 kW. Calculate the power that goes to the sink.

Sections F4–F7

1. **(a)** State Boyle's law.
(b) Explain what is meant by an ideal gas.
2. The temperature of a fixed mass of an ideal gas changes from 27 °C to 87 °C at constant volume. The original pressure was 0.9×10^5 Pa. Calculate the final pressure.
3. Calculate the final pressure for the same temperature change and initial pressure as in question **2** but allowing the volume of the gas to increase by 50%.
4. Calculate the number of moles of gas in a container of volume 2.5×10^{-3} m^3 containing gas at a pressure of 5×10^4 Pa at a temperature of 300 K.
5. **(a)** A gas expands from 2.5×10^{-3} m^3 to 3.8×10^{-3} m^3 at a constant pressure of 1.2×10^5 Pa. Calculate the work done by the gas.
(b) The change is adiabatic. State and explain what happens to the gas as a result of the change.
6. State the difference between evaporation and boiling.
7. Describe briefly an experiment to estimate the size of a molecule.
8. Calculate the kinetic energy of a molecule at a temperature of 300 K. The Boltzmann constant is 1.38×10^{-23} J K^{-1}.
9. Determine the root mean square speed of the molecules of a gas of density 0.12 kg m^{-3} when the gas pressure is $5\ 0 \times 10^5$ Pa.
10. Calculate the rate at which energy is conducted through the base of a copper saucepan of area 0.063 m^2 and thickness 0.0080 m when the bottom surface is at a temperature of 100.5 °C and the liquid in contact with the base of the saucepan is at 100 °C.

Sections G1–G3

1. Write down the symbols for the following particles including their mass and charge numbers.
Proton, electron, alpha particle, neutron, gamma ray photon
2. Describe the atomic structure of $^{206}_{82}$Pb.
3. State what is meant when we say that two nuclides are isotopes.
4. $^{206}_{82}$Pb is formed when a radioactive atom decays by alpha emission. Determine the proton number and mass number of the atom that has decayed.
5. Complete the nuclear equation for the fusion reaction below and identify the particle X.

$$4^1_1\text{p} \rightarrow \alpha + 2\text{X} + 2\nu_e + \gamma$$

6. A small source emits gamma rays. The count rate is 280 s^{-1} when the GM tube is 10 cm from the source. What would you expect it to be when the tube is
(i) 20 cm from the source
(ii) 15 cm from the source.
7. Thorium-234 has a half-life of 24 days. A sample initially contains 6.0×10^{12} atoms. Calculate
(a) the decay constant of thorium in s^{-1}
(b) the initial activity of the thorium-234
(c) the activity after 12 days.
8. Describe briefly one industrial and one medical use of radioactive materials. State the emission and the half-life of the source that is most appropriate for the application you choose.
9. In the fusion reaction in question **5** the proton has a mass 1.007 276 u, the alpha particle has a mass 4.001 506 u, and X has a mass 0.000 548 580 u.
(i) Determine the mass defect in u.
(ii) Calculate the energy in J liberated by the reaction. (1 u = 931 MeV; 1 MeV = 1.6×10^{-13} J.)
10. State the purpose in a nuclear reactor of **(a)** the moderator and **(b)** the coolant.

Sections G4–G5

1. The work function of zinc is 4.31 eV.
(i) Calculate the energy in J of an electron emitted from a zinc target by UV radiation of wavelength 200 nm.
(ii) Calculate the threshold frequency for zinc.
2. Calculate the frequency of the radiation emitted when an electron in a hydrogen atom undergoes a transition from the level at –1.5 eV to the ground state at –13.6 eV.
3. Why do the energy levels have negative values?
4. A carbon dioxide laser emits infrared radiation of wavelength 10.6 μm. Calculate the difference between the energy levels that give rise to this emission.
5. Calculate the wavelength of a photon of energy 2.2 eV.
6. Calculate the photon energy for radiation of wavelength 1.5×10^{-10} m.
7. For lasing action to take place there must be population inversion and a metastable state. Explain what is meant by the italicized terms.
8. An atomic nucleus has a diameter of about 1.0×10^{-14} m. Calculate the minimum energy of an electron that could exist in the nucleus.

Sections G6–G8

1. The rest energy of an electron is 0.511 MeV. Calculate its rest mass. (1 eV = 1.6×10^{-19} J.)
2. Calculate the mass of an electron that is moving at a speed of 0.7c.
3. How much energy is released when an electron and a positron (each of mass 9.1×10^{-31} kg) annihilate each other? What form will the energy take?
4. An electron collides inelastically with an atom. Explain what is meant by an inelastic collision and why a collision with an atom may be inelastic.
5. Calculate the radius of a uranium-238 nucleus given that $R_0 = 1.2 \times 10^{-15}$ m.
6. To what class of particles do the following particles belong?
muon; omega-minus; electron-neutrino; neutron; kaon
7. Write down the quark structure of an antiproton.
8. Use conservation laws to determine which of the following are possible reactions:
(a) $\pi^+ \rightarrow \mu^+ + \nu_\mu$
(b) $p^+ + \pi^- \rightarrow \pi^+ + K^-$
9. Draw a Feynman diagram that shows the exchange of a π^+ between a neutron and a proton.

Sections H4–H6

1. A piano wire made of steel (Young modulus = 2.0×10^{11} Pa) has a length of 1.30 m and a diameter of 2.0 mm. It stretches 2.5 m when it is tightened. Calculate the tension in the wire.
2. Into what classes of materials do the following fall? copper, rubber, nylon, diamond, glass

3. A tendon stretches 3.5 mm when the tension is 14 N. It obeys Hooke's law up to this tension. Calculate the energy stored in the tendon.
4. Draw a diagram showing how the strain varies with stress for rubber. Explain what is happening during the different stages at a molecular level.
5. What makes brittle materials brittle?
6. Marble is strong in compression and has a Young modulus of 5.0×10^{10} Pa. A pillar of length 5.0 m and cross-sectional area 1.5 m^2 supports a load of 20 000 kg. Calculate
   (a) the stress in the pillar
   (b) the strain in the pillar
   (c) the compression of the pillar.
7. Explain at a molecular level what happens when metals such as copper undergo plastic deformation.
8. Explain what is meant by a *composite material.* Give two examples of such a material.
9. Explain using band theory why some materials are electrical conductors while others are electrical insulators.
10. State the difference between a p-type and an n-type semiconductor.

Sections H10–H12

1. A 'square wave' is made up from a sinusoidal frequency equal to that of the square wave together with odd harmonics of this frequency. The amplitude of the harmonics decreases as frequency increases. A square wave of frequency 400 Hz is transmitted using a base bandwidth of 50 Hz to 4 kHz. Draw a frequency spectrum for the transmitted wave.
2. Determine the total bandwidth required for a radio station that is transmitting a base bandwidth of 50 to 4000 Hz.
3. A sinusoidal frequency of 1500 Hz is to be transmitted by amplitude-modulation on a carrier of frequency 200 kHz.
   (a) State the frequencies that will be received for this transmission.
   (b) Draw the frequency spectrum of the received signal.
   (c) Sketch the waveform of the signal received.
4. What is meant by *demodulation*?
5. **(a)** State the frequency band that is used for TV transmission.
   (b) Explain why this band is chosen rather than that used for medium-wave radio.
6. How many colour TV channels could be transmitted using the VHF waveband (30–300 MHz)?
7. The strength of a signal in a cable falls from 500 mW to 100 mW. Calculate the loss in dB.
8. **(a)** How often should a signal be sampled to transmit digitally music that has a base bandwidth from 15 Hz to 15 kHz?
   (b) If each sample uses 8 bits determine the minimum bit rate needed to transmit the information.

Answers to self-assessment questions

Sections A1–A2

1. **(a)** $3.5 \times 10^{6} \Omega$ **(b)** 2.20×10^{-10} F
   (c) 1.5×10^{-3} m s^{-1} **(d)** 2.5×10^{-5} m^2
2. **(a)** 50 Mm **(b)** 3.2 mA **(c)** 39 ns
3. **(a)** $P = F/A$ **(b)** kg m^{-1} s^{-2}
4. A quantity that has no units, e.g. a ratio of two lengths.
5. Random and systematic errors or uncertainties.
6. 5.4%
7. **(a)** 8.3×10^{7} m^3
   (b) 2.3% or absolute uncertainty = $\pm 0.2 \times 10^{3}$ m^3

Sections B1–B3

1. 6.3 m s^{-1}
2. **(a)** 20 m **(b)** 9.3 m s^{-2}
3. Mass is the quantity of matter in a body (scalar). Weight is the force of attraction of the Earth on the mass (vector).
4. **(a)** See B2. **(b)** Total PE and KE of all the particles in the object. **(c)** See B2.
5. **(a)** 4.0×10^{5} J **(b)** 2×10^{5} J
6. 1350 W
7. 4.52 s
8. **(a)** 50.3 m **(b)** 18.8 m s^{-1}
9. **(a)** 0.31 m **(b)** 12 m s^{-1} **(c)** 2.5 m s^{-1}

Sections B4–B5

1. Vector quantity has magnitude (size) and direction. Scalar quantity has magnitude only.
2. **(a)** 5 N to the right **(b)** 3 N to the right
   (c) 9.4 N at 32° to horizontal
   (d) 20.5 N at 15° to horizontal (by scale drawing)
3. No resultant moment (torque) and no resultant force.
4. Horizontal force = 205 N; vertical force = 143 N
5. T_1 = 39 N, T_2 = 22 N, 45 N
6. 2.2 N
7. See B5.
8. A reads 503 N; B reads 197 N.
9. If given a displacement it returns to its original position. Wide base; heavy base (load it with metal) so that centre of gravity is low.

Sections B6–B10

1. 22.9 N
2. Total energy and momentum is conserved in both types. In an elastic collision kinetic energy is conserved.
3. 28.6 m s^{-1} in original direction of motion.
4. 1.7×10^{5} m s^{-1}
5. **(a)** 100 N
   (b) When velocity is constant there is no acceleration. This is the case only when there is no resultant force.
6. (See B9.) Viscosity is a frictional force that opposes relative motion between layers of fluid or between a solid and a fluid.
7. 11.9 mm.
   Radius^4 has to be doubled. Radius therefore has to be increased by 0.95 mm.
8. **(a)** 3.8×10^{-4} N **(b)** 6.7×10^{-3} m s^{-1}
9. **(a)** 5.2 m s^{-2} **(b)** ~26 m

Sections B11–B13

1. **(a)** The angle swept out by a radial line per second.
   (b) rad s^{-1} (radian per second)
2. **(a)** 0.040 s
   (b) 12.6 m s^{-1}
3. 1.7×10^{-14} N toward the centre of the circular orbit.
4. Velocity is a vector, so when the velocity changes, as it does in circular motion, the velocity changes. There is acceleration because this is rate of change of velocity.
5. **(a)** This is equal to the friction between the cyclist and the road.
   (b) The horizontal component of the normal reaction to the track.
   (c) For an Earth satellite, the gravitational force between the Earth and the satellite.

6.

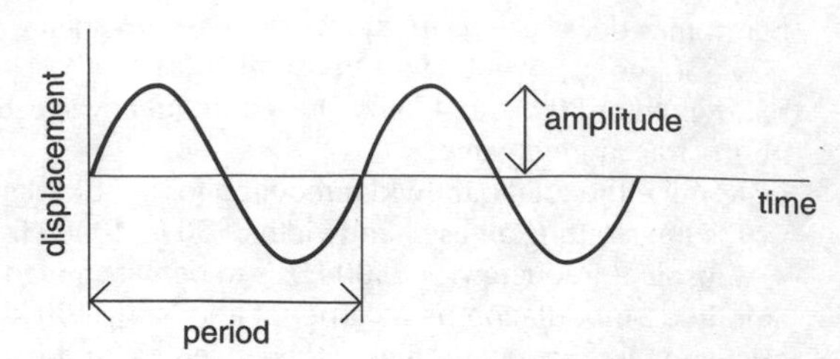

7. **(a)** 0.20 s **(b)** 39.5 m s^{-2} **(c)** 1.25 m s^{-1}
8. See B12.
9. **(a)** 111 kN m^{-1}
(b) 28 kN m^{-1}
10. **(a)** 2.22×10^{-4} J ($A\omega$ = maximum v; $\omega = 2\pi/T$; $KE = \frac{1}{2}mv^2$)
(b) 1.67×10^{-4} J **(c)** 0.055×10^{-4} J
(At half maximum displacement PE is $\frac{1}{4}$ maximum and KE is $\frac{3}{4}$ maximum.)
11. See B13.

Sections C1–C2

1.

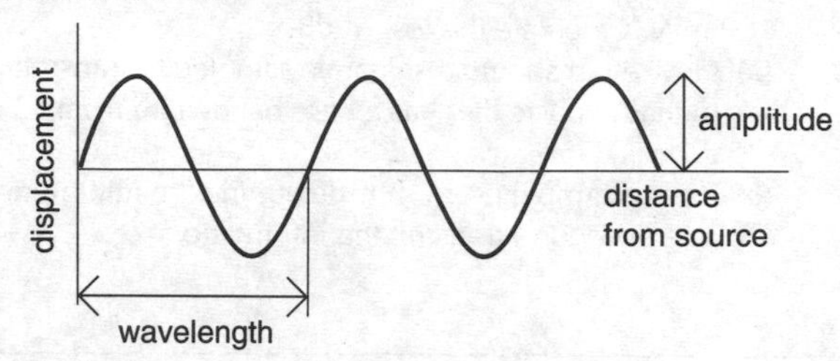

2 **(a)** Refraction: deviation of path of a wave at an interface between two media.
Diffraction: deviation when wave meets an obstruction such as a gap.
(b) Longitudinal: particles of medium oscillate in direction of energy transfer.
Transverse: particles of medium oscillate perpendicular to direction of energy transfer.
(c) Progressive: point of maximum displacement moves in direction of transfer of energy.
Stationary: points of maximum and minimum amplitude are fixed.
(d) Node: point of zero amplitude in a stationary wave.
Antinode: point of maximum amplitude in a stationary wave.

3.

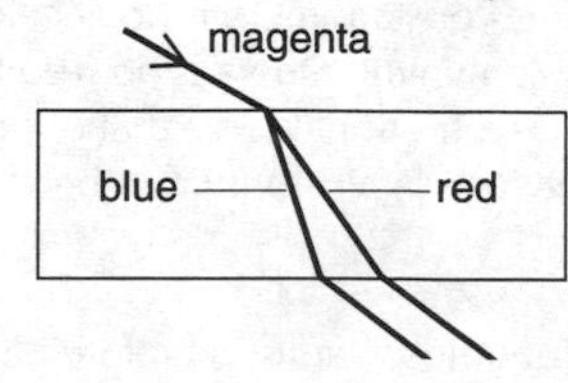

4. UHF radio, microwaves, visible light, ultraviolet, gamma rays.
5. See C2.
6. 15.8°
7. 40°
8. 83°
9. The radiation is in the form of a transverse wave. Unpolarized transverse waves contain vibrations in all planes perpendicular to the direction of propagation of energy. The process of polarization selects one of these planes. Longitudinal waves contain vibrations in the direction of energy so no further selection of oscillations is possible.
10. 9.0 W m^{-2} (using inverse square law)

Section C3

1. **(a)** See C3.
(b) Two sources must be coherent (same frequency and constant phase difference) and have similar amplitudes to produce good contrast between interference maxima and minima.
2. Diffraction spreads light form a single slit so that it illuminates two slits to produce coherent sources. Diffraction spreads out the light from two slits so that the beams overlap and produce interference patterns.
3. 1.1 mm
4. **(a)** At the maximum position the waves are arriving in phase and interfering constructively. By moving 0.8 m the wave from the lower speaker travels half a wavelength further than the other so that the waves are antiphase and interfere destructively.
(b) 0.95 m
5. 7.3°
6. **(a)**

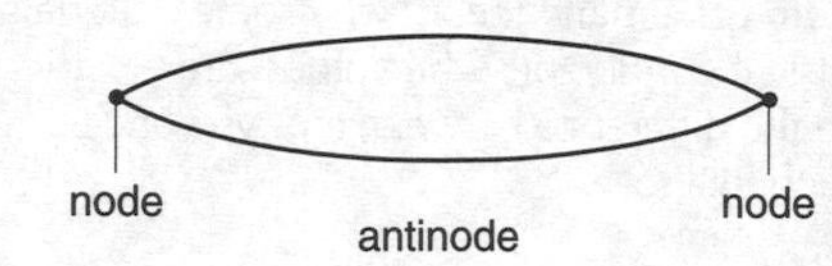

(b) 600 Hz; 0.22 m
(c) four times
7. 342 m s^{-1}

Sections C4–C5

1. A real image can be focused onto a screen, a virtual image cannot. Light rays actually pass through the position of a real image.
2.

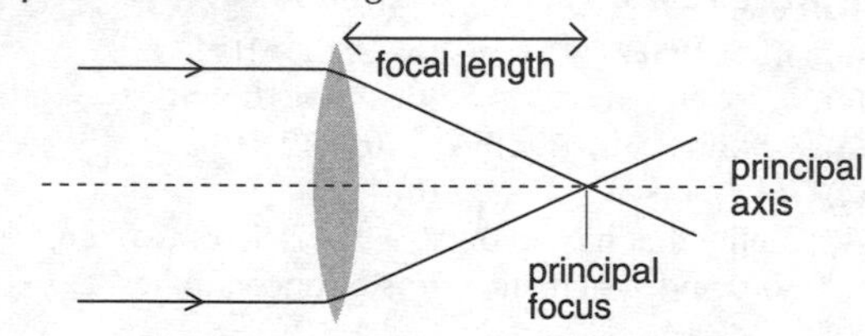

3. 0.18 m from the lens on opposite side to the object; a real image
4. **(a)** 0.038 m from the lens (on the same side as the image)
(b) 6.6 ×
5. Chromatic aberration and spherical aberration (see C4).
6. **(a)** At the principal focus of the objective lens
(b) At 'infinity'
7. **(a)** 30 ×
(b) 0.93 m
8. 2.6×10^{-6} rad or 1.5×10^{-4}°

Sections D1–D4

1. 5.6×10^{18} electrons
2. **(a)** An EMF of 12 V means that 12 J of work is done when 1 coulomb of charge passes around the complete circuit. The resistance of the components inside the supply (chemicals, wires, etc.) is 0.5 Ω.
(b) 2.7 A
(c) 24 A
3. **(a)** 6.7 Ω **(b)** 11.5 Ω **(c)** 38.8 kΩ
(d) 1.5 Ω
4. **(a)** 2.0 A **(b)** 6.0 Ω
5. **(a) (i)** 0.63 A **(ii)** 1.56 W
(b) (i) 2.5A **(ii)** 6.3 W
6. **(a)** More collisions of electrons with the lattice ions
(b) More electrons liberated; has a greater effect than the increased collisions of electrons with lattice ions

7. 2.3×10^{-4} m s^{-1}
8. 4.7×10^{-7} Ω m
9. **(a)** $V_1 = 0$; $V_2 = 3.6$ V; $V_3 = 16$ V; $A_1 = 0.16$ A; $A_2 = 0.08$ A
   (b) 5.2 V
   (c) Some EMF is 'lost' in producing current in the internal resistance of the supply.
10. **(a)** 3.0 V **(b)** 1.26 V
    (c) 0.3 A in **(a)**; 0.38 A in **(b)**
11. **(a)** 17 V **(b)** 18 W **(c)** 36 W
12. **(a)** 32.9:1 step down (more turns on coil connected to 240 V supply)
    (b) **(i)** 0.88 mA **(ii)** 1.04 mA

Sections E1–E4

1. Similarities: for point charges or masses
   Both obey inverse square law for variation of force with distance: both obey inverse *r* law for variation of potential with distance; both act at a distance with no requirement for a material medium.
   Difference: electric fields may produce attraction or repulsion; gravitational fields only produce attraction.
2. 7.3×10^{-9} N
3. 1.8×10^{-18} J
4. 3.2×10^{-19} N in opposite direction to the direction of the field.
5. **(a)** 200 μC of charge is 'stored' for each volt of potential difference between the terminals of the capacitor.
   (b) The area of the plates; the separation of the plates; the permittivity of the medium between the plates
6. **(a)** 0.15 s **(b)** 0.24 s
7. **(a)** 67 μF **(b)** 4.8×10^{-3} J
8. 3.7 m s^{-2}
9. **(a)** Potential at infinity (a very long way from the mass) is zero. Energy has to be supplied to a mass to move it to infinity. Since energy is added to get to zero potential the potential is negative.
   (b)

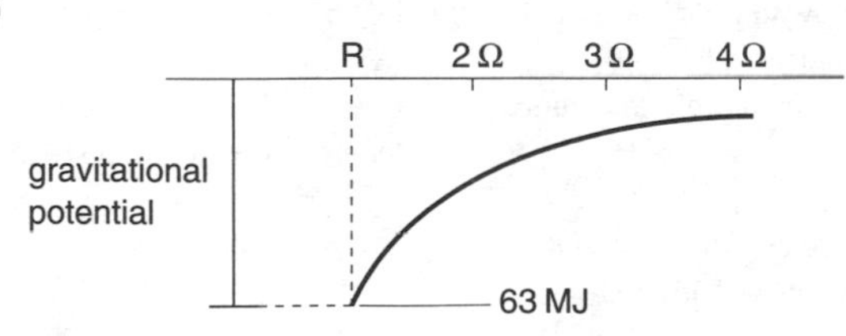

10. Speed = 1670 m s^{-1}; period = 106 min
11. **(a)** mass and distribution of mass
    (b) **(i)** 6 rad s^{-2} **(ii)** 30 rad s^{-1} **(iii)** 11.9 revs (75 rad)

Sections E5–E8

1. See E5
2. A point where there is no magnetic effect. The resultant magnetic flux density is zero.
3. **(a)** The strength of the magnetic field when force of 1 N is exerted on a wire of length 1 m when it caries a current of 1 A.
   (b) 25 mN
4. 2.3×10^{-6} N. The force is a repelling force.
5. See E7
6. 1.0 V
7. (a) 240 A s^{-1} **(b)** 40 mA
8. 4.8×10^{-14} N
9. When an electron moves at right angles to a magnetic field the magnetic force is always perpendicular to the direction of motion of the electron and has constant magnitude. This is the condition for circular motion. An electron that starts at right angles to the field is accelerated in the direction of the field but has constant velocity perpendicular to the field. This leads to a parabolic path.
10. 3.0×10^{7} m s^{-1}
11. 9600 V m^{-1}

Sections F1–F3

1. 3.45×10^{5} Pa
2. See F2
4. 25.6 J mol K^{-1}
5. 958 s
6. **(a)** 40% **(b)** 12 kW

Sections F4–F7

1. See F4
2. 1.08×10^{5} Pa
3. 0.72×10^{5} Pa
4. 0.050 mol
5. **(a)** 156 J
   (b) Work done in expansion so temperature falls
6. See F5
7. See F6
8. 6.2×10^{-21}
9. 350 m s^{-1}
10. 1560 W

Sections G1–G3

1. **(a)** ${}^{1}_{1}\mathrm{p}$ ${}^{0}_{-1}\mathrm{e}$ ${}^{4}_{2}\alpha$ ${}^{0}_{1}\mathrm{n}$ ${}^{0}_{0}\gamma$
2. 82 protons and 124 neutrons in the nucleus with 82 electrons 'orbiting' the nucleus
3. The nuclides have the same number of protons in the nucleus but a different number of neutrons.
4. Proton number 84; mass (nucleon) number 210
5. $4\,{}^{1}_{1}\mathrm{p} \rightarrow {}^{4}_{2}\alpha + 2\,{}^{0}_{1}\mathrm{X} + 2\,{}^{0}_{0}\nu_e + {}^{0}_{0}\gamma$; X is a positron.
6. **(i)** 70 **(ii)** 124
7. **(a)** 8.0×10^{-6} s
   (b) 4.8×10^{8} Bq
   (c) 3.4×10^{8} Bq
8. See G2
9. **(i)** 0.026 501u
   (ii) 3.95×10^{-12} J
10. See G3

Sections G4–G5

1. **(i)** 3.0×10^{-19} J
   (ii) 1.0×10^{15} Hz
2. 2.9×10^{15} Hz
3. A free electron has zero energy. Energy has to be put in to raise an electron to zero energy so the electron energies inside atoms are negative.
4. 1.87×10^{-20} J
5. 5.3×10^{14} Hz
6. 1.32×10^{-15} J
7. See G5
8. 10×10^{-12} J
   (This is much greater than beta particle energies and suggests that there are no electrons in the nucleus.)

Sections G6–G8

1. 9.1×10^{-31} kg
2. 1.27×10^{-30} kg
3. 1.6×10^{-13} J; gamma radiation
4. Inelastic collision – no loss of KE in the collision
   Ionization and excitation (see G7)
5. 7.4×10^{-15} m
6. lepton; baryon; lepton; baryon; meson
   (baryons and mesons are hadrons)
7. $\overline{\mathrm{uud}}$

8. **(a)** This is possible. Lepton and charge numbers are conserved.
   (b) This is not possible. Charge is conserved but baryon number is not. p is a baryon. All the others are mesons.
9.

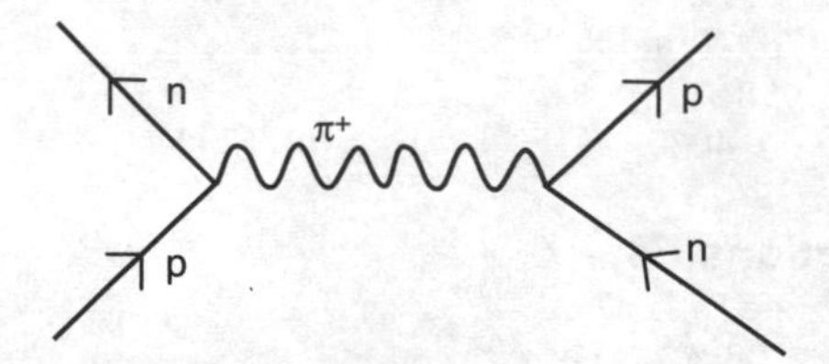

Sections H4–H6

1. 1210 N
2. Polycrystalline; polymer; polymer; crystalline; amorphous
3. 0.025 J
4. See H4.
5. See H4 (stretching glass).
6. **(a)** 1.3×10^5 Pa
   (b) 2.6×10^{-6}
   (c) 1.3×10^{-5} m
7. See H5 (defects).
8. See H6.
9. See H6.
10. See H6.

Sections H10–H12

1.

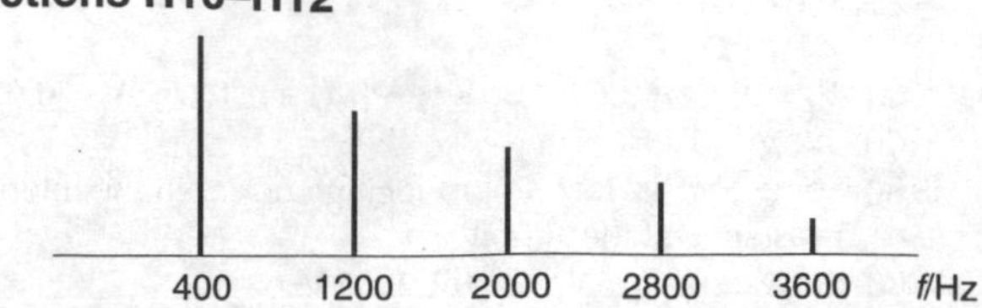

2. 8000 Hz
3. **(a)** 198.5 kHz, 200 kHz, 201.5 kHz

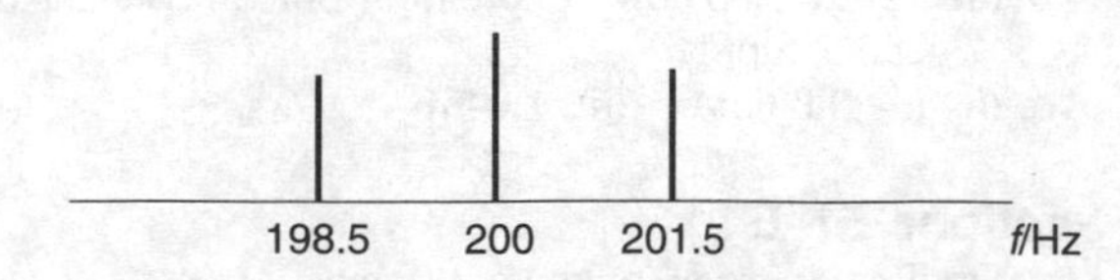

   (b)

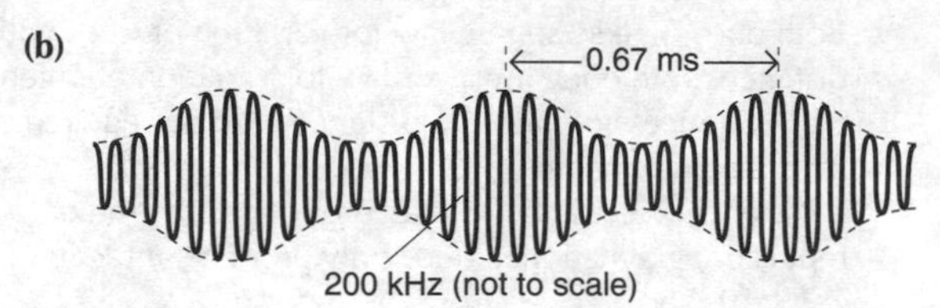

4. See H10.
5. **(a)** UHF
   (b) Large bandwidth (24 MHz for colour TV) required Medium-wave bandwidth has frequency range of less than 3 MHz
6. 10
7. –7 dB
8. **(a)** 30 kHz
   (b) 240 kHz

Physical data

Physical quantity	*Symbol*	*Value*
speed of light in a vacuum	c	2.998×10^8 m s^{-1}
permittivity of free space	ε_0	8.854×10^{-12} F m^{-1}
permeability of free space	μ_0	$4\pi \times 10^{-7}$ H m^{-1}
proton rest mass	m_p	1.673×10^{-27} kg
neutron rest mass	m_n	1.675×10^{-27} kg
electron rest mass	m_e	9.110×10^{-31} kg
proton charge	e	1.602×10^{-19} C
electron charge	$-e$	-1.602×10^{-19} C
	(minus sign often omitted)	
specific charge: electron	e/m_e	1.759×10^{11} C kg^{-1}
Planck constant	h	6.626×10^{-34} J s
gravitational constant	G	6.672×10^{-11} N m^2 kg^{-2}
Avogadro constant	N_A	6.022×10^{23} mol^{-1}
universal molar gas constant	R	8.314 J K^{-1} mol^{-1}
Boltzmann constant	k	1.381×10^{-23} J K^{-1}
absolute zero		0 K, –273.15 °C
standard atmospheric pressure		1.013×10^5 Pa

Physical quantity	*Symbol*	*Value*
kilowatt hour	kW h	3.600×10^6 J
electronvolt	eV	1.602×10^{-19} J
unified atomic mass unit	u	1.661×10^{-27} kg
	(energy equivalent: 931.5 MeV)	
acceleration of free fall (mean, at Earth's surface)	g, g_0	9.807 m s^{-2}
mass of Earth		5.976×10^{24} kg
mass of Sun		1.989×10^{30} kg
mass of Moon		7.350×10^{22} kg
equatorial radius of Earth		6.378×10^6 m
mean distance of Earth from Sun		1.496×10^{11} m
mean distance of Moon from Earth		3.844×10^8 m
solar constant		1.352×10^3 W m^{-2}
astronomical unit	AU	1.496×10^{11} m
parsec	pc	3.086×10^{16} m
light year	ly	9.461×10^{15} m

Equations to learn

You must learn the following equations. They will not be provided for you on your examination formula sheet. All the equations need to be known for A level and most of them need to be known for AS level too. You should check whether there are any that are not required in the AS specification you are using.

Certainly needed at AS and A level

speed = $\frac{\text{distance}}{\text{time}}$ $\qquad v = \frac{\Delta s}{\Delta t}$

acceleration = $\frac{\text{change in velocity}}{\text{time}}$ $\qquad a = \frac{\Delta v}{\Delta t}$

force = mass × acceleration $\qquad F = ma$

density = $\frac{\text{mass}}{\text{volume}}$ $\qquad \rho = \frac{m}{V}$

work done = force × distance moved in direction of force $\qquad \Delta W = F\Delta s$

power = $\frac{\text{energy transferred}}{\text{time taken}} = \frac{\text{work done}}{\text{time taken}}$ $\qquad P = \frac{E}{t}$

weight = mass × acceleration of free fall
= mass × gravitational strength $\qquad$ weight = mg

kinetic energy = $\frac{1}{2}$ mass × velocity^2 $\qquad E_k = \frac{1}{2}mv^2$

change in gravitational potential energy = mass × gravitational field strength × change in height $\qquad \Delta E_p = mg\Delta h$

current = $\frac{\text{charge}}{\text{time}}$ $\qquad I = \frac{\Delta Q}{\Delta t}$

potential difference = current × resistance $\qquad V = IR$
electrical power = potential difference × current $\qquad P = VI$

energy = potential difference × current × time $\qquad E = VIt$

potential difference = $\frac{\text{energy transferred}}{\text{time}}$ $\qquad V = \frac{E}{Q}$

resistance = $\frac{\text{resistivity} \times \text{length}}{\text{cross-sectional area}}$ $\qquad R = \frac{\rho l}{A}$

Possibly needed at AS level

pressure = $\frac{\text{force}}{\text{area}}$ $\qquad P = \frac{F}{A}$

momentum = mass × velocity $\qquad p = mv$

pressure × volume = number of moles × molar gas constant × absolute temperature $\qquad pV = nRT$

wave speed = frequency × wavelength $\qquad v = f\lambda$

Needed at A level only

centripetal force = $\frac{\text{mass} \times \text{velocity}^2}{\text{radius}}$ $\qquad F = \frac{mv^2}{r}$

force between point charges = $\frac{\text{charge (1)} \times \text{charge (2)}}{4\pi \times \text{permittivity} \times \text{(distance between charges)}^2}$ $\qquad F = \frac{Q_1Q_2}{4\pi\varepsilon r^2}$

force between point masses = $\frac{\text{universal gravitational constant} \times \text{mass (1)} \times \text{mass (2)}}{\text{(distance between masses)}^2}$ $\qquad F = \frac{Gm_1m_2}{r^2}$

capacitance = $\frac{\text{charge stored}}{\text{potential difference}}$ $\qquad C = \frac{Q}{V}$

For a transformer:

$\frac{\text{potential difference across coil (1)}}{\text{potential difference across coil (2)}} = \frac{\text{number of turns on coil (1)}}{\text{number of turns on coil (2)}}$ $\qquad \frac{V_1}{V_2} = \frac{N_1}{N_2}$

Needed for OCR Physics A, A and AS level

moment of a force	$T = Fx$
torque of a couple	$T = Fx$
power	$P = Fv$
stress	$\sigma = \frac{F}{A}$
strain	$\varepsilon = \frac{\Delta l}{l}$
the Young modulus	$E = \frac{\sigma}{\varepsilon}$
electric current	$I = \frac{\Delta Q}{\Delta t}$
power	$P = I^2R$ $P = \frac{V^2}{R}$
resistors in series	$R = R_1 + R_2 + \ldots$
resistors in parallel	$\frac{1}{R} = \frac{1}{R_1} + \frac{1}{R_2} + \ldots$
force on a current-carrying conductor	$F = BIl\sin\theta$
photon energy	$E = hf$
photoelectric effect	$hf = \Phi + \frac{1}{2}mv_{max}^2$
de Broglie equation	$\lambda = \frac{h}{mv}$
refractive index	$n = \frac{c_i}{c_r}$ $n = \frac{\sin i}{\sin r}$
wave speed	$v = f\lambda$
double-slit interference	$\lambda = \frac{ax}{D}$
force	$F = \frac{\Delta p}{\Delta t}$

Needed for OCR Physics A, A level only

centripetal acceleration	$a = \frac{v^2}{r}$
simple harmonic motion	$a = -(2\pi f)^2x$ $x = A\sin 2\pi ft$ $x = A\cos 2\pi ft$
gravitational field of strength	$g = \frac{F}{m}$ $g = \frac{GM}{r^2}$
electric field strength	$E = \frac{F}{Q}$ $E = \frac{Q}{4\pi\varepsilon_0 r^2}$ $E = \frac{V}{d}$
capacitance	$C = \frac{Q}{V}$
energy of charged capacitor	$W = \frac{1}{2}QV$
time constant of CR circuit	$\tau = CR$
force on moving charged particle	$F = Bqv$
magnetic flux	$\Phi = BA$
induced e.m.f.	$e = N\frac{\Delta\Phi}{\Delta t}$
thermal energy change	$\Delta Q = mc\Delta\theta$
mass–energy	$\Delta E = \Delta mc^2$
radioactivity	$A = \lambda N$
apparent magnitude	$m = -2.5 \log I + \text{constant}$
apparent/absolute magnitude	$m - M = 5 \log\left(\frac{r}{10}\right)$
Hubble's law	$v = H_0 d$
age of Universe	$t \approx \frac{1}{H_0}$
Doppler formula	$\frac{\Delta\lambda}{\lambda} = \frac{v}{c}$
lens formula	$\text{power} = \frac{1}{f} = \frac{1}{u} + \frac{1}{v}$
X-ray attenuation	$I = I_0 e^{-\mu x}$
Hall voltage	$V_H = Bvd$
inverting amplifier gain	$G = \frac{R_F}{R_{IN}}$
power ratio	no. of decibels (dB) $= 10 \log\left(\frac{P_1}{P_2}\right)$

INDEX

*If a reference is given in **bold**, you should look this up first.*

S

T

U

V

W

X

Y